Biosorbents

This book focuses on the biologically derived adsorbent with numerous applications in wastewater treatment, metal recovery, biosensor development, and so forth. It initiates with the description of biological sources of biosorbents followed by applications of biosorbents, biosorption isotherms, assessment of biosorbents with various tools, pretreatment of biosorbents, and its mode of action. Some less explored areas like separation of radionuclides, biosorption of volatile organic compounds, and animal-based biosorbents are also explained.

Features:

- Focuses on fundamentals, characteristics of flora and fauna-mediated biosorbents used extensively.
- Describes entire aspects of tools and techniques related to assessment and monitoring of biosorbents.
- Includes adsorption kinetics, adsorption isotherm, and mechanism of action of biosorbents.
- Covers advancements in pretreatment methods to enhance the adsorption process of biosorbents.
- Reviews recent applications which include heavy metal removal, dye remediation, and separation of radionuclides and nano-biosorbents.

This book is aimed at graduate students and researchers in bioprocess engineering, microbiology, and biotechnology.

Emerging Materials and Technologies

Series Editor: Boris I. Kharissov

The *Emerging Materials and Technologies* series is devoted to highlighting publications centered on emerging advanced materials and novel technologies. Attention is paid to those newly discovered or applied materials with potential to solve pressing societal problems and improve quality of life, corresponding to environmental protection, medicine, communications, energy, transportation, advanced manufacturing, and related areas.

The series takes into account that, under present strong demands for energy, material, and cost savings, as well as heavy contamination problems and worldwide pandemic conditions, the area of emerging materials and related scalable technologies is a highly interdisciplinary field, with the need for researchers, professionals, and academics across the spectrum of engineering and technological disciplines. The main objective of this book series is to attract more attention to these materials and technologies and invite conversation among the international R&D community.

Wastewater Treatment with the Fenton Process
Principles and Applications
Dominika Bury, Piotr Marcinowski, Jan Bogacki, Michal Jakubczak, and Agnieszka Jastrzebska

Mechanical Behavior of Advanced Materials: Modeling and Simulation
Edited by Jia Li and Qihong Fang

Shape Memory Polymer Composites
Characterization and Modeling
Nilesh Tiwari and Kanif M. Markad

Impedance Spectroscopy and its Application in Biological Detection
Edited by Geeta Bhatt, Manoj Bhatt and Shantanu Bhattacharya

Nanofillers for Sustainable Applications
Edited by N.M Nurazzi, E. Bayraktar, M.N.F. Norrrahim, H.A. Aisyah, N. Abdullah, and M.R.M. Asyraf

Chemistry of Dehydrogenation Reactions and its Applications
Edited by Syed Shahabuddin, Rama Gaur and Nandini Mukherjee

Biosorbents
Diversity, Bioprocessing, and Applications
Edited by Pramod Kumar Mahish, Dakeshwar Kumar Verma, and Shailesh Kumar Jadhav

Principles and Applications of Nanotherapeutics
Imalka Munaweera and Piumika Yapa

For more information about this series, please visit: www.routledge.com/Emerging-Materials-and-Technologies/book-series/CRCEMT

Biosorbents
Diversity, Bioprocessing, and Applications

Edited by
Pramod Kumar Mahish, Dakeshwar Kumar Verma, and Shailesh Kumar Jadhav

CRC Press
Taylor & Francis Group
Boca Raton London New York

CRC Press is an imprint of the
Taylor & Francis Group, an **informa** business

First edition published 2024
by CRC Press
2385 NW Executive Center Drive, Suite 320, Boca Raton FL 33431

and by CRC Press
4 Park Square, Milton Park, Abingdon, Oxon, OX14 4RN

CRC Press is an imprint of Taylor & Francis Group, LLC

ISBN: 9781032399744 (hbk)
ISBN: 9781032431857 (pbk)
ISBN: 9781003366058 (ebk)

DOI: 10.1201/9781003366058

Typeset in Times
by codeMantra

Contents

Editors

Pramod Kumar Mahish is presently serving as Assistant Professor of Biotechnology in Govt. Digvijay Autonomous PG College, Rajnandgaon, Chhattisgarh, India and Chairman Board of Studies Biotechnology at Hemchand Yadav Vishwavidyalaya Durg. He has nine years of teaching and 12 years of research experience. Dr. Mahish has published 17 research papers and nine books. He has also written fifteen book chapters with various other platforms. He has completed three research projects and is presently guiding two Ph.D. scholars. He is a fellow member of the World Researchers Association and a life member of the Aerobiology Association of India and Microbiologists Society India. He was awarded UGC National Fellowship for his Ph.D. research work. Dr. Mahish is engaged in the editing of two books on the De-Gruyter and ACS platform.

Dakeshwar Kumar Verma is an Assistant Professor of Chemistry at Govt. Digvijay Autonomous Postgraduate College, Rajnandgaon, Chhattisgarh, India. His research is mainly focused on the preparation and designing of sustainable and eco-friendly corrosion inhibitors for metals and alloys that are useful for several industrial applications. Dr. Verma is the author of several research papers, review articles, and book chapters in peer-reviewed international journals. He is also an editor/author of various books that will be published by various publications. Dr. Verma has an H index of 22 and an i-10 index of 30 with a total of 1500 plus citations. He has received the Council of Scientific and Industrial Research Junior Research Fellowship national award in 2013. He also availed Ministry of Human Resource Development (MHRD) fellow during his Ph.D.

Shailesh Kumar Jadhav is presently serving as Senior Professor of Biotechnology in the School of Studies in Biotechnology, Pt. Ravishankar Shukla University, Raipur, Chhattisgarh, India. He has 34 years of teaching and 29 years of research experience. Prof. Jadhav has published more than 125 research papers and 12 books. His area of interest is bioprocessing, green energy, and microbial technologies. He has awarded 26 Ph.D. and completed seven research projects of different Government agencies of India. He has received various awards in the fields of teaching, research, and administration. Prof. Jadhav has an H index of 20 and an i10 index of 39.

Contributors

Adewale George Adeniyi is in the Department of Chemical Engineering, University of Ilorin, Ilorin, Nigeria.

Shreanshi Agrahari is in the Department of Chemistry (Centre of Advanced Study), Institute of Science, Banaras Hindu University, Varanasi, India.

Shahid Ahmad is in the School of Physical Sciences, University of the Punjab, Lahore, Pakistan.

Míriam Cristina Santos Amaral is in the Department of Sanitation and Environmental Engineering, School of Engineering, Universidade Federal de Minas Gerais, Campus Pampulha, MG, Brazil.

Godwin Anywar is in the Department of Plant Sciences, Microbiology and Biotechnology, Makerere University, Kampala, Uganda.

Sonia Bajaj is in the Department of Zoology, Shri Shankaracharya Mahavidyalaya, Bhilai Chhattisgarh, India.

Khasan Berdimuradov is in the Faculty of Industrial Viticulture and Food Production Technology, Shahrisabz Branch of Tashkent Institute of Chemical Technology, Shahrisabz, Uzbekistan.

Elyor Berdimurodov is in the Department of Chemical & Materials Engineering, New Uzbekistan University, Tashkent, Uzbekistan; Medical School, Central Asian University, Tashkent, Uzbekistan; Faculty of Chemistry, National University of Uzbekistan, Tashkent, Uzbekistan.

Thais Girardi Carpanez is in the Department of Sanitation and Environmental Engineering, School of Engineering, Universidade Federal de Minas Gerais, Campus Pampulha, MG, Brazil.

Giovanni Souza Casella is in the Department of Sanitation and Environmental Engineering, School of Engineering, Universidade Federal de Minas Gerais, 6627 Antônio Carlos Avenue, Campus Pampulha, MG, Brazil.

Nagendra Kumar Chandrawanshi is in the School of Studies in Biotechnology, Pt. Ravishankar Shukla University Raipur, Chhattisgarh, India.

Ravishankar Chauhan is in the Department of Botany, Pandit Ravishankar Tripathi Government College Bhaiyathan, Surajpur, India.

Omar Dagdag is at the Department of Mechanical Engineering, Gachon University, Seongnam 13120, Republic of Korea.

Eduardo Coutinho de Paula is in the Department of Sanitation and Environmental Engineering, School of Engineering, Universidade Federal de Minas Gerais, Campus Pampulha, MG, Brazil.

Kanika Dulta is in the Department of Food Technology, School of Applied and Life Sciences, Uttaranchal University, Dehradun, Uttarakhand, India.

Muhammad Ehsan is in the School of Physical Sciences, University of the Punjab, Lahore, Pakistan.

Ilyos Eliboyev is in the Faculty of Chemistry, National University of Uzbekistan, Tashkent, Uzbekistan.

Ebuka Chizitere Emenike is in the Department of Pure and Industrial Chemistry, Nnamdi Azikiwe University, Awka, Nigeria.

Ravindra Kumar Gautam is in the Department of Chemistry (Centre of Advanced Study), Institute of Science, Banaras Hindu University, Varanasi, India.

Nisha Gupta is in the School of Studies in Biotechnology, Pt. Ravishankar Shukla University, Raipur, Chhattisgarh, India.

Muhammad Imran is at the Centre for Inorganic Chemistry, School of Chemistry, University of the Punjab, Lahore, Pakistan.

Kingsley O. Iwuozor is in the Department of Pure and Industrial Chemistry, Nnamdi Azikiwe University, Awka, Nigeria.

Shailesh Kumar Jadhav is in the School of Studies in Biotechnology, Pt. Ravishankar Shukla University, Raipur, Chhattisgarh, India.

Akanksha Jain is in the Department of Biotechnology, Shri Shankaracharya Mahavidyalaya, Bhilai, Chhattisgarh, India.

Deepali Koreti is in the School of Studies in Biotechnology, Pt. Ravishankar Shukla University Raipur, Chhattisgarh, India.

Anjali Kosre is in the School of Studies in Biotechnology, Pt. Ravishankar Shukla University Raipur, Chhattisgarh, India.

Shoomaila Latif is in the School of Physical Sciences, University of the Punjab, Lahore, Pakistan.

Pramod Kumar Mahish is in the Postgraduate Department and Research Centre of Biotechnology, Govt. Digvijay (Autonomous) P.G. College Rajnandgaon, Chhattisgarh, India.

Madhu Manikpuri is at the Centre for Basic Sciences (CBS), Pt. Ravishankar Shukla University, Raipur, Chhattisgarh, India.

Varsha Meshram is in the School of Studies in Biotechnology, Pt. Ravishankar Shukla University, Raipur, Chhattisgarh, India.

Victor Rezende Moreira is in the Department of Sanitation and Environmental Engineering, School of Engineering, Universidade Federal de Minas Gerais, Campus Pampulha, MG, Brazil.

Julius Mulindwa is in the Department of Biochemistry, Makerere University, Kampala, Uganda.

Jane Namukobe is in the Department of Chemistry, Makerere University, Kampala, Uganda.

Jéssica Mesquita do Nascimento is at the Center for Exact, Natural and Technological Sciences – CCENT, State University of the Tocantina Region of Maranhão, Imperatriz, Maranhão, Imperatriz, MA, Brazil.

Moses Okol is in the Department of Biochemistry, Makerere University, Kampala, Uganda.

Johnson O. Oladele is in the Department of Chemical Sciences, Faculty of Science, Kings University, Ode-Omu, Osun State, Nigeria.

Suchitra Kumari Panigrahy is in the Department of Biotechnology, Kalinga University, Raipur, Chhattisgarh, India.

Alka Patle is in the Department of Chemistry, Govt. Agrasen College Bilha, Bilaspur, Chhattisgarh, India.

Tarun Kumar Patle is in the Department of Chemistry, Pt. Sundarlal Sharma (Open) University, Bilaspur, Chhattisgarh, India.

Jai Shankar Paul is in the School of Studies in Biotechnology, Pt. Ravishankar Shukla University, Raipur, Chhattisgarh, India.

Timothy D. Phillips is in the Department of Veterinary Physiology and Pharmacology, College of Veterinary Medicine and Biomedical Sciences, Texas A&M University, College Station, TX, USA.

Khemraj Sahu is in the School of Studies in Biotechnology, Pt. Ravishankar Shukla University, Raipur, Chhattisgarh, India.

Lucilaine Valéria de Souza Santos is in the Department of Sanitation and Environmental Engineering, School of Engineering, Universidade Federal de Minas Gerais, Campus Pampulha, MG, Brazil.

N. Saranya is in the Department of Bioengineering, Institute of Biotechnology, SIMATS School of Engineering, Chennai, Tamil Nadu, India.

Rangabhashiyam Selvasembian is in the Department of Biotechnology, School of Chemical and Biotechnology, SASTRA Deemed University, Thanjavur, Tamil Nadu, India.

M. Hassan Siddique is at Hi Tech Blending Pvt Ltd, ZIC Motor Oil, Lahore, Pakistan.

Ankit Kumar Singh is in the Department of Chemistry (Centre of Advanced Study), Institute of Science, Banaras Hindu University, Varanasi, India.

Divyanshi Singh is in the Faculty of Applied Sciences and Biotechnology, Shoolini University, Solan, Himachal Pradesh, India.

Kirti Singh is in the Faculty of Applied Sciences and Biotechnology, Shoolini University, Solan, Himachal Pradesh, India.

Swati Singh is in the Faculty of Applied Sciences and Biotechnology, Shoolini University, Solan, Himachal Pradesh, India.

Jamilu Ssenku is in the Department of Plant Sciences, Microbiology & Biotechnology, Makerere University, Kampala, Uganda.

Swati Tanti is in the Department of Biotechnology, Faculty of Science, Kalinga University, Raipur, Chhattisgarh, India.

Veena Thakur is at Government Pt. Shyam Shankar Mishra College Deobhog, Gariyaband, Chhattisgarh, India.

Ida Tiwari is in the Department of Chemistry (Centre of Advanced Study), Institute of Science, Banaras Hindu University, Varanasi, India.

Shubhra Tiwari is in the School of Studies in Biotechnology, Pt. Ravishankar Shukla University, Raipur, Chhattisgarh, India.

Patience Tugume is in the Department of Plant Sciences, Microbiology & Biotechnology, Makerere University, Kampala, Uganda.

Dakeshwar Kumar Verma is in the Department of Chemistry, Government Digvijay Autonomous Postgraduate College, Rajnandgaon, Chhattisgarh, India.

Dristi Verma is in the School of Studies in Biotechnology, Pt. Ravishankar Shukla University, Raipur, Chhattisgarh, India.

Tikendra Kumar Verma is at Laxman Prasad Baidh Govt. Girls College, Bemetara, Chhattisgarh, India.

Meichen Wang is in the Department of Veterinary Physiology and Pharmacology, College of Veterinary Medicine and Biomedical Sciences, Texas A&M University, College Station, TX, USA.

1 Introduction and Characteristics of Biosorbents

Kanika Dulta, Ebuka Chizitere Emenike, Kingsley O. Iwuozor, Divyanshi Singh, Swati Singh, Kirti Singh, Adewale George Adeniyi, Saranya N., and Rangabhashiyam Selvasembian

1.1 INTRODUCTION

It is generally acknowledged that the rise in global population has prompted fast urbanization and industrial development, which in turn result in the production of a high volume of effluents that negate the ecosystem and human health. Release of effluents from several industries like leather tanning, dyeing, pharmaceutical, electroplating, and chemical processing without proper treatment deteriorates water bodies. The environmental impact studies revealed the existence of numerous toxins, which ultimately cause dangerous effects on the lives of people, animals, and plants (Khan et al., 2022; Sivaraman et al., 2022). Before being released into the environment, industrial effluents are frequently treated using a variety of physicochemical and biological techniques. Chemisorption, flotation, ultrasound oxidation, photocatalytic oxidation, steam reforming process oxidation, wet air oxidation, electrochemical oxidation, hydrogen peroxide treatment, nano-filtration, adsorption, and bioremediation are some of the effective methods utilized for contaminant alleviation. Biosorption being a simple, cost-effective process attracted researchers in wastewater treatment. Biomass from various natural resources has attained a prime focus as biosorbents which are tremendously explored for contaminant removal from wastewater. The raw forms of plant biomass like fallen leaves, stems, bark, shells, pods, and aquatic plants which are prone to specific metal-binding are explored as biosorbents for the decontamination of metal-bearing untreated wastewater.

For the intended degradation of organic pollutants, a diverse variety of biomolecules found in bio-based sources have been used as biosorption. All forms of microbial, plant, but also animal biomass, as well as its derivatives have drawn a lot of attention in a number of contexts and in connection with a number of different compounds (Volesky & Holan, 1995; Volesky, 2001; Al-Masri et al., 2010). In latest years, research on polysaccharides, industrial sludge, agricultural residues have gained momentum in wastewater treatment (Blázquez et al., 2011; Witek-Krowiak et al., 2011; Reddy et al., 2012; Witek-Krowiak & Reddy, 2013; Rangabhashiyam et al., 2022).

DOI: 10.1201/9781003366058-1

1.2 BIOSORPTION

Biosorption may be defined as a physical operation in which the pollutant molecules are concentrated upon surface of biosorbent material under optimum conditions (Gadd, 2009). The physicochemical process of biosorption includes ion exchange, surface complexation, adsorption, absorption, and dissolution. Due to its effectiveness and simplicity, biosorption has long been considered a potential process for the elimination of pollutants from solutions and perhaps the recovery of pollutants. Biosorbents include living or dead biomass. Adsorption's subclass of biosorption uses a biological substrate as just the sorbent. Presence of diverse functional groups on the surface of the biomass components serves as a sorption site for the attachment of pollutants (Davis et al., 2003). The operational costs are reasonable because this procedure does not require a significant investment. The biological elements can also be derived via industrial effluents, agriculture, and therefore are frequently cheap. Advantages of biosorption include less operational and merely no maintenance costs, cheap and abundantly available resources, less sludge generation, the possibility of desorption, and regeneration of biosorbents (Kratochvil & Volesky, 1998; Malleswari et al., 2022).

The mechanism of pollutant biosorption using various biomass-based adsorbents was found to be either adsorption or adsorption coupled reduction. The mechanism and scheme of biosorption highly depend on the type of biosorbent used, its origin, chemical composition, environmental conditions, speciation of pollutants in water, etc. (Torres, 2020). Other promising characteristics of biosorbents may be of higher surface area, porosity, structure, and composition of membranes in case of microbial sorbents which highly influence the biosorption capacity of the biosorbents either in its native form or modified form (de Freitas et al., 2019a,b; Wang et al., 2019; Yücel et al., 2020).

1.2.1 Characteristics of Biosorbents

Parameters to be considered in the selection of biosorbents in removing contaminants from aqueous solutions include analyzing the availability, nontoxicity, reusability, chemical nature of the biomass to be utilized as biosorbents (Das, 2010; Lim & Aris, 2014). Biomass obtained from plant sources are basically comprised of lignocelluloses including lignins, celluloses, and hemicelluloses (Zoghlami & Paës, 2019). Apart from the lignocellulosic components as their structural materials, plants produce primary and secondary metabolites which include pigments, phytochemicals like polyphenols, alkaloids, terpenoids, and saponins which are rich in anionic and cationic functional groups that make the plant-derived biosorbents superior options for pollutant sequestration. According to the nature of the pollutant, properties like net charge, speciation, degradation ability, optimum conditions in which the pollutants are stable, modifications of the biosorbents can be made. Biomass is the primary component of a biosorption procedure. The biomass may be physical, chemical, or biological modifications that would impart specific functional groups and net charge to the biosorbents, increases surface area, exposes active groups, and increases porosity and accessibility to pollutants (Elgarahy et al., 2021). Also, live and dead cells of microorganisms like bacteria, fungi and yeast in free form or immobilized forms were utilized as biosorbents. Promising results have been obtained with the cells of bacteria, fungi, and yeast species (Jaafari & Yaghmaeian, 2019; Moghazy et al., 2019; Rogowska et al., 2019; Santaeufemia et al., 2019; Contreras-Cortés et al., 2020; Jin et al., 2020).

Given that many of the biosorption techniques are reversed, it is possible to recover the components that have been biosorbed while renewing the biosorbent. Toxicant affinity (elevated under optimal circumstances), renewal, and reuse (strong likelihood of biosorbent rejuvenation, with probable usage more than a sequence of cycles), plus toxicant restoration are influencing features of biosorption. Recovery of toxicant is achievable with suitable elutant administration. Acidic or alkaline treatments have frequently proven to be effective recovery methods for toxicants.

1.3 SOURCES OF BIOSORBENT

Biosorbent materials are sourced from algae, plants, bacteria, fungi, and animals. Biosorbents can be divided into high-cost and low-cost sorbents based on their availability, preparation procedures. Low-cost sorbents comprise natural biomass that are sourced from the environment such as plant wastes, agro wastes, vegetable wastes, fruit wastes, grain husks, pods, aquatic organisms like seaweeds, and waste or by-products from industries such as fermentation products and waste sludge. Biosorbent materials, which also are particularly manufactured for biosorption purposes, are among the more expensive sorbents. Such materials ought to be readily recoverable and reusable and ought to have a strong biosorption characteristic.

Biosorption is a property of nearly every biological material, particularly macroalgae (seaweeds), plant, animal biomass, and products produced from those sources (like chitosan). Different types of microbial, plant, and animal biomass, as well as their derived products, have been extensively studied in the hunt for highly effective and affordable biosorbents and new potential for pollution management and resource recovery (Fomina and Gadd, 2014; Iwuozor et al., 2022). The main groups of organisms make up the majority of biosorbents: bacterium, fungal, algae, industrial effluent, agricultural by-products, environmental by-products, etc. (Table 1.1). The biosorption technique is mature and widely used as a result of the simple accessibility and affordable conversion of these natural sources into biosorbents, and expense, reliability, as well as affordability are the three key considerations when choosing biosorbents for large-scale industrial purpose. The elimination of heavy metals, dyes, PPCPs, and other emerging contaminants using natural elements is a topic of research that is getting more and more attention. Native and dead cells of microbes are also used as biosorbents owing to their higher selectivity, degradation ability and effectiveness.

For commercial applications, the biosorbent can come from (i) free industrial waste; (ii) naturally occurring organisms that are widely accessible; and (iii) quickly growable biomass, particularly cultivated for biosorption (Park et al., 2010).

TABLE 1.1
Biomass subjected for biosorption is characterized in different patterns

S. no.	Classification	Illustration	References
1.	Bacterium	Gram-positive microorganism (*Bacillus* sp., etc.), gram-negative microorganism (*Pseudomonas* sp., etc.).	Tinyiro et al., 2011
2.	Fungi	Molds (*Rhizopus* sp., etc.) yeast (*Saccharomyces cerevisiae*).	Martha et al., 2019; Rossi et al., 2020
3.	Algae	Micro-algae (*Chlorella* sp., etc.), macro-algae (Chlorophyceae, Rhodophyceae, Phaeophyceae).	Prosenc et al., 2021; Kumar et al., 2023
4.	Industrial effluents	Food/beverage residue, anaerobic by-product, stimulated sludges, as well as fermented waste product.	Liu et al., 2021; Pap et al., 2017
5.	Agricultural by-products	Waste from fruits and vegetables, rice husks, oat bran, etc.	Sridhar et al., 2022; Ma et al., 2020
6.	Environmental by-products	Weeds, woodchips, tree trunks, and plant remnants.	Das et al., 2023; Saha et al., 2013
7.	Additional ingredients	Compounds derived from cellulose as well as chitin.	Jung et al., 2023

1.4 METHODOLOGIES FOR EVALUATING BIOSORBENTS

Comprehending the reaction conditions as well as its implications forms a major part of the categorization. Characterization methods that are frequently employed include scanning electron microscopy (SEM)–energy dispersive X ray spectrometry, transmission electron microscopy (TEM), Fourier transform infrared (FTIR) spectroscopy, atomic absorption spectroscopy, Brunauer–Emmett–Teller (BET), mercury intrusion porosimetry, CHNS analyzer, X-ray diffraction (XRD). These methods are helpful for figuring the surface area, crystalline nature, pore volume, functional group analysis, thermal stability, etc. The relevant analytical methods are outlined in detail below and are utilized to evaluate the properties of biosorbents. Some of these techniques give clues about the mechanism of biosorption (Park et al., 2010).

1.4.1 Particle Size Analyzer (Particle Analyzers)

This device is a powerful analytical tool used to describe the physical attributes of tiny particles. The technique can be applied as a particle size analyzer based on its design or purpose. For evaluating both sizes and distribution of the particles that compose a material, particle analyzers are utilized. Particle size analyzers are utilized in various industries for quality checks, quality assurance, production, as well as research and advancement. Laser diffraction particle size analyzers just use angles of light scattered by such a flow of particles as they travel through some kind of laser beam to determine the dimensions of the particles. This method permits the measurement of bulk materials continuously over a large size zone (10 nm–3 mm). Laser diffraction particle analyzer's size restrictions but also responsiveness affects the quantity and positioning of its sensors. Particles in solutions are mostly examined using dynamic light scattering particle analyzers (size range: 1 nm to 6 um). By measuring the variations in scattered laser light intensity caused either by particles' Brownian motion, dynamic light scattering can estimate dimension. Through electronically aligning the particles and thereafter analyzing their dispersion, driven diffraction particle size analyzers may affect the quantity even of smaller particles (0.5 nm–200 nm) in liquids.

1.4.2 X-Ray Diffraction Analysis

X-ray diffraction (XRD) is a highly adaptable technique used for determining the crystalline structure of a compound. XRD is an important non-destructive technique for analyzing a wide range of materials and powders (Roychand et al., 2021). X-rays are high-energy electromagnetic waves with a wavelength between 103 and 101 nm (Spieß et al., 2009). Biosorbents are analyzed with XRD for the determination of crystal structures and structural modifications, chemical interactions between sorbent and sorbate before and after adsorption from the XRD patterns at different angles. Although XRD is a well-established non-destructive technique, it still requires further improvements in its characterization capabilities, especially when dealing with complex mineral structures (Ali et al., 2022).

Besides chemical characterization, XRD is quite helpful for measuring strain but also analyzing appearance. Waves generated by a diffractometer have a specific frequency that is determined by their source. Since no other type of light has the proper wavelength for inter-atomic-scale diffraction, x-rays are frequently the source. The pattern's atoms behave exactly like something of a diffraction grating whenever these waves enter it, producing bright spots at specific angles. Bragg's law can be employed to calculate the separation of the diffraction grating by determining the angle at which these high points appear.

1.4.3 Scanning Electron Microscopy Analysis

Scanning electron microscopy (SEM) was used to investigate the microstructure of the material. SEM is currently a well-developed method that is used extensively in various scientific applications. SEM is a potent tool for studying materials and is used in metallurgical, geology, biology, as well as healthcare, among other fields. This technique involves forcing an electron beam from an electron cannon to strike the sample holder. Whereas a vacuum is established in the container and an electron gun is positioned on top of the apparatus, the electrons beam is directed toward the sample incident. The instrument's lenses and electromagnetic field regulate the electron beams trajectory. Whenever a material is struck by an accelerating electron beam, some of the incident electrons are return dispersed, some are passed through the sample, and some produce secondary electrons. These secondary and backscattered particles are monitored by the detector. Ultimately, the appearance of the sample can indeed be documented using a computer. SEM is normally used to observe bulk samples at ~1 nm resolution extending from micrometer scale of whole cells to labeled molecules (Abrams et al., 1944; Swift & Brown, 1970; Peckys et al., 2009; Ross et al., 2015).

1.4.4 Transmission Electron Microscopy Analysis

The transmission electron microscopy (TEM) device is frequently used mostly for particle analysis. Particle size and physical properties are precisely measured by TEM. The thermoelectric filament or the field emission filament emits electrons as the device's electron source in TEM. The cathode or electrode accelerates the filament-produced electron to increase its energy. The optical device that produces images of the samples is an electromagnetic lens. With the aid of TEM, the crystalline plane, miller indices, and interatomic distance can be investigated. The diffraction pattern in TEM provides information on the sample's crystalline and amorphous characteristics. The regular placement of the points but also circumferential rings reveals the sample's crystallinity and provides information about just the sample's amorphous state (Nie, 2012; Jiao et al., 2017; Kong & Liu, 2018; Xiong et al., 2021).

1.4.5 Fourier Transform Infra-Red (FTIR Spectroscopy

Fourier transform infra-red (FTIR) spectroscopy was used to look into the existence of various molecules on the surface of the biosorbent, and that these groups are located where the biosorbent binds to metal ions. Typical chemical bonds vibration was also confirmed by FTIR spectrometer (Wibawa et al., 2020) and high-spectral-resolution information are concurrently gathered over a broad spectral band using an FTIR spectrometer. In comparison to a diffraction spectrophotometer FTIR typically analyzes intensities over a limited variety of wavelengths at a time that offers a substantial edge. Since a Fourier transform (a mathematical operation) is necessary to turn the raw data into the real spectrum, the name FTIR spectroscopy was coined. A recorded spectrum indicates the position of bands associated with bond strength and nature, as well as specific functional groups, providing information about molecular structures and interactions (Mourdikoudis et al., 2018).

1.4.6 Brunauer–Emmett–Teller

Nanoscale materials are characterized using the Brunauer-Emmett-Teller (BET) method. It is based on the actual adsorption of a gas on a solid surface and is called just after beginning of its authors' last names, Brunauer, Emmett, and Teller. Due to its relative accuracy, speed, and simplicity, it is frequently used to calculate the surface area of nanostructures. The most popular way for describing a certain surface area is BET. The implementation of the BET technique involves two steps. The first step is to convert a physisorption isotherm into a "BET plot," from which the BET monolayer capacity, nm, can be calculated. The BET area is estimated from nm in the second phase using an adequate molecule cross-sectional surface value. The BET theory is a key analytical tool for the measurement of the specific surface area of materials. It seeks to explain the physical adsorption of gas molecules on a solid surface. BET theory uses probing gases that don't chemically react with material surfaces as adsorbates to measure a specific surface area in systems with multilayer adsorption (Nasrollahzadeh et al., 2019).

1.4.7 CHNS

CHNS elemental analyzers provide a means for the rapid determination of carbon, hydrogen, nitrogen, and sulfur in organic matrices and other types of materials (Roychand et al., 2021). Determining the elemental makeup of biosorbents is aided by CHNS analysis. Comprehending the composition of the material and its capacity to absorb particular contaminants is crucial. Different biosorbents have varying contents of carbon, hydrogen, nitrogen, and sulfur, which affects how successful they are in biosorption processes. The proportion content of every component in a sample is provided by CHNS analysis. Particulates are simple to spot. To guarantee the uniformity and purity of biosorbents, CHNS analysis is employed for quality control purposes. Researchers and practitioners can evaluate the quality of biosorbent materials and preserve the effectiveness of biosorption processes by figuring out the elemental makeup. The information generated by CHNSO elemental analysis aids in identifying an organic compound's structure and chemical make-up. Elemental analysis is a quick, easy, and low-cost method for figuring the chemical components.

1.5 DISTINCTIVE BIOSORPTION PARAMETERS

The biosorption of toxic substances is influenced by a variety of factors which include pH, temperature, contact time, agitation speed, initial pollutant concentration, dose of the biosorbent utilized, and nature of the biosorbents.

1.5.1 pH

The propensity of biomass to absorb a solution is significantly impacted by the pH. The technique can be utilized in a variety of pH settings, though. Plant-derived biosorbents in their native form have diverse functional groups in accordance with their chemical constituents and hence render a net positive, negative, or either neutral charge on the surface. If the biosorbent has positively charged functional groups, they attract anionic pollutants and if it has more number of negatively charged groups, then they tend to attract cationic pollutants. This can be ascertained by determining the point of zero charge (PZC) of the biosorbent. PZC of the biomass or biosorbent is usually determined by contacting the biosorbent in different pH ranges (say from 1.0 to 10.0) of KNO_3 solution for a particular period of time. After a time period, the pH of the solution was determined. A plot is made between initial pH of the solution and the difference between the initial and final pH of the solution. The point at which the difference in pH is zero is considered PZC. Below the PZC value, the biosorbent is positively charged and be used for catching anionic ligands. Above the PZC value, the biosorbent is more negatively charged and can attract cationic contaminants. Maintenance of optimum pH of the solution is also very much important for the sorption to takes place. The removal percentage of pollutants gradually increases till optimum pH, where it reaches maximum and then gradually decreases above optimum pH values. The biosorption process halts beyond optimum pH, and there is a higher chance of desorption of the adsorbed pollutants. Biosorption has an advantage of alleviating pollutants with a wide pH range (pH 3–9) as well as a large array of temperatures (4°C–90°C).

1.5.2 Temperature

Temperature is one of the influencing parameters in adsorption experiments. Increase in temperature increases the collision and kinetic energy of the interacting molecules (Sedlakova-Kadukova et al., 2019). The mechanism of biosorption is unaffected by temperature since the biomass remains dormant. In contrast, a number of researchers noted that intake improved as temperature goes up. Temperature raises typically improve biosorptive clearance of adsorptive pollutants by raising the adsorbate's surface activities and kinetic energy, but they might also harm the biosorbent's physical structure. Temperature can affect the concentration and stability of complex ligands, ligand forms, particularly metal complexes. Regardless, temperature can have a favorable or negative impact on the kinetics of the reaction, the efficiency of biosorption, or perhaps the characteristics of the biosorbent. Increased porous structure in the biosorbent, greater flexibility of heavy metal ions, plus successful interactions could all result in a rise in biosorption as temperature rises. The temperature doesn't really alter the biosorption process, which suggests that perhaps the flow separation is not an absorption major bottleneck, according to certain studies.

1.5.3 Contact Period

The ideal contact duration between biosorbent and also the adsorbate is crucial for achieving maximal biosorption of pollutants from aqueous phase. Optimal contact time between biosorbent and adsorbate at which maximum removal attains has to be determined. Removal percentage of pollutants gradually increases with the contact time till equilibrium after which no biosorption takes place (Priyadarshanee & Das, 2020). The equilibrium time of biosorption is also very important in analyzing the kinetics of the adsorption process, whether the process is following pseudo first order, pseudo second order, or diffusion limited.

1.6 BENEFITS OF BIOSORPTION

Further drawback that biosorption must deal with is the fact that its potential for use on an industrial level has not yet been fully realized. However, the majority of biosorption applications are focused on lab tests. All of this research enables the current understanding of biosorption, which is sufficient to give a firm foundation for its expanded usage. This method isn't frequently employed in industry, though. Live biomass had a higher biosorption capability than non-living things biomass. A cheap, simple, economical, and effective way to get rid of contaminants is biosorption. The ability to be using waste products with no obvious utility is one of the major benefits of the biosorption process since the sorbent material can be highly diverse. Another of the major benefits of the biosorption process is the wide range of sorbent materials available, making it feasible to utilize waste products with no obvious utility and among the benefits of biosorption are.

Good selectivity but also recovering of particular toxic substances; numerous different toxic metals therapeutic interventions as well as mixed wastes; utilization inexpensive and plentiful biomaterials; treatment of a large quantity of wastewater due to rapid kinetics; comparatively low operating cost as well as inadequate capital investment; temperature, pH, and coexisting charged particles as a broad range of environment conditions; substantially lower volume of hazardous materials formed.

Biosorption has been defined as the property of certain biomolecules (or types of biomasses) to bind and concentrate selected ions or other molecules from aqueous solutions (Volesky, 2007). Biosorption is a quick occurrence of non-growing biomass/adsorbents passively storing metals. It provides benefits over traditional methods, among which are as follows: cheap cost; energy accuracy; limitation of chemical and/or biological sludge; absence of the need for extra nutrients; regeneration of biosorption; and potential for metal recovery. The rate of adsorption species is drawn to the adsorbent and bound there by various mechanisms as a result of the adsorbent's increased propensity for that species. Generally speaking, in comparison to traditional heavy metal removal techniques, the biosorption may offer the following benefits:

- Usage readily available, sustainable energy biomaterials that can be generated at a minimal price;
- Capability to recognize massive amounts of wastewater because of the rapid kinetics;
- High specificity in aspects of extraction and recovery of precise toxic substances;
- Capability to handle multiple heavy metals and mixed wastes;
- Fewer requirements for additional costly reactants that would otherwise be required;
- Most significantly, the metal removal capacity of biological biomass is a good or better than other conventional adsorbents (Abbas et al., 2014).

1.7 CONCLUSION

Years of biosorption study have shown how complicated the procedure is, how dependent it is on physicochemical and biological elements, and how unclear the processes are. The conventional use of biosorption as a limited, ecologically benign form of pollutant treatment has not been commercially effective, so this approach needs to be re-evaluated. The employment of physicochemical and biotic manipulations to enhance biosorption (capacity, selectivity, kinetics, and re-use) increases costs and may impact the environment. Biosorption researches have been tremendously encountered with more diverse fields like nanotechnology that incorporates nanoparticles with the biosorbents to have a novel framework of materials. Biosorption is becoming a competitive field in which raw, dead, live, and immobilized materials and organisms were exploited for the enhanced biosorption capacity. To conclude that the new material is a superior option, it is required to compare its properties to those of other potential materials that have previously been established as sorbents (commercial sorbents). For biosorbents to be fully accepted, more actions must be followed. The development of large-scale methods, increased commercialization, and, generally, its use in actual settings are some of the issues facing biosorption at the moment. Few biosorbents are currently being marketed for their usage, despite the fact that the benefits of this form of sorbent are obvious (de Freitas et al., 2019a,b). The insufficient effectiveness of biosorbents would necessitate their modification in order to increase their efficiency. If the biosorbents could at least match the efficiency of commercial ones, that would be preferable. The use of nanomaterials, among other alternatives like chemical or physical alterations, can increase the potency of biosorbents (Benis et al., 2020; Giese et al., 2020; Qin et al., 2020). Further drawback that biosorption must deal with is the fact that its potential for use on an industrial level has yet to be fully realized. However, the preponderance of biosorption applications is focused upon lab tests. All of this research enables the current understanding of biosorption, which is sufficient to give a firm foundation for its expanded usage. This method isn't frequently employed in industry, though.

For economically eliminating harmful metallic ions from compromised fluids, biosorption is an option. A novel class of biosorbent compounds is focused on biomass resources that are perennial or leftover from other processes and only require a minimal amount of processing before being used. The ability to regenerate biosorption materials as well as the elevated tiny effluent stream that enables practical metal recycling/recovery from it both support the operational economy. Design engineers are accustomed to the layout and machinery utilized in the biosorption because it is predicated toward well sorption concepts. Biosorbents can be made from a variety of source materials, including microbial cells, commercial sludge, food scraps, including agricultural residues (bacteria, fungi, and yeast). The possibility of using biodegradable and agricultural refuse as biosorbents is especially interesting since it has previously been shown in a number of semi-industrial scale studies that these materials are effective at attaching toxic substances. The use of biologically significant nanomaterials for the removal of contaminants of toxic substances is one of the latest technological innovations. The main technique for lowering running costs at room temperature is biosorption techniques. The majority of biosorption reactions are temperature dependent, meaning that raising the reaction temperature will cause the leading edge to thin out while also increasing the mass transfer coefficient and the number of effective collisions. The biosorption will reveal chemical sorption if its temperature goes up and mechanical sorption if something decreases. As biosorption continues to develop, possible improvements in both price and performance might be anticipated. To fully comprehend the mechanics of biosorption and what governs the specificity of biosorptive, fundamental study must be carried out.

The conventional use of biosorption as a low-cost, ecologically benign approach of pollutants treatment has not been commercially effective, and this should be reconsidered. The economic strength of biosorption as a technique would hinge on a deeper comprehension of this process, which will be guided by a practical justification of its industrialization but also application areas. Throughout this chapter, we attempt to describe the background of biosorption as well as its numerous characteristics, which typically show applications for strong heavy substances in industry and environment protection.

REFERENCES

Salman H. Abbas, Ibrahim M. Ismail, Tarek M. Mostafa, & Abbas H. Sulaymon, Biosorption of heavy metals: a review, *Journal of Chemical Science and Technology*. October 2014, 3(4), 74–102.

I. M. Abrams & J. W. McBain, A closed cell for electron microscopy, *Journal of Applied Physics*. 1944, 100(2595): 273–274.

M. S. Al-Masri, Y. Amin, B. Al-Akel & T. Al-Naama, Biosorption of cadmium, lead, and uranium by powder of poplar leaves and branches, *Applied Biochemistry and Biotechnology*. February 2010, 160(4), 976–87 [https://doi.org/10.1007/s12010-009-8568-1] [PubMed].

Asif Ali,Yi Wai Chiang, & Rafael M. Santos, X-ray diffraction techniques for mineral characterization: a review for engineers of the fundamentals, applications, and research directions, *Minerals*. 2022, 12(2), 205 [https://doi.org/10.3390/min12020205].

Khaled Zoroufchi Benis, Ali Motalebi Damuchali, Kerry N. Mc Phedran, Jafar Soltan, Treatment of aqueous arsenic – a review of biosorbent preparation methods, *Journal of Environmental Management*. 1 November 2020, 273, 111126 [https://doi.org/10.1016/j.jenvman.2020.111126].

G. Blázquez, M. A. Martín-Lara, G. Tenorio & M. Calero, Batch biosorption of lead (II) from aqueous solutions by olive tree pruning waste: equilibrium, kinetics and thermodynamic study, *Chemical Engineering Journal*. 2011, 168, 170–177 [https://doi.org/10.1016/j.cej.2010.12.059].

Ana Gabriela Contreras-Cortés, Francisco Javier Almendariz-Tapia, Mario Onofre Cortez-Rocha, Armando Burgos-Hernández, Ema Carina Rosas-Burgos, Francisco Rodríguez-Félix, Agustín Gómez-Álvarez, Manuel Ángel Quevedo-López, & Maribel Plascencia-Jatomea, Biosorption of copper by immobilized biomass of *Aspergillus australensis*. Effect of metal on the viability, cellular components, polyhydroxyalkanoates production, and oxidative stress, *Environmental Science and Pollution Research*. 2020, 27, 28545–28560 [https://doi.org/10.1007/s11356-020-07747-y].

Nilanjana Das, Recovery of precious metals through biosorption-a review, *Hydrometallurgy*. 2010, 103, 180–189 [https://doi.org/10.1016/j.hydromet.2010.03.016].

Thomas A. Davis, Bohumil Volesky, & Alfonso Mucci, A review of the biochemistry of heavy metal biosorption by brown algae, *Water Research*. 2003, 37, 4311–4330 [https://doi.org/10.1016/S0043-1354(03)00293-8].

Geovani Rocha de Freitas, Meuris Gurgel Carlos da Silva, & Melissa Gurgel Adeodato Vieira, Biosorbents and patents, *Environmental Science and Pollution Research*. 2019a, 26, 19097–19118 [https://doi.org/10.1007/s11356-019-05330-8].

Geovani Rocha de Freitas, Melissa Gurgel Adeodato Vieira, & Meuris Gurgel Carlos da Silva, Fixed bed biosorption of silver and investigation of functional groups on acidified biosorbent from algae biomass, *Environmental Science and Pollution Research International*. 2019b, 26, 36354–36366 [https://doi.org/10.1007/s11356-019-06731-5].

A. M. Elgarahy, K. Z. Elwakeel, S. H. Mohammad & G. A. Elshoubaky, A critical review of biosorption of dyes, heavy metals and metalloids from wastewater as an efficient and green process, *Cleaner Engineering and Technology*. 2021, 4, 100209 [https://doi.org/10.1016/j.clet.2021.100209].

Fomina, Marina, & Geoffrey Michael Gadd, Biosorption: current perspectives on concept, definition and application, *Bioresource Technology*. 2014, 160, 3–14 [http://dx.doi.org/10.1016/j.biortech.2013.12.102].

Geoffrey Michael Gadd, Biosorption: critical review of scientific rationale, environmental importance and significance for pollution treatment, *Journal of Chemical Technology & Biotechnology*. 2009, 84, 13–28 [https://doi.org/10.1002/jctb.1999].

Ellen C. Giese, Debora D. V. Silva, Ana F. M. Costa, Sâmilla G. C. Almeida, & Kelly J. Dussán, Immobilized microbial nanoparticles for biosorption, *Critical Reviews in Biotechnology*. 2020, 40, 653–666 [https://doi.org/10.1080/07388551.2020.1751583].

Kingsley O. Iwuozor, Ebuka Chizitere Emenike, Joshua O. Ighalo, Steve Eshiemogie, Patrick E. Omuku, & Adewale George Adeniyi, Valorization of sugar industry's by-products: a perspective, *Sugar Tech*. 2022, 1–27 [https://doi.org/10.1007/s12355-022-01143-1].

Jalil Jaafari, & Kamyar Yaghmaeian, Optimization of heavy metal biosorption onto freshwater algae (*Chlorella coloniales*) using response surface methodology (RSM), *Chemosphere*. February 2019, 217, 447–455 [https://doi.org/10.1016/j.chemosphere.2018.10.205].

Z. B. Jiao, J. H. Luan, M. K. Miller, Y. W. Chung & C. T. Liu, Co-precipitation of nanoscale particles in steels with ultra-high strength for a new era, *Materials Today*. April 2017, 20(3), 142–154 [https://doi.org/10.1016/j.mattod.2016.07.002].

Chang-Sheng Jin, Ren-Jian Deng, Bo-Xhi Ren, Bao-Lin Hou, & Andrew S. Hursthouse, Enhanced biosorption of Sb (III) onto living rhodotorula mucilaginosa strain DJHN 070401: optimization and mechanism, *Current Microbiology*. 2020, 77, 2071–2083 [https://doi.org/10.1007/s00284-020-02025-z].

Adnan Khan, Sumeet Malik, Nisar Ali, Yong Yang, Mohammed Salim Akhter, & Muhamma Bila, Chapter 2 - Introduction to nano-biosorbents, *Micro and Nano Technologies*. 2022, 29–43 [https://doi.org/10.1016/B978-0-323-90912-9.00002-2].

Hao Jie Kong, & Chain Tsuan Liu, A review on nano-scale precipitation in steels, *Technologies*. 2018, 6(1), 36 [https://doi.org/10.3390/technologies6010036].

Kratochvil, David & Bohumil Volesky, Biosorption of Cu from ferruginous wastewater by algal biomass, *Water Research*. 1998, 32, 2760–2768 [https://doi.org/10.1016/S0043-1354(98)00015-3].

Ai Phing Lim, & Ahmad Zaharin Aris, A review on economically adsorbents on heavy metals removal in water and wastewater, *Reviews in Environmental Science and Bio-Technology*. 2014, 13, 163–181 [https://doi.org/10.1007/s11157-013-9330-2].

P. V. N. Malleswari, S. Swetha, G. B. Jegadeesan & S. Rangabhashiyam, Biosorption study of amaranth dye removal using *Terminalia chebula* shell, *Peltophorum pterocarpum* leaf and *Psidium guajava* bark, *International Journal of Phytoremediation*. 2022, 24(10), 1081–1099 [https://doi.org/10.1080/15226514.2021.2002261].

Reda M. Moghazya, A. Labena, & Sh. Husienc, Eco-friendly complementary biosorption process of methylene blue using micro-sized dried biosorbents of two macro-algal species (Ulva fasciata and Sargassum dentifolium): full factorial design, equilibrium, and kinetic studies, *International Journal of Biological Macromolecules*. 2019, 134, 330–343 [https://doi.org/10.1016/j.ijbiomac.2019.04.207].

Stefanos Mourdikoudis, Roger M. Pallares, & Nguyen T. K. Thanh, Characterization techniques for nanoparticles: comparison and complementarity upon studying nanoparticle properties, *Nanoscale*. 2018, 10, 12871–12934 [https://doi.org/10.1039/C8NR02278J].

Mahmoud Nasrollahzadeh, Monireh Atarod, Mohaddeseh Sajjadi, S. Mohammad Sajadi, & Zahra Issaabadi, Plant-mediated green synthesis of nanostructures: mechanisms, characterization, and applications, *Interface Science and Technology*. 2019, 28, 199–322 [https://doi.org/10.1016/B978-0-12-813586-0.00006-7].

Jian-Feng Nie, Precipitation and hardening in magnesium alloys, *Metallurgical and Materials Transactions A*. 2012, 43, 3891–3939 [https://doi.org/10.1007/s11661-012-1217-2].

Donghee Park, Yeoung-Sang Yun, & Jong Moon Park, The past, present, and future trends of biosorption, *Biotechnology and Bioprocess Engineering*. 2010, 151, 86–102 [https://doi.org/10.1007/s12257-009-0199-4].

Diana B. Peckys, Gabriel M. Veith, David C. Joy, & Niels de Jonge, Nanoscale imaging of whole cells using a liquid enclosure and a scanning transmission electron microscope, *PLoS One*. 2009. [https://doi.org/10.1371/journal.pone.0008214].

Priyadarshanee, Monika & Surajit Das, Biosorption and removal of toxic heavy metals by metal tolerating bacteria for bioremediation of metal contamination: a comprehensive review, *Journal of Environmental Chemical Engineering*. 2020 [https://doi.org/10.1016/j.jece.2020.104686].

Huaqing Qin, Tianjue Hua, Yunbo Zhai, Ningqin Lu, & Jamila Aliyeva, The improved methods of heavy metals removal by biosorbents: a review, *Environmental Pollution*. March 2020, 258, 113777 [https://doi.org/10.1016/j.envpol.2019.113777].

S. Rangabhashiyam, Pollyanna V. dos Santos Lins, Leonardo M. T. de Magalhães Oliveira, Pamela Sepulveda, Joshua O.Ighalo, Anushka Upamali Rajapaksha, & Lucas Meili, Sewage sludge-derived biochar for the adsorptive removal of wastewater pollutants: A critical review, *Environmental Pollution*. January 2022, 293, 118581.

Desireddy Harikishore Kumar Reddy, Seung-Mok Lee, & Kalluru Seshaiah, Biosorption of toxic heavy metal ions from water environment using honeycomb biomass—an industrial waste material, *Water, Air, and Soil Pollution*. 2012, 223, 5967–5982 [https://doi.org/10.1007/s11270-012-1332-0].

Rogowska, Agnieszka, et al., A study of zearalenone biosorption and metabolisation by prokaryotic and eukaryotic cells, *Toxicon*. November 2019, 169, 81–90 [https://doi.org/10.1016/j.toxicon.2019.09.008].

Ross, Frances M., Opportunities and challenges in liquid cell electron microscopy, *Science*. 2015, 18, 350(6267).

Rajeev Roychand, Savankumar Patel, Pobitra Halder, Sazal Kundu, James Hampton, David Bergmann, Aravind Surapaneni, Kalpit Shah, & Biplob Kumar Pramanika, Recycling biosolids as cement composites in raw, pyrolyzed and ashed forms: a waste utilization pproach to support circular economy, *Journal of Building Engineering*. June 2021, 102199 [https://doi.org/10.1016/j.jobe.2021.102199].

Sergio Santaeufemia, & Julio Abalde Enrique Torres, Eco-friendly rapid removal of triclosan from seawater using biomass of a microalgal species: kinetic and equilibrium studies, *Journal of Hazardous Materials*. 5 May 2019, 369, 674–683 [https://doi.org/10.1016/j.jhazmat.2019.02.083].

J. Sedlakova-Kadukova, A. Kopcakova, L. Gresakova, A. Godany & P. Pristas, Bioaccumulation and biosorption of zinc by a novel Streptomyces K11 strain isolated from highly alkaline aluminium brown mud disposal site, *Ecotoxicology and Environmental Safety*. 15 January 2019, 167, 204–211 [https://doi.org/10.1016/j.ecoenv.2018.09.123].

Subramaniyasharma Sivaraman, Nithin Michael Anbuselvan, Ponnusami Venkatachalam, Saravanan Ramiah Shanmugam, & Rangabhashiyam Selvasembian, Waste tire particles as efficient materials towards hexavalent chromium removal: characterisation, adsorption behaviour, equilibrium, and kinetic modelling, *Chemosphere*. May 2022, 295, 133797.

Lothar Spieß, Gerd Teichert, Robert Schwarzer, Herfried Behnken, & Christoph Genzel, *Moderne R€ontgenbeugung*, second ed. 2009. Teubner Verlag, Wiesbaden.

J. A. Swift & A. C. Brown, An environmental cell for the examination of wet biological specimens at atmospheric pressure by transmission scanning electron microscopy, *Journal of Physics*. 1970, 3(11), 924–926.

Enrique Torres, Biosorption: a review of the latest advances, *Processes*. 2020, 8(12), 1584 [https://doi.org/10.3390/pr8121584].

Bohumil Volesky, Detoxification of metal-bearing effluents: biosorption for the next century, *Hydrometallurgy*. 2001, 59, 203–216.

Bohumil Volesky, Biosorption and me, *Water Research*. October 2007, 41(18), 4017–4029 [https://doi.org/10.1016/j.watres.2007.05.062].

Bohumil Volesky & Z. R. Holan, Biosorption of heavy metals, *Biotechnology Progress*, 1995-Wiley Online Library [https://doi.org/10.1021/bp00033a001].

Nana Wang, Yuyin Qiua, Tangfu Xiaoa, Jianqiao Wanga, Yuxiao Chena, Xingjian Xu, Zhichao Kang, Lili Fan, & Hong Wen Yu, Comparative studies on Pb (II) biosorption with three spongy microbe-based biosorbents: high performance, selectivity and application, *Journal of Hazardous Materials*. 2019, 373, 39–49 [https://doi.org/10.1016/j.jhazmat.2019.03.056].

Pratama Jujur Wibawa, Muhammad Nu, Mukhamad Asy'ari, & Hadi Nur, SEM, XRD and FTIR analyses of both ultrasonic and heat generated activated carbon black microstructures, *Heliyon*. March 2020, 6(3), e0354 [https://doi.org/10.1016/j.heliyon.2020.e03546].

Anna Witek-Krowiak, & D. Harikishore Kumar Reddy, Removal of micro elemental Cr (III) and Cu (II) by using soybean meal waste – unusual isotherms and insights of binding mechanism, *Bioresource Technology*. January 2013, 127, 350–357 [https://doi.org/10.1016/j.biortech.2012.09.072].

Anna Witek-Krowiak, Roman G. Szafran, & Szymon Modelski, Biosorption of heavy metals from aqueous solutions onto peanut shell as a low-cost bio-sorbent, *Desalination*. 2011, 265, 126–134 [https://doi.org/10.1016/j.desal.2010.07.042].

Zhiping Xiong, Ilana Timokhina, & Elena Pereloma, Clustering, nano-scale precipitation and strengthening of steels, *Progress in Materials Science*. May 2021, 100764 [https://doi.org/10.1016/j.pmatsci.2020.100764].

Hande Günan Yücel, Zümriye Aksu, Gülşah Büşra Yalçınkaya, Sevgi Ertuğrul Karatay, & Gönül Dönmez, A comparative investigation of lithium(I) biosorption properties of aspergillus versicolor and *Kluyveromyces marxianus*, *Water Science and Technology*. 2020, 81, 499–507 [https://doi.org/10.2166/wst.2020.126].

2 Bacterial and Fungal Biosorbents

Jéssica Mesquita do Nascimento

ACRONYMS

EPS: Extracellular Polymeric Substances
FTIR: Fourier Transform Infrared Spectroscopy
RSM: Response Surface Methodology
POPs: Persistent Organic Pollutants
TET: Tetracycline
CIP: Ciprofloxacin
SDZ: Sulfadiazine
SMX: Sulfamethoxazole
AB62: Acid Blue 62
AB25: Acid Blue 25
AB40: Acid Blue 40
RB19: Reactive Blue 19
MB: Methylene Blue
RR198: Reactive Red 198
PCBs: Polychlorinated Biphenyls
DDT: Dichlorodiphenyltrichloroethane

2.1 INTRODUCTION

Microorganisms are beings that existed on planet Earth billions of years before the emergence of plants and animals. In early classification systems, microorganisms were included as belonging to the Kingdoms Plantae and Animalia. This classification system was introduced in 1735 by Carolus Linnaeus who divided organisms into only two kingdoms [1, 2]. Subsequently, the classification developed by Carl Woese (1978) based on cellular organization grouped organisms into three domains (Bacteria, Archaea, and Eukarya) [2]. Microorganisms have representatives in all three domains, as there are microorganisms with eukaryotic cells (fungi), prokaryotic cells (bacteria), and without peptidoglycan in the cell wall (archaea) [1, 2].

Prokaryotic organisms, from the Greek (pre-nucleus), and eukaryotes, from the Greek (true nucleus), are distinguished based on their cellular structures [3, 4]. The main difference between prokaryotic and eukaryotic cells is that prokaryotic cells do not have a nucleus and other membrane-bound structures called organelles – organelles are specialized cellular structures that have specific, therefore simpler functions, whereas eukaryotic cells have a nucleus and have cell

DOI: 10.1201/9781003366058-2

organelles [3]. Another characteristic of prokaryotic cells is that they have varied morphologies such as bacilli, cocci, and spirilla [1].

Prokaryotic and eukaryote organisms share the same constituents, such as nucleic acids, proteins, lipids, and carbohydrates. They also carry out the same types of chemical reactions to metabolize food, form proteins, and store energy. It is especially the structure of cell walls and membranes and the absence of organelles that differentiate prokaryotes from eukaryotes [2]. Several microorganisms play roles in industrial processes – field of activity of industrial microbiology and environmental processes – field of action of environmental microbiology. Numerous microbial strains were evaluated in bioremediation processes and proved to have the ability to remove organic and inorganic pollutants [1, 2, 5, 6].

Among these strains, bacteria and fungi deserve to be highlighted, since they have good removal capacities and efficiencies, fast growth rate (bacteria), production of secondary metabolites that facilitate the decontamination process (bacteria), and presence of chitin that acts in capturing the species to be removed (fungi), easy handling and safe – non-pathogenic (most of the fungi studied are not pathogenic) in addition to being by-products of various industrial processes, with their biomass being reused in decontamination processes (fungi) [7–11].

2.1.1 Microorganisms and Their Role in the Balance of Life

Microscopic organisms with animal and plant characteristics were discovered at the end of the 17th century. This classification system was introduced in 1735 by Carolus Linnaeus who divided organisms into just two kingdoms – Plantae and Animalia [2]. Over the decades, there was a need for a new classification system that would remove the barely distinguishable classification that was only in the animal and plant kingdoms [1, 2].

In 1978, Carl Woese developed a classification system based on cellular organization and grouped organisms into three domains (Bacteria, Archaea, and Eukarya), this system is still adopted today [2]. In this classification, animals, plants, and fungi are kingdoms of the Eukarya domain; pathogenic prokaryotes, and many non-pathogenic prokaryotes, as well as photoautotrophic prokaryotes, make up the Bacteria domain; and the last domain, the Archaea which is composed of prokaryotes that do not have peptidoglycan in their cell walls [2]. The group of microorganisms includes bacteria, fungi (yeasts and molds), protozoa, microscopic algae, and viruses (acellular organisms that many authors consider the limit between the living and non-living), consequently, microorganisms are present in all three domains [2] (Figure 2.1).

Microorganisms are tiny living beings that, in general, are individually too small to be visualized with the naked eye, requiring the use of a microscope [2]. Overall, we associate microorganisms with infections, food spoilage, or with pandemic diseases such as SARS-CoV-2 [12].

However, these microscopic beings play an important role in maintaining the balance of life [2]. Certain microorganisms play an essential role in photosynthesis: the microbiota present in the soil helps in the degradation of waste and the incorporation of nitrogen gas from the air into organic compounds; marine and freshwater microorganisms are the basis of the food chain in these ecosystems; in addition to

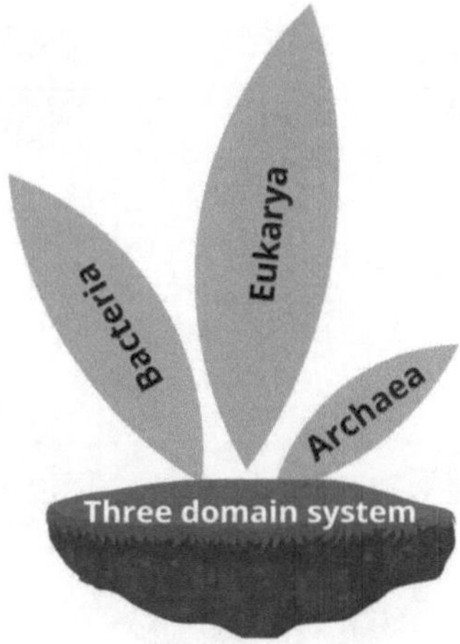

FIGURE 2.1 Three-domain classification system

the microflora present in our intestine, responsible for the digestion and synthesis of some vitamins required by our body [2].

Microorganisms are prominent in commercial applications, such as the synthesis of chemicals (vitamins, organic acids, enzymes, alcohols, and drugs) and fermentation processes such as the production of beer, bread, yogurt, cheeses, wines, and others [2]. The main microorganisms that stand out in these processes are fungi, yeasts, and bacteria such as the genus *Saccharomyces* sp., *Streptomyces sp.*, and *Lactobacillus* sp. [1].

In addition to industrial microbiology (a field of microbiology that uses microorganisms on a large scale for the synthesis of commercial products or for carrying out important chemical transformations) [1], environmental microbiology is another field of action for microorganisms. Environmental bioremediation processes are widely known and studied, where microorganisms act in the decontamination of organic and inorganic contaminants [1, 2, 13, 14] (Figure 2.2).

Numerous microbial strains are studied and evaluated in bioremediation processes (biosorption and bioaccumulation) of pollutants, whether organic or inorganic [13, 15–18]. Biosorption is defined as the ability of certain biomasses to bind and concentrate metal ions or other molecules from aqueous solutions as organic compounds on their surface (surface process). Whereas bioaccumulation is based on active metabolic transport, that is capture and trapping of the toxic species inside the cell (intracellular process). Biosorption can occur through dead biomass (or by some molecules and/or their active groups) as it is a metabolically passive process and based mainly on the "affinity" between the biosorbent and the adsorbate (species to be adsorbed) [7, 8, 19].

In the mid-1980s, the first reports emerged of the ability of some microorganisms to accumulate metallic elements [20]. Other studies have indicated the ability to bind via metabolism independent of microbial biomass, whether living or dead, to potentially toxic metals. As a result, biosorption studies using microorganisms have grown over the years [20–24].

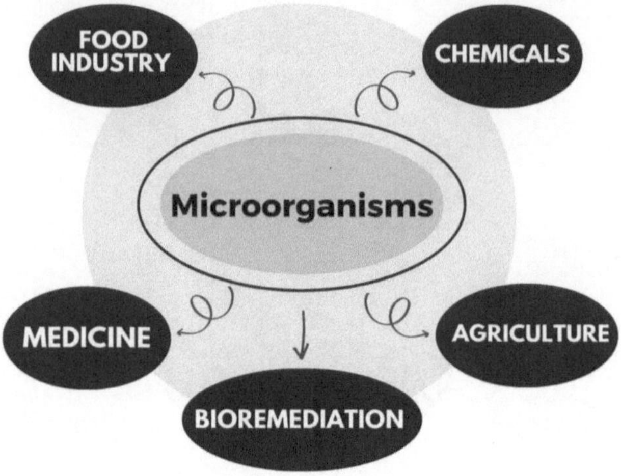

FIGURE 2.2 Fields of action of microorganisms

Microorganisms are reported as potential biosorbents of toxic metals and organic compounds such as dyes and pesticides [25, 26]. Since the cell wall composition of these organisms is full of functional groups that will become active capture sites for the species to be removed [16, 27–30].

However, the composition of the microbial wall differs between different microbial groups, therefore the mechanism and efficiency of biosorption using microorganisms will depend on the active binding sites present in the cellular components [31]. Therefore, in the literature, there are reports of biosorption using fungi, microalgae, and bacteria [31, 32].

2.2 BACTERIA AS BIOSORBENTS

Prokaryotes are grouped into two domains: Bacteria and Archaea. The Bacteria domain is composed of a diversity of prokaryotic strains. In this group are pathogenic as well as non-pathogenic bacteria [1]. Bacteria are relatively simple and unicellular organisms [2].

Although bacteria and archaea have similarities, their chemical composition is different. Bacterial strains are distinguished by many factors, such as morphology, chemical composition, nutritional requirements, biochemical activities, and energy source. The size of the bacteria ranges from 0.2 to 2 μm in diameter and from 2 to 8 μm in length. In terms of morphology, they can be shaped like a sphere (cocos, which means fruiting), rod (bacilli, which means rod), and spiral [1, 2]. It is estimated that 99% of bacteria in nature exist in the form of biofilms [2].

Many prokaryotes secrete a substance called a glycocalyx on their cell surface. The bacterial glycocalyx is a viscous polymer of gelatinous consistency that is situated external to the cell wall and is generally composed (composition varies between bacterial species) of polysaccharide, polypeptide, or both. The glycocalyx is a very important constituent of biofilms [2].

Biofilms are defined as a natural form of cellular immobilization formed by an agglomeration of microbial cells attached to solid supports resulting from microbial attachment arising from a matrix of extracellular polymeric substances (EPS) [33, 34]. One of the main functions of EPS is to protect microorganisms against the toxic effects of metallic species; this can occur through several mechanisms such as immobilization (capture of the toxic species) and, therefore, preventing its entry into the intracellular environment (bioaccumulation). The composition of EPS is nucleic acids, proteins, lipids, and complex carbohydrates, which play an important role in the biosorptive process [35, 36].

The peptidoglycan component present in bacterial walls is a polymer containing peptide bond where the glycan units are cross-linked with meso-diaminopimelic acid, D-alanine, and D-glutamic acid. The cell wall differential is the cross-linking between the peptide chains [31]. In Gram-negative bacteria, the cell wall contains an average of 10% peptidoglycan, in addition to lipoproteins, phospholipids, glycoproteins, lipopolysaccharides, and enzymes. These components are described as active in the capture of metals [31].

In general, Gram-positive bacteria have a cell wall composed of many layers of peptidoglycan, forming a rigid and thick structure, whereas Gram-negative cell walls have only a thin layer of peptidoglycan [2].

Bacteria are considered the most abundant microorganisms on planet Earth and have the ability to grow in a wide range of environmental conditions. Due to its small size, fast growth rate, high surface-to-volume ratio, ease of cultivation, and excellent performance in the biosorption process, it is one of the most evaluated biosorbents in this process [35, 37]. In the literature, several authors have studied the performance of bacteria as biosorbents of metallic species (Table 2.1) and organic contaminants (Table 2.2) [5, 38–40].

2.2.1 Potentially Toxic Metals

Potentially toxic metals are persistent contaminants that accumulate along the food chain, in addition to contaminating soil, subsoil, and groundwater [26, 41]. With the advance of industrial and population growth, the number of solid waste and industrial effluents has increased greatly over the years, resulting in greater environmental contamination [42–44].

Among the numerous decontamination technologies, biosorption appears as a cheap and environmentally friendly technology [45, 46]. Microorganisms are widely studied as biosorbents of metals; in the literature, several authors have investigated the potential of bacteria as biosorbents of potentially toxic metals (Table 2.1) [38, 46].

Most studies have evaluated bacteria that have been isolated from contaminated environments, thus, strains that are isolated and identified have resistance to potentially toxic metals [5, 30, 47, 48].

The bacterium *Parapedobacter* sp. ISTM3 recognized as an EPS-producing strain was isolated from a cave and its EPS was characterized and purified to be used in Cr^{6+} biosorption studies. The EPS biosorption profile produced by *Parapedobacter* sp. in mixed solutions with metallic concentration (Zn^{+2}, Cu^{+2}, Pb^{+2}, Cr^{6+}, Fe^{2+}, and Cd^{2+}) of 20 mg L^{-1} revealed that EPS showed the best efficiency and biosorption capacity for Cr^{6+} compared to the other potentially toxic metals studied. Biosorption for Cr^{6+} was favorable under acidic conditions (pH 5.0) with a biosorption capacity of 19.03 mg g^{-1} and a removal rate of 95.10% [49].

Lead-contaminated groundwater samples were used as a source to isolate bacteria with distinct morphological characteristics that were identified as *Bacillus toyonensis* SCE1, *Bacillus anthracis* SCE2, *Acinetobacter baumannii* SCE3, *Bacillus toyonensis* SCE4, and *Bacillus toyonensis* SCE5 which were tested for their biosorption capabilities [50].

The results indicated that the biosorption efficiency observed for the different bacterial strains followed the order of *Bacillus toyonensis* SCE1 > *Acinetobacter baumannii* SCE3 > *Bacillus toyonensis* SCE4 > *Bacillus toyonensis* SCE5 > *Bacillus anthracis* SCE2, and the percentage of biosorption was 84.6%, 76.6%, 65.3%, 62.6%, and 46%, respectively [50].

At the same time, the maximum lead biosorption capacity is 15 mg g^{-1} for the species *Bacillus toyonensis* SCE1, followed by 12 mg g^{-1} for the species *Acinetobacter baumannii* SCE3. For all systems studied, the pseudo-second-order kinetic model provided the best correlation of the experimental data, thus indicating that the metal–biomass interaction is via chemisorption [50].

The evaluation of 21 bacterial strains *Bifidobacterium* (n = 11), *Lactobacillus* (n = 4), *Escherichia* (n = 2), *Bacteroides* (n = 1), *Clostridium* (n = 1), *Enterococcus* (n = 1), and *Providencia alcalifaciens* (n = 1) isolated from the human intestine as biosorbents and As^{5+} metabolizers showed that seven strains showed high As^{5+} biosorption, ranging from 20.1 to 29.8%, which may be related to functional groups on bacterial surfaces, such as groups hydroxyl, amino, and carboxyl. Furthermore, six of these seven strains were versatile as they also had roles in reducing As^{5+} to As^{3+}, which is mainly regulated by the arsC gene [51].

The Gram-negative strain *Delftia lacustris* MS3 was isolated from a lead and zinc mine for Pb^{2+} biosorption studies. The results showed that the studied process obtained a better fit to the pseudo-second-order model, which indicates that the removal rate control step is a function of the chemical interaction between functional groups present on the surface of the biomass and the Pb^{2+} ions [48].

The genus of Gram-positive bacteria *Bacillus sp.* is known to have the ability to form dormant cells (endospores) at times of nutritional need (nutritional depletion). This defense mechanism causes cells to become resistant to extreme temperatures, lack of water, and exposure to toxic chemicals [2]. Many authors have isolated, identified, and evaluated strains of *Bacillus sp.* as biosorbents of potentially toxic metals [52].

The biosorptive profile of the *Bacillus subtilis* KC6 strain that was isolated from a pyrite mine was studied for Cd^{2+} removal. The assays showed that the KC6 bacterium showed a great capacity for resistance and removal of Cd^{2+} in solution. The highest biosorption capacity of Cd^{2+} was found at concentrations of 10 and 40 mg L^{-1}, being about 86.33% and 65.33%, respectively [52].

Bacillus amyloliquefaciens was isolated from marine soils and showed high resistance to Chromium Cr^{6+} ions. The biosorptive process was fast, during the 60 minutes of contact time, the bacteria removed 79.90% Cr^{6+}. According to the Langmuir isotherm model, the maximum adsorption capacity was 48.44 mg g^{-1} [53].

Bacillus pumilus SWU7-1, a Gram-positive bacterium, was isolated from soil not contaminated by Sr^{2+} and evaluated for Sr^{2+} biosorption assays. Tests with lower initial concentrations showed that living biomass can resist and capture Sr^{2+} ions. The best biosorption efficiencies (> 90%) were found at an initial metallic concentration of 54–130 mg L^{-1}, whereas the best biosorptive capacities were found at higher metallic concentrations, ($q = 275$ mg g^{-1}) at 700 mg L^{-1} of Sr^{2+} concentration. The biosorption process was better adjusted to the Langmuir model than Freundlich, showing a maximum biosorption capacity of 299.4 mg g^{-1} [37].

Lactic acid bacteria (LAB) present a morphology of cocci and Gram-positive bacilli that synthesize lactic acid as the main or only product via the fermentation process. LAB are anaerobic, but many are not sensitive to O_2 and can therefore grow in its presence [1]. This genus of bacteria is also widely studied in the biosorption of metals [54].

The study of LAB – *Weissella viridescens* ZY-6 – for removal of Cd^{2+} ions in an aqueous solution revealed that the binding capacity of Cd^{2+} to bacterial biomass was influenced by Cd^{2+} concentration, biomass dosage, incubation time, temperature, and pH. In binary metallic systems, the rate of Cd^{2+} removal by ZY-6 decreased in the presence of other metallic cations such as Mg^{2+}, Mn^{2+}, and Fe^{2+}. Binding experiments using different cellular components showed that Cd^{2+} was mainly biosorbed on bacterial cell walls, and only a small amount accumulated inside the cells [54].

Twelve strains of LAB were evaluated as biosorbents of Cd^{2+} and Pb^{2+} ions. A LAB strain was able to efficiently biosorb the studied metals in apple juice samples. The selected strain of LAB was functionalized by Fe_3O_4 magnetic nanoparticles (MNPs) and modified by ethylenediaminetetraacetic acid dianhydride (EDTAD) to generate the adsorbent EDTAD–MNPs–LAB (EMB) [55].

The maximum adsorption capacity of the EMB adsorbent for Cd^{2+} and Pb^{2+} in apple juice was 0.57 mg g^{-1} and 0.17 mg g^{-1}, respectively. The EMB adsorbent can be quickly separated from the sample solution with the help of the magnetic field, in addition to the fact that the adsorption process had no significant impact on the quality of the fruit juice [55].

The biosorption of Ni^{2+} and Zn^{2+} in a bubble column reactor showed that the removal of zinc was 96% and 54% for nickel using isolated bacteria as biosorbent. A comparison between an unaerated and aerated column indicated a higher percentage of removal with the same contact time. The contact time study also confirmed that the longer the process takes place, the removal efficiency increases, while the ability to biosorb metal ions decreases. Maximum removals were reached in the first seven hours of mixing in the reactor. Equilibrium time was visualized at 27 hours for Zn^{2+} and 96 hours for Ni^{2+} [56].

The inactivated and lyophilized *Pseudomonas putida* strain showed to be able to biosorb Cd^{2+}, Cu^{2+}, Pb^{2+}, and Zn^{2+} ions in addition to having considerably high capacities, with a removal efficiency of 80% and fast kinetics – the metallic species were removed in less than five minutes contact time [57].

TABLE 2.1
Biosorption of potentially toxic metals by bacteria

Bacterium	Ion	Metal (mg L^{-1})	q (mg g^{-1})	R (%)	Reference
***Ochrobactrum* sp. GDOS**	Cd^{2+}	200	34.36	–	[58]
***Pseudomonas azotoformans* JAW1**	Cd^{2+}	25	–	44.67	[47]
	Cu^{2+}			63.32	
	Pb^{2+}			78.23	
***Klebsiella* sp. 3S1**	Ag^{+}	1,000	99.24	–	[59]
***Ralstonia solanacearum* KTSMBNL 13**	Pb^{2+}	100	–	90	[60]
***Tepidimonas fonticaldi* AT-A2**	Au^{3+}	15	1.45	71	[61]
***Bacillus pumilus* sp. AS1**	Pb^{2+}	50	0.6	–	[62]
Arthrospira (spirulina) platensis	^{137}Cs			±60	[63]
	^{233}U			03	
	^{241}Am			77	
	^{237}Np			±10	
	^{239}Pu			67	
	^{90}Sr			62	
***Pseudomonas* sp. live**	Cd^{2+}	–	92.59	89.1	[64]
***Pseudomonas* sp. dead**			63.29	98.5	
Bacillus subtilis KC6	Cd^{2+}	40	–	65.33	[52]
***Delftia lacustris* MS3**	Pb^{2+}	50		59.41	[48]
***Bacillus pumilus* SWU7-1**	Sr^{2+}	70	±50	94.69	[37]
***Pseudomonas aeruginosa* ASU 6A**	Zn^{2+}	50	±45	–	[65]
***Bacillus cereus* AUMC B52**			±40		
Lactobacillus acidophilus	Pb^{2+}	0.035	–	65.61	[66]
	Cd^{2+}			71.95	
Vibrio alginolyticus PBR1	Cd^{2+}	50	–	59.78	[67]
	Pb^{2+}			82.20	
Parapedobacter sp. ISTM3	Cr^{6+}	20	19.032	95.10	[49]
Bifidobacterium breve	As^{5+}			20.1– 29.8	[51]
Bacteroides vulgatus	As^{3+}				
Clostridium butyricum					
Enterococcus avium					
Escherichia coli1					
Escherichia coli2					
Lactobacillus rhamnosus					
Providencia alcalifaciens					
Bacillus amyloliquefaciens	Cr^{6+}	500		79.9	[53]
Weissella viridescens ZY-6	Cd^{2+}	10		±100	[54]
Bacillus toyonensis SCE1	Pb^{2+}	15	12.7	84.6	[50]
Bacillus anthracis SCE2			6.9	46	
Acinetobacter baumannii SCE3			11.5	76.6	
Bacillus toyonensis SCE4			9.8	65.3	
Bacillus toyonensis SCE5			9.4	62.6	

(*Continued*)

TABLE 2.1 (*Continued*)
Biosorption of potentially toxic metals by bacteria

Bacterium	Ion	Metal (mg L^{-1})	q (mg g^{-1})	R (%)	Reference
Lactobacillus rhamnosus	Cd^{2+}	10	0.57	20.67	[55]
functionalized by MNPs of Fe_3O_4	Pb^{2+}		0.17	62	
Bacillus sp.	Ni^{2+}			54	[56]
Pseudomonas sp.	Zn^{2+}			96	
Klebsiella sp.					
Escherichia sp.					
Bacillus xiamenensis viva	Pb^{2+}	500	216.75	43.35	[68]
Bacillus xiamenensis morta			207.4	41.48	
Bacillus badius AK	Cd^{2+}	100	131.58		[69]
Pseudomonas putida	Cu^{2+}	0.1	6.6		[57]
	Cd^{2+}		8.0	80	
	Zn^{2+}		6.9	80	
	Pb^{2+}		56.2	80	

Where metal represents the metal concentration studied, q the biosorption capacity, and R the removal efficiency.

2.2.2 Organic Pollutants

Persistent organic pollutants (POPs) have this name because they are resistant to degradation by chemical, physical, and biological means. POPs are used in various applications such as flame retardants, fire suppressants, heat transfer agents, surfactants, cosmetics, and pesticides mainly due to their lower reactivity and stability [70].

POPs cause numerous undesirable effects on human health and the environment. These pollutants can be distributed in soil, air, and water [71]. Like potentially toxic metals, these contaminants are also persistent and can bioaccumulate living organisms. The main ones (POPs) are polychlorinated biphenyls, dichlorodiphenyltrichloroethane, and dioxins [72]. Dyes and drugs are also another class of organic contaminants that impair the quality of life [73, 74].

Bioremediation in general involves three main methods, which are biosorption, bioaccumulation, and biodegradation. The processes of bioaccumulation and biodegradation occur via active transport, that is, the microorganism needs to be alive to perform it, while biosorption occurs via passive transport, not requiring maintenance of the microorganism [72, 74].

The metabolically active bioremediation processes performed by microorganisms are biodegradation and bioaccumulation [75, 76]. Biodegradation refers to the process carried out by microorganisms to reduce complex chemical compounds into a simpler metabolism/immobilization pathway [35, 74]. Bioaccumulation is the process of removing chemical species through accumulation and metabolism. Biosorption is a process that may involve physical and/or chemical interactions between the surface of the microorganism and the species to be removed, this binding occurs without the need for energy expenditure. These processes can occur at the same time, so some authors study them concomitantly [13, 41, 77].

Bacteria can remove POPs through mechanisms such as biodegradation, biosorption, volatilization, hydrolysis, and mineralization. Several authors have studied bacterial strains as biosorbents of organic contaminants, there are reports of biosorption using live or dead biomass (Table 2.2) [72, 73].

Microorganisms can bioremediate organic pollutants, but some are reported to be more efficient, especially strains of *Bacillus*, *Corynebacterium*, *Staphylococcus*, *Streptococcus*, *Shigella*, *Alcaligenes*, *Acinetobacter*, *Escherichia*, *Klebsiella*, and *Enterobacter* [74].

The bacterium *Rhodopseudomonas palustris* 51ATA was evaluated as a biosorbent of the azo salt dye Fast Black K. The results showed that the strain has the potential to remove the dye studied at various concentrations (25–400 mg L^{-1}). Regarding temperature, the dye was better biosorbed on the bacterial surface at lower temperatures 25–35°C [25].

The isolation and characterization of bacterial strains capable of discoloring and/or degrading azo dyes that are applied in textile production (monoazo dye Reactive Orange 16 and diazo dye Reactive Green 19) were investigated from activated sludge systems used in the treatment of (textile) residual waters. After the pre-screening step of 125 isolated strains, a strain belonging to the genus *Acinetobacter* (ST16.16/164) and another belonging to *Klebsiella* (ST16.16/034) outperformed the other strains tested in the studied process. Both strains exhibited good decolorization ability (>80%) over a wide temperature range (20°C–40°C) and maintained good decolorization activity at temperatures as low as 10°C (especially the *Klebsiella* strain – ST16.16/034) [78].

The bacterial strain, *Penicillium simplicissimum* (isolate 10, KP713758), was investigated for biosorption and biodegradation activities for triphenylmethane (TPM) dyes. The bacteria showed good decolorization activities in relation to methyl violet (MV, 100 mg L^{-1}), crystal violet (CV, 100 mg L^{-1}), and cotton blue (CB, 50 mg L^{-1}), with 98%, 95%, and 82% removal rates at 13, 14, and 1 day(s). Malachite green (MG, 100 mg L^{-1}), the most recalcitrant dye, was partially discolored (54%) at 14 days [79].

The biosorption capacity of the bacterial strain *Acidithiobacillus thiooxidans* was investigated to remove sulfur blue dye 15 (SB15) from water samples. This bacterium is reported to have the ability to oxidize sulfur compounds to sulfuric acid, in addition to promoting cell adhesion to the surface of sulfide particles, and is therefore considered a potential biosorbent. The data had a better fit to the Langmuir model, according to the kinetics of the process, the best fit was for the pseudo-second-order model. At pH 8.3 and dye concentrations of up to 2,000 mg L^{-1}, the bacterium had a biosorption efficiency of 87.5% [80].

The potential of *Streptomyces bacillaris* as an efficient biological agent for the removal of TPM dyes was analyzed. The strain showed the ability to effectively decolorize MG, MV, CV, and CB. The tests revealed that the high decolorization activity for the dyes MG (94.7%), MV (91.8%), CV (86.6%), CB (68.4%) *S. bacillaris* is due to biosorption and biodegradation processes [81].

Pesticides are widely used in crops, mainly to reduce crop losses from pest proliferation. The bioremediation of organic pesticide residues that include organophosphates and organochlorines is carried out via techniques such as adsorption, which is currently one of the most adopted methodologies to remove these pollutants from the environment [82].

The isolation of bacterial species (*Pseudomonas stutzeri*) from soil contaminated by pesticides and their use as a mixed biosorbent composed of *P. stutzeri* and *Delonix regia* seeds modified with sulfuric acid for biosorption of pesticides present in the soil and aquatic environment showed that mixed biosorbent may be an effective alternative in removing the pesticide chlorpyrifos [82].

The evaluation of the effect of pH and temperature on the biosorption process of chlorpyrifos using the mixed biosorbent (*P. stutzeri* + *Delonix regia*) revealed that the increase in temperature decreased the removal rate of the investigated pesticide, the best working temperature was 30°C, about pH, the best removal rate was at neutral pH [82].

Biosorption of veterinary medicinal products (VMP) – tetracycline (TET), ciprofloxacin (CIP), sulfadiazine (SDZ), and sulfamethoxazole (SMX) – through a consortium of dead and dry biomass of microalgae – *Scenedemus almeriensis* bacteria was studied. The data showed that the relative removal of antibiotics was greater at low equilibrium concentrations [83].

After 96 hours of contact, antibiotic removal via biosorption using an antibiotic concentration of 1 mg L^{-1} was 75% TET, 43% CIP, and 12% SDZ; SMX did not show biosorption removal [83].

TABLE 2.2
Biosorption of organic pollutants by bacteria

Bacterium	Item analyzed	Ci (mg L^{-1})	R (%)	Reference
***Rhodopseudomonas palustris* 51ATA**	Fast black K dye	400	30.57	[25]
Penicillium simplicissimum	MV	100	98	[79]
	CV	100	95	
	CB	50	82	
	MG	100	54	
Acidithiobacillus thiooxidans	SB15	2,000	87.5	[80]
P. stutzeri and seeds of *Delonix regia*	Chlorpyrifos	25	95.29	[82]
The dry consortium of microalgae–bacteria *Scenedemus almeriensis*	TET	1	75	[83]
	CIP		43	
	SDZ		12	
	SMX		0	
Indigenous microbial consortium via MFCs	Acid Blue 25 (AB62)	50	4.92	[84]
	Acid blue 25 (AB25)		45.94	
	Acid blue 40 (AB40)		37.22	
	Reactive blue 19 (RB19)		1.78	

Where Ci represents the concentration studied and R the removal efficiency.

The use of an indigenous microbial consortium with acclimatization implementation to degrade four anthraquinone dyes was researched. In addition to biodiscoloration, the authors also studied the interactive effects of biosorption and biotoxicity via microbial fuel cells (MFCs). At the concentration of 50 mg L^{-1}, biodegradation predominated in contrast to biosorption at this concentration [84]. Although biosorption and bioaccumulation processes are not dominant mechanisms in antibiotic removal, they are indispensable prior processes for biodegradation to occur [85].

2.3 FUNGI AS BIOSORBENTS

Fungi are part of the Eukarya domain and therefore are organisms that have eukaryotic cells. Fungi can have a single cell (unicellular) and several cells (multicellular). Multicellular fungi, such as mushrooms, can be compared to plants; however, they cannot carry out the process of photosynthesis [86].

Fungi have cell walls composed mainly of chitin, a substance that makes up 80%–90% of the cell wall. Chitin is composed of polysaccharides that contain nitrogen, inorganic ions, polyphosphates, lipids, and proteins [31, 87]. The unicellular form of fungi, called yeasts, which are often confused as a group apart from fungi, is oval that are larger than bacteria. Filamentous fungi have visible biomasses, the mycelia, which are composed of long filaments (hyphae) that branch and intertwine [86].

Fungi are microorganisms that are widely abundant in nature, as are bacteria. Filamentous fungi are known as decomposers and also parasites of animals and plants. Therefore, fungi are essential for recycling dead and decaying plants and animals [88].

Fungi are easy-to-grow microorganisms, have a high biomass yield, and have been evaluated as metal biosorbents in aqueous solution [87], as they can live in environments with high metallic concentration and accumulate micronutrients and toxic metals exhibiting high removal capacity [35]. In the literature, several studies have evaluated the biosorption of metals via fungal biomass (Table 2.3) [6, 41, 89–91].

In the literature, there are reports of studies that evaluated the fungal biomass after chemical and/or physical modification, as a nanoadsorbent composition, and in multielement systems to assess the competitiveness of metallic species. Yeasts are also highly valued in biosorption processes because they are easy to handle and cultivate, in addition to being a very common by-product of the food industry (genus *Saccharomyces* sp.) [13, 14, 92].

2.3.1 Potentially Toxic Metals

Fungi are microorganisms highly valued as biosorbents of potentially toxic metals. In the literature, several strains had their potential studied for numerous metallic species [93–95].

Fungal chitosan and nano-chitosan were produced using the mycelia of the fungus *Cunninghamella elegans* for the biosorption of potentially toxic metals, Pb^{2+} and Cu^{2+}. The results showed that chitosan and nano-chitosan had high biosorption capacity. Chitosan nanoparticles were more effective than fungal chitosan for biosorption of the metallic species under study [96].

The fungus *Pycnoporus sanguineus* was evaluated in a multielement system to remove Cd^{2+}, Pb^{2+}, and Cr^{3+}. The tests showed that the fungal biomass showed good removal efficiency for the metallic species under study [97].

The fungal biomass of *P. sanguineus* was also evaluated in its natural form and with chemical modifications of NaOH and HCl to evaluate the biosorption potential in a multielement system of Cd^{2+}, Pb^{2+}, and Cr^{3+} ions. The results showed that the modification influenced the biosorption process [98].

The fungus *Trichoderma harzianum* was used in the biosorption and consequent biosynthesis of gold nanoparticles. Numerous studies have shown that many biosorbents also can reduce, stabilize, and agglomerate metal ions in the form of nanoparticles. The results revealed that the fungus *T. harzianum* evaluated had a high gold biosorption capacity, 1,340 mg g^{-1} in 180 minutes of contact, and removal efficiency above 60%, in addition to biologically synthesizing gold nanoparticles with an average diameter of 30 nanometers [90].

The yeast *Saccharomyces cerevisiae* Perlage® BB, a commercial strain destined for the production of sparkling wines, was studied as a Cu^{2+} biosorbent. The biomass showed good removal capacity and efficiency. The study of isotherms revealed that the biosorption of Cu^{2+} was adjusted to the Langmuir model, which showed a maximum biosorption capacity of 4.73 mg g^{-1} [99].

The evaluation of the kinetics and equilibrium of Ni^{2+} biosorption from an aqueous solution was investigated using dead biomass of *Mucor hiemalis* physicochemically treated in a batch system. Biosorption equilibrium was established at a contact time of 150 minutes. The pseudo-second-order kinetic equation was the model that best fitted the data [100].

The white rot fungus, *Phanerochaete chrysosporium*, is a strong degrader of several xenobiotics. In addition to having a good performance in the removal of potentially toxic metals by biosorption, the investigation of the fungus *P. chrysosporium* after optimization of the process via functionalization with $CaCO_3$ as a Pb^{2+} and Cd^{2+} biosorbent showed high biosorption efficiency for the metallic species under study [101].

The study of the fungus *Ganoderma lucidum* as a biosorbent of Pb^{2+} and Cd^{2+} ions, from fungus residues and culture substrate, after pretreatment with three different chemical procedures (0.5 M NaOH, 10% (v/v) H_2O_2, and 0.5 M NaCl) revealed that the chemical modification positively influenced the process, as the removal efficiency increased from 87% to 93% for Pb^{2+} and 84% to 97% for Cd^{2+} using pretreatment with NaCl which was chosen during the study [102].

The fungus *Trichoderma* sp. BSCR02 was evaluated on Cr^{6+} biosorption. The results of the dead biomass mediated biosorption process after optimization showed that the maximum removal was observed at pH 5, with an initial metal concentration of 200 mg L^{-1}, the temperature of 35°C, supplemented with 1.6 mg mL^{-1} of biosorbent, and with a contact time of 120 minutes. The fungal biomass was active for five cycles of biosorption [103].

Where metal represents the metal concentration studied, q is the biosorption capacity, and R is the removal efficiency.

TABLE 2.3
Biosorption of potentially toxic metals by fungi

Fungus	Ion	Metal (mg L^{-1})	q (mg g^{-1})	R (%)	Reference
Cunninghamella elegans fungal nano-chitosan	Pb^{2+}	300	87.51	–	[96]
C. elegans fungal nano-chitosan	Cu^{2+}	300	59.62	–	
C. elegans fungal chitosan	Pb^{2+}	300	89.12	–	
C. elegans fungal chitosan	Cu^{2+}	300	57.23	–	
Penicillium piscarium	U	100	3.96	79.2	[91]
Pycnoporus sanguineus	Cd^{2+}	25	–	±70	[97]
	Pb^{2+}	25	–	±80	
	Cr^{3+}	25	–	±78	
Pleurotus ostreatus	Cu^{2+}	10	–	86	[104]
Aspergillus niger PTN31	Pb^{2+}	100	20.7	±85	[105]
***A. niger* PTN31 (EPS)**	Pb^{2+}	100	291.6	±80	
***Pycnoporus sanguineus* immobilized in calcium alginate**	Cu^{2+}	300	2.76	–	[106]
Fusarium solani	Zn^{2+}	600	5.81	–	[107]
F. solani treatment with NaOH	Zn^{2+}	600	12.5	–	
Trichoderma harzianum	Au^{3+}	400	1340	±60	[44]
Pycnoporus sanguineus	Cd^{2+}	40	±5.9	±75	[98]
	Pb^{2+}		±3.2	±40	
	Cr^{3+}		±5.5	±70	
P. sanguineus **modified with NaOH**	Cd^{2+}	40	±6.2	±80	
	Pb^{2+}		±3.8	±48	
	Cr^{3+}		±6.1	±76	
***P. sanguineus* modified with HCl**	Cd^{2+}	40	±6.5	±80	
	Pb^{2+}		±4.1	±49	
	Cr^{3+}		±6.3	±77	
Papiliotrema huenov living biomass	Mn^{2+}	110	16.6	75.5	[108]
	Cu^{2+}	128	15.7	70.5	
Papiliotrema huenov dead biomass	Mn^{2+}	110	27.1	60.3	
	Cu^{2+}	128	26.4	56.5	
Cryptococcus laurentii (AL65)	Pb^{2+}	20	9.1	46	[109]
***Pichia pastoris* X33 complete cell**	Cu^{2+}	100	6.2	41.1	[110]
P. pastoris cell wall	Cu^{2+}	100	11.53	21.2	
P. pastoris cell membrane	Cu^{2+}	100	10.97	20.7	
P. pastoris cytoplasm	Cu^{2+}	100	8.87	18.5	

(*Continued*)

TABLE 2.3 (*Continued*)
Biosorption of potentially toxic metals by fungi

Fungus	Ion	Metal (mg L⁻¹)	q (mg g⁻¹)	R (%)	Reference
***P. pastoris* treated with Proteinase -K (30 U)**	Cu^{2+}	100		18.1	
***P. pastoris* treated with β-mannanase (50 U)**	Cu^{2+}	100		28.2	
P. pastoris manana	Cu^{2+}	100		34	
P. pastoris glucan	Cu^{2+}	100	–	12	
Kodamaea transpacifica	Cr^{6+}	100	–	±38	[111]
***Kodamaea transpacifica* modified with (BZK)**	Cr^{6+}	100	476.19	85.8	
Saturnispora quitensis	Cr^{6+}	100	–	±40	
***Saturnispora quitensis* modified with (BZK)**	Cr^{6+}	100	416.67	85.4	
Kazachstania yasuniensis	Cr^{6+}	100	–	±38	
Kazachstania yasuniensis modificada com (BZK)	Cr^{6+}	100	114.94	80.7	
Saccharomyces cerevisiae	Cr^{6+}	100	–	±39	
***Saccharomyces cerevisiae* modified with (BZK)**	Cr^{6+}	100	120.48	75.8	
S. cerevisiae	Pb^{2+}	5,5	160.14	–	[112]
***S. cerevisiae* immobilized on chitosan MNPs**	Cu^{2+}	400	140	50	[113]
S. cerevisiae in the continuous bioreactor system	Cu^{2+} Pb^{2+}	180	29.9	20	[114]
S. cerevisiae	Mn^{2+}	5,6		3.4	[94]
***S. cerevisiae* immobilized in sawdust**	Cd^{2+}	100	1.2	–	[92]
S. cerevisiae Perlage® BB	Cu^{2+}	25		76	[99]
S. cerevisiae	Pb^{2+}	0,0528	–	91.6	[28]
	Cd^{2+}	0,0528	–	95.3	
Mucor hiemalis	Ni^{2+}	50	21.49		[100]
Phanerochaete chrysosporium	Cd^{2+}	50	28.6	22	[101]
	Pb^{2+}		47.69	45	
Ganoderma lucidum untreated	Pb^{2+}	1		87	[102]
	Cd^{2+}			84	
Ganoderma lucidum treated with NaCl	Pb^{2+}			93	
	Cd^{2+}			97	
Trichoderma sp. BSCR02 live	Cr^{6+}	50		77.36	[103]
Trichoderma sp. BSCR02 dead				83.26	

2.3.2 Organic Pollutants

Fungi are widely reported as bioremediation of organic pollutants [115]. Most dye-degrading fungi are white-rot basidiomycetes, with the *Phanerochaete chrysosporium* strain being one of the most studied species [81, 116].

The study of the fungus *Aspergillus niger* as a biosorbent of the Procion Red MX-5B dye showed an efficiency of 30%, at pH 4, dye concentration of 200 mg L^{-1}, the temperature of 30°C, and contact time of three hours [117].

Inactive spheres of immobilized carboxy methyl cellulose (CMC) from *Aspergillus fumigatus* that were isolated from an activated sludge system for treating wastewater from a dyeing plant were used to evaluate the biosorption of azo dye (reactive bright red K-2BP). The results showed that the maximum biosorption capacity was at pH 2 (31.5 mg g^{-1}) and that the data fit the Freundlich isothermal model [118].

The selection of fungi isolated from contaminated soils and dye wastewater showed that of the 58 fungal isolates, *Mucor circinelloides* performed best with 94% Congo red removal capacity at a concentration of 150 mg L^{-1}. In this study, the authors evaluated some conditions, such as the difference between wet and dry biomass. The results showed that dye biosorption was 39% higher in wet biomass. The maximum adsorption capacity according to the Langmuir model was 169.49 mg g^{-1} [119].

The characterization of the biosorbent by Fourier transform infrared spectroscopy, pretreatment (NaOH was the one with the best response, 19% increase in efficiency), and Zeta potential analysis revealed that the dye removal was performed using the biosorption process and that the hydroxyl and amine groups present in the fungal cell wall polymers played the main role in the dye biosorption [119].

The evaluation of dead biomass of the fungus *Aspergillus nidulans* G, which was able to use tobacco wastewater as the only substrate, as a biosorbent of Congo red dye, showed that the kinetics and biosorption equilibrium were well described by the model pseudo-second-order and by Langmuir, respectively. The maximum biosorption capacity of Congo red (CR) was 357.14 mg g^{-1} at 30°C and pH 6.8 [120].

An isolated fungus, *Aspergillus foetidus*, was studied as a biosorbent of black reactive dye 5. The tests showed that the maximum value of the biosorption capacity according to the Langmuir model was 106 mg g^{-1} at 50°C for pretreated fungal biomass with 0.1 M NaOH. Thermodynamic studies showed that the biosorption process was favorable, spontaneous, and endothermic in nature [121].

The use of dead fungal biomass of *Aspergillus fumigatus* as a biosorbent for the methylene blue (MB) dye and the consequent optimization of the process conditions showed that the best conditions are an initial concentration of 12 mg L^{-1} of MB, buffered at alkaline pH, contact time of 120 minutes, and room temperature showed an efficiency of 93.5% [122].

The filamentous fungus *Thamnidium elegans* was studied as a biosorbent in batch and dynamic mode to remove the Reactive Red 198 (RR198) dye. The biosorbent dosage effect study indicated that increasing the dosage from 0.2 to 0.8 g L^{-1} resulted in a significant increase in dye biosorption (40–90%). This improvement is due to the increase in the number of active sites for biosorption [123].

The yeast *Saccharomyces cerevisiae* was used to biosorb the Basic Blue 41 dye. The Surface and Response methodology was used to optimize the process. The results showed maximum biosorption of ±94% under static conditions, within 14 hours of contact, and at a temperature of 20°C using 0.6% biosorbent with dye solution with an initial concentration of 150 mg L^{-1} [124].

The removal of Congo red anionic dye from aqueous solution by the fungi *Aspergillus carbonarius* and *Penicillium glabrum* was analyzed. The best conditions of the process indicated that the maximum biosorption capacity was 99.01 mg g^{-1}

for *Aspergillus carbonarius* and 101.01 mg g^{-1} for *Penicillium glabrumem* at pH 4.5, biosorbent dosage of 0.33 g L^{-1}, contact time of 180 minutes, and initial dye concentration of 50 mg L^{-1} [125].

A submerged membrane fungal (*Coriolus versicolor*) reactor was used for biosorption/adsorption tests on powdered activated carbon (PAC), membrane retention, and biodegradation. Preliminary results with only fungal cultures, which can secrete extracellular enzymes, as in vitro assays using enzyme solution confirmed the much slower degradation rate, but higher biosorption of the polymeric dye (Poly S119) – removal rate of 99% compared to the other azo dye studied (Acid Orange II). The high biosorption rate of Poly S119 dye in the bioreactor condition favored excellent retention by the cake layer on the microfiltration membrane inside the bioreactor [126].

The fungus *Penicillium oxalicum* was investigated to evaluate the biosorption of eight aromatic compounds with different functional groups. The affinity of the biosorbent for the eight compounds at pH 6.0 follows the following trend: 1-naphthalenamine > naphthol > benzoic acid > p-toluidine > p-cresol > p-toluic acid > phenol > p-toluenesulfonic acid [127].

TABLE 2.4
Biosorption of organic pollutants by fungi

Fungus	Item analyzed	Ci (mg L^{-1})	R (%)	Reference
Aspergillus niger	Azo procion red MX-5B	200	30	[117]
Inactive CMC spheres immobilized from *Aspergillus fumigatus*	Azo dye (reactive bright red K-2BP)	96,6		[118]
Mucor circinelloides	Congo red dye	150	94	[119]
Aspergillus nidulans G	Congo red dye	20	96,40	[120]
Aspergillus foetidus	Black reactive dye 5	100	>99	[121]
Aspergillus fumigatus	MB	12	93,5	[122]
Thamnidium elegans	RR198	100	>90	[123]
Saccharomyces cerevisiae	Basic blue dye 41	150	94	[124]
Aspergillus carbonarius M333	Congo red dye	50		[125]
Penicillium glabrum Pgl				
Coriolus versicolor	Poly S119	0,1	99	[126]
	Acid Orange II		94	
Penicillium oxalicum	1-naphthalenamine	10–100		[127]
	Naphthol			
	Benzoic acid			
	p-toluidine			
	p – cresol			
	p-toluic acid			
	phenol			
	p-toluenesulfonic acid			
Champignon mushroom stem residues (*Agaricus bisporus*) and the shiitake (*Lentinula edodes*)	Paracetamol	2,000	98	[128]
	17 α-ethinyl estradiol (EE2)	2	100	

The potential of stem residues from champignon mushrooms (*Agaricus bisporus*) and shiitake mushrooms (*Lentinula edodes*) as biosorbents for pharmaceutical products (paracetamol and 17 α-ethinyl estradiol (EE2)) was evaluated. The kinetic study showed that (EE2) shiitake and champignon stalks showed 100% removal, in 20 and 30 minutes of contact, respectively [128] (Table 2.4).

Where Ci represents the concentration studied and R the removal efficiency.

2.4 CONSIDERATIONS

The growing population and industrial development promote a greater production of solid waste and industrial effluents contaminated by organic and inorganic species. The implementation of decontamination techniques becomes, therefore, essential, intending to promote the correct disposal within the governmental limits of these contaminant species.

Biosorption emerges as a promising methodology for the removal of potentially toxic metals and organic pollutants. Numerous biomasses are studied to carry out this process. With the emphasis on microorganisms that have a high capacity to retain these toxic contaminants. Bacteria and fungi deserve to be highlighted because they have numerous advantages in this bioremediation technique.

Bacteria have a fast growth rate and can produce EPS and remove pollutants by various mechanisms, as well as fungi, which, despite growing slower compared to bacteria, are considered easier to handle biomass, in addition to being very common by-products of industrial processes, becoming a potential residue for reuse in biosorption.

GLOSSARY

Bioremediation: Technique where microorganisms are used to remove or metabolize toxic substances of organic and inorganic nature.

Biosorption: Physical–chemical process of superficial removal of chemical species via live or dead biomass, as it is metabolically passive, many dead biomasses are studied. This technique involves the affinity between the biosorbent and the species to be removed.

Bioaccumulation: Process of removal of chemical species by living biomass through accumulation and metabolism, this technique occurs via active metabolism.

Biodegradation: Process carried out by live microorganisms (metabolically active process) where the reduction of complex chemical compounds in simpler pathways of metabolism/immobilization occurs.

Persistent Organic Pollutants (POPs): They receive this denomination because they are resistant to degradation by physical, chemical, or biological means.

Dyes: These are substances capable of transmitting their color to other substrates. Dyes can be classified into three main groups: (1) anionic; (2) cationic; and (3) non-ionic. The cationic or anionic azo dye has one or more azo bonds (N = N).

REFERENCES

[1] Madigan, Michael T; Martinko, John M.; Parker, Jack: *Microbilogia de Brock*. 12. Aufl.: Jones & Bartlett, 2010—ISBN 8536320931.

[2] Tortora, Gerard J.; Funke, Berdell R.; Case, Christine L.: *Microbiology – An Introduction*. 10 ed.: Pearson Education, 2010—ISBN 0-321-55007-2; ISBN 13: 978-0-321-55007-1.

[3] Singh, Bijender; Satyanarayana, Tulasi.: *1- Basic Microbiology*. 3. Aufl. : Elsevier B.V., 2017—ISBN 9780444636683

[4] Kwaasi, Aaron A.: Contents classification of microorganisms detection of foodborne pathogens and their toxins classification of microorganisms. Caballero, Benjamin; Trugo, Luiz, and Finglas, Paul M. (Hrsg.). *Encyclopedia of Food Sciences and Nutrition*. Elsevier, (2003), 3877–3885.

[5] Madakka, Mekapogu; Jayaraju, Nadimikeri; Rajesh, Nambi; Subhosh Chandra, Muni Ramanna Gari: *Development in the Treatment of Municipal and Industrial Wastewater by Microorganism*. Amsterdã: Elsevier Inc., 2018—ISBN 9780128163283.

[6] Shoaib, Amna; Aslam, Nida; Aslam, Nabila: Trichoderma harzianum: Adsorption, desorption, isotherm and FTIR studies. In: *Journal of Animal and Plant Sciences* Bd. 23, Nr. 5 (2013), 1460–1465—ISBN 1018-7081.

[7] Vijayaraghavan, K.; Balasubramanian, R.: Is biosorption suitable for decontamination of metal-bearing wastewaters? A critical review on the state-of-the-art of biosorption processes and future directions. In: *Journal of Environmental Management* Bd. 160 (2015), 283–296.

[8] Alluri, Hima Karnika; Ronda, Srinivasa Reddy; Settalluri, Vijaya Saradhi; Jayakumar Singh, Bondili; Suryanarayana, Visweswra; Venkateshwar, Pathapalli: Biosorption: An eco-friendly alternative for heavy metal removal. In: *African Journal of Biotechnology* Bd. 6, Nr. 25 (2007), 2924–2931.

[9] Phian, Sonika; Nagar, Shilpi; Kaur, Jasleen; Rawat, Charu Dogra: Emerging issues and challenges for microbes-assisted remediation. In: *Microbes and Microbial Biotechnology for Green Remediation* 28 (2022), 47–89.

[10] Rana, Neha; Gupta, Piyush: Microbes: An eco-friendly tool in wastewater treatment. In: *Synergistic Approaches for Bioremediation of Environmental Pollutants : Recent Advances and Challenges* (2022), 161–183.

[11] Sharif, Nadia; Bibi, Ayesha; Zubair, Naila; Munir, Neelma: Heavy metal accumulation potential of aquatic fungi. In: *Freshwater Mycology* 11 (2022), 193–208.

[12] Solomon, Magan; Liang, Chen: Human coronaviruses: The emergence of SARS-CoV-2 and management of COVID-19. In: *Virus Research* Bd. 319 (2022), 198882.

[13] Massoud, Ramona; Hadiani, Mohammad Rasoul; Hamzehlou, Pegah; Khosravi-Darani, Kianoush: Bioremediation of heavy metals in food industry: Application of Saccharomyces cerevisiae. In: *Electronic Journal of Biotechnology* Bd. 37 (2019), 56–60—ISBN 07173458.

[14] Talukdar, Daizee; Jasrotia, Teenu; Sharma, Rohit; Jaglan, Sundeep; Kumar, Rajeev; Vats, Rajeev; Kumar, Raman; Mahnashi, Mater H.; u. a.: Evaluation of novel indigenous fungal consortium for enhanced bioremediation of heavy metals from contaminated sites. In: *Environmental Technology and Innovation* Bd. 20 (2020), 101050.

[15] Bhargava, Arpit; Jain, Navin; Khan, Mohd Azeem; Pareek, Vikram; Dilip, R. Venkataramana; Panwar, Jitendra: Utilizing metal tolerance potential of soil fungus for efficient synthesis of gold nanoparticles with superior catalytic activity for degradation of rhodamine B. In: *Journal of Environmental Management* Bd. 183 (2016), 22–32.

[16] Jiang, Xincheng; Zhou, Xudong; Li, Caiyun; Wan, Zhenjia; Yao, Lunguang; Gao, Pengcheng: Adsorption of copper by flocculated Chlamydomonas microsphaera

microalgae and polyaluminium chloride in heavy metal-contaminated water. In: *Journal of Applied Phycology* Bd. 31 (2019), 1143–1151.
[17] Medfu Tarekegn, Molalign; Zewdu Salilih, Fikirte; Ishetu, Alemitu Iniyehu: Microbes used as a tool for bioremediation of heavy metal from the environment. In: *Cogent Food and Agriculture* Bd. 6, 1 (2020) 1–19.
[18] Ren, Binqiao; Jin, Yu; zhao, Luyang; Cui, Chongwei; Song, Xiaoxiao: Enhanced Cr (VI) adsorption using chemically modified dormant Aspergillus niger spores: Process and mechanisms. In: *Journal of Environmental Chemical Engineering* Bd. 10, Nr. 1 (2022), 106955.
[19] Naja, Ghinwa, Murphy, Vanessa; Volesky, Bohumil: Biosorption, metals. *Encyclopedia of Industrial Biotechnology: Bioprocess, Bioseparation, and Cell Technology*, (2010), 1–28
[20] Vijayaraghavan, Krishnaswamy; Yun, Yeoung-sang: Bacterial biosorbents and biosorption. In: *Biotechnology Advances* Bd. 26, Nr. 3 (2008), 266–291.
[21] Roane, Timberley M; Pepper, Ian L; Gentry, Terry J: *Microorganisms and Metal Pollutants.* 2. Aufl.: Elsevier Inc., 2015—ISBN 9780123946263.
[22] Vendruscolo, Francielo; da Rocha Ferreira, Glalber Luiz; Antoniosi Filho, Nelson Roberto: Biosorption of hexavalent chromium by microorganisms. In: *International Biodeterioration and Biodegradation* Bd. 119 (2017), 87–95.
[23] Sharma, Rohit; Talukdar, Daizee; Bhardwaj, Shefali; Jaglan, Sundeep; Kumar, Rajeev; Kumar, Raman; Akhtar, M. Shaheer; Beniwal, Vikas; u. a.: Bioremediation potential of novel fungal species isolated from wastewater for the removal of lead from liquid medium. In: *Environmental Technology and Innovation* Bd. 18 (2020), 100757.
[24] Abdel-Raouf, Neveen; Sholkamy, Essam Nageh; Bukhari, Nagat; Al-Enazi, Nouf Mohammed; Alsamhary, Khawla Ibrahim; Al-Khiat, Soad Humead A.; Ibraheem, Ibraheem Borie M.: Bioremoval capacity of Co^{+2} using Phormidium tenue and Chlorella vulgaris as biosorbents. In: *Environmental Research* Bd. 204 (2022), 111630.
[25] Öztürk, Ayten; Bayol, Emel; Abdullah, Meysun I.: Characterization of the biosorption of fast black azo dye K salt by the bacterium Rhodopseudomonas palustris 51ATA strain. In: *Electronic Journal of Biotechnology* Bd. 46 (2020), 22–29.
[26] Singh, Simranjeet; Kumar, Vijay; Datta, Shivika; Dhanjal, Daljeet Singh; Sharma, Kankan; Samuel, Jastin; Singh, Joginder: Current advancement and future prospect of biosorbents for bioremediation. In: *Science of the Total Environment* Bd. 709 (2020), 135895.
[27] Fawzy, Mustafa A.: Biosorption of copper ions from aqueous solution by Codium vermilara: Optimization, kinetic, isotherm and thermodynamic studies. In: *Advanced Powder Technology* Bd. 31, Nr. 9 (2020), 3724–3735.
[28] Hadiani, Mohammad Rasoul; Darani, Kianoush Khosravi; Rahimifard, Nahid; Younesi, Habibollah: Biosorption of low concentration levels of Lead (II) and Cadmium (II) from aqueous solution by Saccharomyces cerevisiae: Response surface methodology. In: *Biocatalysis and Agricultural Biotechnology*, 15(2018), 25–34.
[29] Huang, Fei; Li, Kai; Wu, Ren Ren; Yan, Yu Jian; Xiao, Rong Bo: Insight into the Cd^{2+} biosorption by viable Bacillus cereus RC-1 immobilized on different biochars: Roles of bacterial cell and biochar matrix. In: *Journal of Cleaner Production* Bd. 272 (2020), 122743.
[30] Parsania, Somayeh; Mohammadi, Parisa; Soudi, Mohammad Reza: Biotransformation and removal of arsenic oxyanions by Alishewanella agri PMS5 in biofilm and planktonic states. In: *Chemosphere* Bd. 284 (2021), 131336.
[31] Singh, Simranjeet; Kumar, Vijay; Singh Dhanjal, Daljeet; Datta, Shivika; Singh, Satyender; Singh, Joginder: Biosorbents for heavy metal removal from industrial effluents. In: *Bioremediation for Environmental Sustainability* (2021), 219–233.
[32] Vijayaraghavan, Krishnaswamy; Yun, Yeoung Sang: Bacterial biosorbents and biosorption. In: *Biotechnology Advances* Bd. 26, Nr. 3 (2008), 266–291.

[33] Tatsaporn, Todhanakasem; Kornkanok, Ketbumrung: Using potential lactic acid bacteria biofilms and their compounds to control biofilms of foodborne pathogens. In: *Biotechnology Reports* Bd. 26 (2020), e00477.

[34] Priyadarshanee, Monika; Das, Surajit: Biosorption and removal of toxic heavy metals by metal tolerating bacteria for bioremediation of metal contamination: A comprehensive review. In: *Journal of Environmental Chemical Engineering* Bd. 9, Nr. 1 (2021), 104686.

[35] Yin, Kun; Wang, Qiaoning; Lv, Min; Chen, Lingxin: Microorganism remediation strategies towards heavy metals. In: *Chemical Engineering Journal* Bd. 360, August 2018 (2018), 1553–1563.

[36] Gupta, Pratima; Diwan, Batul: Bacterial exopolysaccharide mediated heavy metal removal: A review on biosynthesis, mechanism and remediation strategies. In: *Biotechnology Reports* Bd. 13 (2017), 58–71.

[37] Dai, Qunwei; Zhang, Ting; Zhao, Yulian; Li, Qiongfang; Dong, Faqin; Jiang, Chunqi: Potentiality of living Bacillus pumilus SWU7-1 in biosorption of strontium radionuclide. In: *Chemosphere* Bd. 260 (2020), 127559.

[38] Huang, Haojie; Jia, Qingyun; Jing, Weixin; Dahms, Hans Uwe; Wang, Lan: Screening strains for microbial biosorption technology of cadmium. In: *Chemosphere* Bd. 251 (2020), 126428.

[39] Girijan, Sudeeptha; Kumar, Mathava: Immobilized biomass systems: An approach for trace organics removal from wastewater and environmental remediation. In: *Current Opinion in Environmental Science and Health* Bd. 12 (2019), 18–29.

[40] Akhigbe, Lulu; Ouki, Sabeha; Saroj, Devendra: Disinfection and removal performance for Escherichia coli and heavy metals by silver-modified zeolite in a fixed bed column. In: *Chemical Engineering Journal* Bd. 295 (2016), 92–98—ISBN 1385-8947.

[41] Ignatova, Lyudmila; Kistaubayeva, Aida; Brazhnikova, Yelena; Omirbekova, Anel; Mukasheva, Togzhan; Savitskaya, Irina; Karpenyuk, Tatyana; Goncharova, Alla; u. a.: Characterization of cadmium-tolerant endophytic fungi isolated from soybean (Glycine max) and barley (Hordeum vulgare). In: *Heliyon* Bd. 7, Nr. 11 (2021), e08240.

[42] Nascimento, Jéssica M.; de Oliveira, Jorge Diniz; Rizzo, Andrea C.L.; Leite, Selma G.F.: Biosorption Cu (II) by the yeast Saccharomyces cerevisiae. In: *Biotechnology Reports* Bd. 21, 2018 (2019), e00315.

[43] Baltazar, Marcela dos Passos Galluzzi; Gracioso, Louise Hase; Avanzi, Ingrid Regina; Karolski, Bruno; Tenório, Jorge Alberto Soares; do Nascimento, Claudio Augusto Oller; Perpetuo, Elen Aquino: Copper biosorption by Rhodococcus erythropolis isolated from the Sossego Mine – PA – Brazil. In: *Journal of Materials Research and Technology* 8 (2018), 2–10.

[44] do Nascimento, Jéssica Mesquita; Cruz, Nildo Duarte; de Oliveira, Gabriel Rodrigues; Sá, Waldeemeson Silva; de Oliveira, Jorge Diniz; Ribeiro, Paulo Roberto S.; Leite, Selma G.F.: Evaluation of the kinetics of gold biosorption processes and consequent biogenic synthesis of AuNPs mediated by the fungus Trichoderma harzianum. In: *Environmental Technology and Innovation* Bd. 21 (2021), 101238.

[45] Nascimento, Jéssica Mesquita do; Oliveira, Jorge Diniz de; Leite, Selma Gomes Ferreira: Chemical characterization of biomass flour of the babassu coconut mesocarp (Orbignya speciosa) during biosorption process of copper ions. In: *Environmental Technology & Innovation* Bd. 16 (2019), 100440.

[46] Qin, Huaqing; Hu, Tianjue; Zhai, Yunbo; Lu, Ningqin; Aliyeva, Jamila: *The Improved Methods of Heavy Metals Removal by Biosorbents: A Review.* Bd. 258: Elsevier, 2020—ISBN 8673188822829.

[47] Choińska-Pulit, Anna; Sobolczyk-Bednarek, Justyna; Łaba, Wojciech: Optimization of copper, lead and cadmium biosorption onto newly isolated bacterium using a Box-Behnken design. In: *Ecotoxicology and Environmental Safety* Bd 149, (2018), 275–283.

[48] Samimi, Mohsen; Shahriari-Moghadam, Mohsen: Isolation and identification of Delftia lacustris Strain-MS3 as a novel and efficient adsorbent for lead biosorption: Kinetics and thermodynamic studies, optimization of operating variables. In: *Biochemical Engineering Journal* Bd. 173 (2021), 108091.

[49] Tyagi, Bhawna; Gupta, Bulbul; Thakur, Indu Shekhar: Biosorption of Cr (VI) from aqueous solution by extracellular polymeric substances (EPS) produced by Parapedobacter sp. ISTM3 strain isolated from Mawsmai cave, Meghalaya, India. In: *Environmental Research* Bd. 191, April (2020), 1–9.

[50] Mathew, Blessy Baby; Krishnamurthy, Nideghatta Beeregowda: Screening and identification of bacteria isolated from industrial area groundwater to study lead sorption: Kinetics and statistical optimization of biosorption parameters. In: *Groundwater for Sustainable Development* Bd. 7 (2018), 313–327.

[51] Wang, Pengfei; Du, Huili; Fu, Yaqi; Cai, Xiaolin; Yin, Naiyi; Cui, Yanshan: Role of human gut bacteria in arsenic biosorption and biotransformation. In: *Environment International* Bd. 165 (2022), 107314.

[52] Xie, Yanluo; He, Nan; Wei, Mingyang; Wen, Tingyao; Wang, Xitong; Liu, Huakang; Zhong, Shiqiang; Xu, Heng: Cadmium biosorption and mechanism investigation using a novel Bacillus subtilis KC6 isolated from pyrite mine. In: *Journal of Cleaner Production* Bd. 312, April (2021), 127749.

[53] Ramachandran, Govindan; Chackaravarthi, Gnanasekaran; Rajivgandhi, Govindan Nadar; Quero, Franck; Maruthupandy, Muthuchamy; Alharbi, Naiyf S.; Kadaikunnan, Shine; Khaled, Jamal M.; u.a.: Biosorption and adsorption isotherm of chromium (VI) ions in aqueous solution using soil bacteria Bacillus amyloliquefaciens. In: *Environmental Research* Bd. 212 (2022), 113310.

[54] Li, Wen; Chen, Yinyuan; Wang, Tao: Cadmium biosorption by lactic acid bacteria Weissella viridescens ZY-6. In: *Food Control* Bd. 123 (2021), 107747.

[55] Li, Xueke; Ming, Qiaoying; Cai, Rui; Yue, Tianli; Yuan, Yahong; Gao, Zhenpeng; Wang, Zhouli: Biosorption of Cd^{2+} and Pb^{2+} from apple juice by the magnetic nanoparticles functionalized lactic acid bacteria cells. In: *Food Control* Bd. 109 (2020), 106916.

[56] Arjomandzadegan, Mohammad; Rafiee, Poorya; Moraveji, Mostafa Keshavarz; Tayeboon, Maryam: Efficacy evaluation and kinetic study of biosorption of nickel and zinc by bacteria isolated from stressed conditions in a bubble column. In: *Asian Pacific Journal of Tropical Medicine* Bd. 7, Nr. S1 (2014), S194–S198.

[57] Pardo, Rafael; Herguedas, Mar; Barrado, Enrique; Vega, Marisol: Biosorption of cadmium, copper, lead and zinc by inactive biomass of Pseudomonas Putida. In: *Analytical and Bioanalytical Chemistry* Bd. 376, Nr. 1 (2003), 26–32.

[58] Khadivinia, Elnaz; Sharafi, Hakimeh; Hadi, Faranak; Zahiri, Hossein Shahbani; Modiri, Sima; Tohidi, Azadeh; Mousavi, Amir; Salmanian, Ali Hatef; u. a.: Cadmium biosorption by a glyphosate-degrading bacterium, a novel biosorbent isolated from pesticide-contaminated agricultural soils. In: *Journal of Industrial and Engineering Chemistry* Bd. 20, Nr. 6 (2014), 4304–4310—ISBN 1226-086X.

[59] Muñoz, Antonio Jesús; Espínola, Francisco; Ruiz, Encarnación: Biosorption of Ag(I) from aqueous solutions by Klebsiella sp. 3S1. In: *Journal of Hazardous Materials* Bd. 329 (2017), 166–177—ISBN 1873-3336 (Electronic) 0304-3894 (Linking).

[59] Pugazhendhi, Arivalagan; Boovaragamoorthy, Gowri Manogari; Ranganathan, Kuppusamy; Naushad, Mu; Kaliannan, Thamaraiselvi: New insight into effective biosorption of lead from aqueous solution using Ralstonia solanacearum: Characterization and mechanism studies. In: *Journal of Cleaner Production* Bd. 174 (2018), 1234–1239—ISBN 0959-6526.

[60] Han, Yin Lung; Wu, Jen Hao; Cheng, Chieh Lun; Nagarajan, Dillirani; Lee, Ching Ray; Li, Yi Heng; Lo, Yung Chung; Chang, Jo Shu: Recovery of gold from industrial wastewater by extracellular proteins obtained from a thermophilic bacterium Tepidimonas fonticaldi AT–A2. In: *Bioresource Technology* Bd. 239 (2017), 160–170.

[61] Sayyadi, Shayan; Ahmady-Asbchin, Salman; Kamali, Kasra; Tavakoli, Nadia: Thermodynamic, equilibrium and kinetic studies on biosorption of $Pb^{+2}$from aqueous solution by Bacillus pumilus sp. AS1 isolated from soil at abandoned lead mine. In: *Journal of the Taiwan Institute of Chemical Engineers* Bd. 80 (2017), 701–708.
[62] Zinicovscaia, Inga; Safonov, Alexey; Zelenina, Daria; Ershova, Yana; Boldyrev, Kirill: Evaluation of biosorption and bioaccumulation capacity of cyanobacteria Arthrospira (spirulina) platensis for radionuclides. In: *Algal Research* Bd. 51, September (2020), 102075.
[63] Xu, Shaozu; Xing, Yonghui; Liu, Song; Hao, Xiuli; Chen, Wenli; Huang, Qiaoyun: Characterization of Cd^{2+} biosorption by Pseudomonas sp. strain 375, a novel biosorbent isolated from soil polluted with heavy metals in Southern China. In: *Chemosphere* Bd. 240 (2020), 124893.
[64] Joo, Jin-ho; Hassan, Sedky H.A.; Oh, Sang-eun: Comparative study of biosorption of Zn^{2+} by Pseudomonas aeruginosa and Bacillus cereus. In: *International Biodeterioration & Biodegradation* Bd. 64, Nr. 8 (2010), 734–741.
[65] Afraz, Vahideh; Younesi, Habibollah; Bolandi, Marzieh; Hadiani, Mohammad Rasoul: Optimization of lead and cadmium biosorption by Lactobacillus acidophilus using response surface methodology. In: *Biocatalysis and Agricultural Biotechnology* Bd. 29, June (2020), 101828.
[66] Parmar, Paritosh; Shukla, Arpit; Goswami, Dweipayan; Patel, Baldev; Saraf, Meenu: Optimization of cadmium and lead biosorption onto marine Vibrio alginolyticus PBR1 employing a Box-Behnken design. In: *Chemical Engineering Journal Advances* Bd. 4, October (2020), 100043.
[67] Mohapatra, Ranjan Kumar; Parhi, Pankaj Kumar; Pandey, Sony; Bindhani, Birendra Kumar; Thatoi, Hrudayanath; Panda, Chitta Ranjan: Active and passive biosorption of Pb(II)using live and dead biomass of marine bacterium Bacillus xiamenensis PbRPSD202: Kinetics and isotherm studies. In: *Journal of Environmental Management* Bd. 247 (2019), 121–134.
[68] Vishan, Isha; Saha, Biswanath; Sivaprakasam, Senthilkumar; Kalamdhad, Ajay: Evaluation of Cd(II) biosorption in aqueous solution by using lyophilized biomass of novel bacterial strain Bacillus badius AK: Biosorption kinetics, thermodynamics and mechanism. In: *Environmental Technology & Innovation* Bd. 14 (2019), 100323.
[70] Tufail, Muhammad Aammar; Iltaf, Jawaria; Zaheer, Tahreem; Tariq, Leeza; Amir, Muhammad Bilal; Fatima, Rida; Asbat, Ayesha; Kabeer, Tahira; u. a.: Recent advances in bioremediation of heavy metals and persistent organic pollutants: A review. In: *Science of the Total Environment* Bd. 850 (2022), 157961.
[71] Adithya, Srikanth; Jayaraman, Ramesh Sai; Krishnan, Abhishek; Malolan, Rajagopal; Gopinath, Kannappan Panchamoorthy; Arun, Jayaseelan; Kim, Woong; Govarthanan, Muthusamy: A critical review on the formation, fate and degradation of the persistent organic pollutant hexachlorocyclohexane in water systems and waste streams. In: *Chemosphere* Bd. 271 (2021), 129866.
[72] Chan, Sook Sin; Khoo, Kuan Shiong; Chew, Kit Wayne; Ling, Tau Chuan; Show, Pau Loke: Recent advances biodegradation and biosorption of organic compounds from wastewater: Microalgae-bacteria consortium – A review. In: *Bioresource Technology* Bd. 344 (2022), 126159.
[73] Elgarahy, Ahmed M.; Elwakeel, Khalid Z.; Mohammad, Samya H.; Elshoubaky, Gihan Ahmed El: A critical review of biosorption of dyes, heavy metals and metalloids from wastewater as an efficient and green process. In: *Cleaner Engineering and Technology* Bd. 4, Nr. June (2021), 100209.
[74] Gaur, Nisha; Narasimhulu, Korrapati; PydiSetty, Y.: Recent advances in the bioremediation of persistent organic pollutants and its effect on environment. In: *Journal of Cleaner Production* Bd. 198(2018), 1602–1631.

[75] Cheng, Zhuowei; Zhang, Xiaomin; Kennes, Christian; Chen, Jianmeng; Chen, Dongzhi; Ye, Jiexu; Zhang, Shihan; Dionysiou, Dionysios D.: Differences of cell surface characteristics between the bacterium Pseudomonas veronii and fungus Ophiostoma stenoceras and their different adsorption properties to hydrophobic organic compounds. In: *Science of the Total Environment* Bd. 650 (2019), 2095–2106.
[76] Vélez, Jessica M. Bedoya; Martínez, José Gregorio; Ospina, Juliana Tobón; Agudelo, Susana Ochoa: Bioremediation potential of Pseudomonas genus isolates from residual water, capable of tolerating lead through mechanisms of exopolysaccharide production and biosorption. In: *Biotechnology Reports* Bd. 32 (2021), 1–10.
[77] Bano, Amna; Hussain, Javaid; Akbar, Ali; Mehmood, Khalid; Anwar, Muhammad; Hasni, Muhammad Sharif; Ullah, Sami; Sajid, Sumbal; u. a.: Biosorption of heavy metals by obligate halophilic fungi. In: *Chemosphere* Bd. 199 (2018), 218–222.
[78] Meerbergen, Ken; Willems, Kris A.; Dewil, Raf; Van Impe, Jan; Appels, Lise; Lievens, Bart: Isolation and screening of bacterial isolates from wastewater treatment plants to decolorize azo dyes. In: *Journal of Bioscience and Bioengineering* Bd. 125, Nr. 4 (2018), 448–456.
[79] Chen, Si Hui; Yien Ting, Adeline Su: Biosorption and biodegradation potential of triphenylmethane dyes by newly discovered Penicillium simplicissimum isolated from indoor wastewater sample. In: *International Biodeterioration & Biodegradation* Bd. 103 (2015), 1–7.
[80] Nguyen, Thai Anh; Fu, Chun Chieh; Juang, Ruey Shin: Biosorption and biodegradation of a sulfur dye in high-strength dyeing wastewater by Acidithiobacillus thiooxidans. In: *Journal of Environmental Management* Bd. 182 (2016), 265–271.
[81] Adenan, Nurul Hidayah; Lim, Yau Yan; Ting, Adeline Su Yien: Removal of triphenylmethane dyes by Streptomyces bacillaris: A study on decolorization, enzymatic reactions and toxicity of treated dye solutions. In: *Journal of Environmental Management* Bd. 318 (2022), 115520.
[82] Saravanan, Anbalagan; Kumar, Ponnnusamy Senthil; Jeevanantham, Sathasivam; Harikumar, P.; Bhuvaneswari, V.; Indraganti, Sravya: Identification and sequencing of bacteria from crop field: Application of bacteria—agro-waste biosorbent for rapid pesticide removal. In: *Environmental Technology & Innovation* Bd. 25 (2022), 102116.
[83] Zambrano, Johanna; García-Encina, Pedro Antonio; Hernández, Félix; Botero-Coy, Ana M.; Jiménez, Juan J.; Irusta-Mata, Rubén: Removal of a mixture of veterinary medicinal products by adsorption onto a Scenedesmus almeriensis microalgae-bacteria consortium. In: *Journal of Water Process Engineering* Bd. 43 (2021), 102226.
[84] Reyes, Kim Rafaelle E.; Tsai, Po Wei; Tayo, Lemmuel L.; Hsueh, Chung Chuan; Chen, Bor Yann: Biodegradation of anthraquinone dyes: Interactive assessment upon biodecolorization, biosorption and biotoxicity using dual-chamber microbial fuel cells (MFCs). In: *Process Biochemistry* Bd. 101 (2021), 111–127.
[85] Lv, Mengyu; Zhang, Dongqing; Niu, Xiaojun; Ma, Jinling; Lin, Zhang; Fu, Mingli: Insights into the fate of antibiotics in constructed wetland systems: Removal performance and mechanisms. In: *Journal of Environmental Management* Bd. 321 (2022), 116028.
[86] Tortora, Christine L. Case; Berdell R. Funke; Gerard, J.: *Micro Biologia.* 12ª. Porto Alegre: Artmed, 2017—ISBN 9788582713549.
[87] Abdel-Ghani, Nour T.; El-Chaghaby, Ghadir A.: Biosorption for metal ions removal from aqueous solutions: A review of recent studies. In: *International Journal of Latest Research in Science and Technology* Bd. 3, Nr. 1 (2014), 24–42—ISBN 2278-5299.
[88] Sugiharto, Sugiharto: A review of filamentous fungi in broiler production. In: *Annals of Agricultural Sciences* Bd. 64, Nr. 1 (2019), 1–8.
[89] Yin, Kun; Wang, Qiaoning; Lv, Min; Chen, Lingxin: Microorganism remediation strategies towards heavy metals. In: *Chemical Engineering Journal* Bd. 360, Nr. October 2018 (2019), 1553–1563.

[90] Nascimento, Jéssica Mesquita; Cruz, Nildo Duarte; de Oliveira, Gabriel Rodrigues; Sá, Waldeemeson Silva; de Oliveira, Jorge Diniz; Ribeiro, Paulo Roberto S.; Leite, Selma G.F.: Evaluation of the kinetics of gold biosorption processes and consequent biogenic synthesis of AuNPs mediated by the fungus Trichoderma harzianum. In: *Environmental Technology and Innovation* Bd. 21 (2021), 101238.
[91] Coelho, Ednei; Reis, Tatiana Alves; Cotrim, Marycel; Mullan, Thomas K.; Corrêa, Benedito: Resistant fungi isolated from contaminated uranium mine in Brazil shows a high capacity to uptake uranium from water. In: *Chemosphere* Bd. 248 (2020), 1–9.
[92] Anagnostopoulos, Vasileios A.; Vlachou, Athina; Symeopoulos, Basil D.: Immobilization of Saccharomyces cerevisiae on low-cost lignocellulosic substrate for the removal of Cd^{2+} from aquatic systems In: *Jounal of Environment & Biotechnology Research* Bd. Nr. 1 (2015), 23–29.
[93] Sağ, Yesims: Biosorption of heavy metals by fungal biomass and modeling of fungal biosorption: A review. In: *Separation and Purification Methods* Bd. 30, Nr. 1 (2001), 1–48.
[94] Fadel, Hoda M.; Hassanein, Naziha M.; Elshafei, Maha M.; Mostafa, Amr H.; Ahmed, Marwy A.; Khater, Hend M.: Biosorption of manganese from groundwater by biomass of Saccharomyces cerevisiae. In: *HBRC Journal* Bd. 13, Nr. 1 (2017), 106–113.
[95] Bazrafshan, Edris; Zarei, Amin Allah; Mostafapour, Ferdos Kord: Biosorption of cadmium from aqueous solutions by Trichoderma fungus: kinetic, thermodynamic, and equilibrium study. In: *Desalination and Water Treatment* Bd. 57, Nr. 31 (2016), 14598–14608—ISBN 9113544861.
[96] Alsharari, Sultan F.; Tayel, Ahmed A.; Moussa, Shaaban H.: Soil emendation with nano-fungal chitosan for heavy metals biosorption. In: *International Journal of Biological Macromolecules* Bd. 118 (2018), 2265–2268.
[97] Nascimento, Jéssica M.; Oliveira, J.D.: Biossorção de metais potencialmente tóxicos (Cd^{2+}, Pb^{2+} E Cr^{3+}) em biomassa seca de Pycnoporus sanguineus. In: *Ecletica Quimica* Bd. 39, Nr. 1 (2014), 151–163.
[98] Nascimento, Jéssica M.; Juliermerson, Jonas; Dos Santos, S.; De Oliveira, Jorge D.: *Competitive biosorption of Cd(II), Pb(II) and Cr(III) Using Fungal Biomass Pycnoporus Sanguineus. In: Jounal of Environment & Biotechnology Research* Bd. Nr. 6, (2017), 123–127.
[99] do Nascimento, Jéssica M.; de Oliveira, Jorge Diniz; Rizzo, Andrea C.L.; Leite, Selma G.F.: Biosorption Cu (II) by the yeast Saccharomyces cerevisiae. In: *Biotechnology Reports* Bd. 21, Nr. 2018 (2019), e00315.
[100] Shroff, Kshama A.; Vaidya, Varsha K.: Kinetics and equilibrium studies on biosorption of nickel from aqueous solution by dead fungal biomass of Mucor hiemalis. In: *Chemical Engineering Journal* Bd. 171, Nr. 3 (2011), 1234–1245.
[101] Lu, Ningqin; Hu, Tianjue; Zhai, Yunbo; Qin, Huaqing; Aliyeva, Jamila; Zhang, Hao: Fungal cell with artificial metal container for heavy metals biosorption: Equilibrium, kinetics study and mechanisms analysis. In: *Environmental Research* Bd. 182, Nr. December 2019 (2020), 109061.
[102] Rozman, Ula; Kalčíková, Gabriela; Marolt, Gregor; Skalar, Tina: Potential of waste fungal biomass for lead and cadmium removal : Characterization, biosorption kinetic and isotherm studies. In: *Environmental Technology & Innovation* Bd. 18 (2020), 100742.
[103] Benila Smily, John Rose Mercy; Sumithra, Pasumalai Arasu: Optimization of chromium biosorption by fungal adsorbent, trichoderma sp. BSCR02 and its desorption studies. In: *HAYATI Journal of Biosciences* Bd. 24, Nr. 2 (2017), 65–71.
[104] Buratto, Ana Paula; Costa, Raquel Dalla; Ferreira, Edilson da Silva: Application of fungal biomass of Pleurotus ostreatus in the process of biosorption of copper ions (II) Aplicação de biomassa fúngica de Pleurotus ostreatus em processo de biossorção de íons cobre (II). In: *Eng Sanit Ambient*, Nr. 17 (2012), 413–420.

[105] Dang, Chenyuan; Yang, Zhenxing; Liu, Wen; Du, Penghui; Cui, Feng; He, Kai: *Role of Extracellular Polymeric Substances in Biosorption of Pb^{2+} by a High Metal Ion Tolerant Fungal Strain Aspergillus Niger PTN31*. Bd. 6: Elsevier B.V., 2018—ISBN 1062744799.
[106] Yahaya, Yus Azila; Mat Don, Mashitah; Bhatia, Subhash: Biosorption of copper (II) onto immobilized cells of Pycnoporus sanguineus from aqueous solution: Equilibrium and kinetic studies. In: *Journal of Hazardous Materials* Bd. 161, Nr. 1 (2009), 189–195.
[107] El Sayed, Manal T.; El-Sayed, Ashraf S.A.: Bioremediation and tolerance of zinc ions using Fusarium solani. In: *Heliyon* Bd. 6, Nr. 9 (2020), e05048.
[108] Van, Phu Nguyen; Truong, Hai, Thi Hong; Pham, Tuan Anh; Cong, Tuan Le; Le, Tien; Nguyen, Kim Cuc Thi: Removal of manganese and copper from aqueous solution by yeast papiliotrema huenov. In: *Mycobiology* Bd. 49, Nr. 5 (2021), 507–520.
[109] Rusinova-Videva, Snezhana; Nachkova, Stefka; Adamov, Aleksandar; Dimitrova-Dyulgerova, Ivanka: Antarctic yeast Cryptococcus laurentii (AL65): Biomass and exopolysaccharide production and biosorption of metals. In: *Journal of Chemical Technology and Biotechnology* Bd. 95, Nr. 5 (2019), 1372–1379.
[110] Chen, Xinggang; Tian, Zhuang; Cheng, Haina; Xu, Gang; Zhou, Hongbo: Adsorption process and mechanism of heavy metalions by different components of cells, using yeast (Pichia pastoris) and Cu^{2+}as biosorption models. In: *RSC Advances* Bd. 11 (2021), 17080.
[111] Campaña-pérez, Juan Fernando; Barahona, Patricia Portero; Martín-ramos, Pablo; Javier, Enrique; Barriga, Carvajal; Campaña-pérez, Juan Fernando: Ecuadorian yeast species as microbial particles for Cr (VI) biosorption. In: *Environmental Science and Pollution Research* Bd. 26 (2019), 28162–28172.
[112] Ferreira, Joelma Morais; Luiz, Flávio; Leonor, Odelsia; Alsina, Sanchez; Conrado, Líbia De Sousa; Bezerra, Eliane; Costa, Wolia: Estudo do equilíbrio e cinética da biossorção do Pb^{2+} por Saccharomyces cerevisiae. In: *Química Nova* Bd. 30, Nr. 5 (2007), 1188–1193.
[113] Peng, Qingqing; Liu, Yunguo; Zeng, Guangming; Xu, Weihua; Yang, Chunping; Zhang, Jingjin: Biosorption of copper(II) by immobilizing Saccharomyces cerevisiae on the surface of chitosan-coated magnetic nanoparticles from aqueous solution. In: *Journal of Hazardous Materials* Bd. 177, Nr. 1–3 (2010), 676–682—ISBN 1873-3336 (Electronic) r0304-3894 (Linking).
[114] Amirnia, Shahram; Ray, Madhumita B.; Margaritis, Argyrios: Heavy metals removal from aqueous solutions using Saccharomyces cerevisiae in a novel continuous bioreactor-biosorption system. In: *Chemical Engineering Journal* Bd. 264 (2015), 863–872.
[115] Bouras, Hadj Daoud; Isik, Zelal; Arikan, Ezgi Bezirhan; Yeddou, Ahmed Réda; Bouras, Noureddine; Chergui, Abdelmalek; Favier, Lidia; Amrane, Abdeltif; u. a.: Biosorption characteristics of methylene blue dye by two fungal biomasses. In: *International Journal of Environmental Studies* Bd. 78, Nr. 3 (2021), 365–381.
[116] Isik, Birol; Ugraskan, Volkan; Cankurtaran, Ozlem: Effective biosorption of methylene blue dye from aqueous solution using wild macrofungus (Lactarius piperatus). In: *Separation Science and Technology (Philadelphia)* Bd. 57, Nr. 6 (2022), 854–871.
[117] Almeida, Érica Janaína Rodrigues; Corso, Carlos Renato : Comparative study of toxicity of azo dye Procion Red MX-5B following biosorption and biodegradation treatments with the fungi aspergillus niger and aspergillus terreus. In: *Chemosphere* Bd. 112 (2014), 317–322.
[118] Wang, Bao E.; Hu, Yong You; Xie, Lei; Peng, Kang: Biosorption behavior of azo dye by inactive CMC immobilized Aspergillus fumigatus beads. In: *Bioresource Technology* Bd. 99, Nr. 4 (2008), 794–800.
[119] Azin, Ehsan; Moghimi, Hamid: Efficient mycosorption of anionic azo dyes by Mucor circinelloides: Surface functional groups and removal mechanism study. In: *Journal of Environmental Chemical Engineering* Bd. 6, Nr. 4 (2018), 4114–4123.

[120] Xi, Yu; Shen, Yong Fang; Yang, Fan; Yang, Gao Ju; Liu, Chang; Zhang, Zheng; Zhu, Da Heng: Removal of azo dye from aqueous solution by a new biosorbent prepared with Aspergillus nidulans cultured in tobacco wastewater. In: *Journal of the Taiwan Institute of Chemical Engineers* Bd. 44, Nr. 5 (2013), 815–820.

[121] Patel, Rachna; Suresh, Sumathi: Kinetic and equilibrium studies on the biosorption of reactive black 5 dye by Aspergillus foetidus. In: *Bioresource Technology* Bd. 99, Nr. 1 (2008), 51–58.

[122] Kabbout, Rana; Taha, Samir: Biodecolorization of textile dye effluent by biosorption on fungal biomass materials. In: *Physics Procedia* Bd. 55 (2014), 437–444.

[123] Akar, Tamer; Arslan, Sercan; Akar, Sibel Tunali: Utilization of Thamnidium elegans fungal culture in environmental cleanup: A reactive dye biosorption study. In: *Ecological Engineering* Bd. 58 (2013), 363–370.

[124] El-Gendy, Nour Sh; El-Salamony, Radwa A.; Amr, Salem S.Abu; Nassar, Hussein N.: Statistical optimization of basic blue 41 dye biosorption by Saccharomyces cerevisiae spent waste biomass and photo-catalytic regeneration using acid TiO2 hydrosol. In: *Journal of Water Process Engineering* Bd. 6 (2015), 193–202.

[125] Bouras, Hadj Daoud; Yeddou, Ahmed Réda; Bouras, Noureddine; Hellel, Djamila; Holtz, Michael D.; Sabaou, Nasserdine; Chergui, Abdelmalek; Nadjemi, Boubekeur: Biosorption of Congo red dye by Aspergillus carbonarius M333 and Penicillium glabrum Pg1: Kinetics, equilibrium and thermodynamic studies. In: *Journal of the Taiwan Institute of Chemical Engineers* Bd. 80 (2017), 915–923.

[126] Hai, Faisal Ibney; Yamamoto, Kazuo; Nakajima, Fumiyuki; Fukushi, Kensuke: Removal of structurally different dyes in submerged membrane fungi reactor—Biosorption/PAC-adsorption, membrane retention and biodegradation. In: *Journal of Membrane Science* Bd. 325, Nr. 1 (2008), 395–403.

[127] Zhang, Yu; Wei, Dongbin; Huang, Rongde; Yang, Min; Zhang, Shujun; Dou, Xiaomin; Wang, Dongsheng; Vimonses, Vipasiri: Binding mechanisms and QSAR modeling of aromatic pollutant biosorption on Penicillium oxalicum biomass. In: *Chemical Engineering Journal* Bd. 166, Nr. 2 (2011), 624–630.

[128] Menk, Josilene de Jesus; do Nascimento, Ashiley Ingrid Soares; Leite, Fernanda Gomes; de Oliveira, Renan Angrizani; Jozala, Angela Faustino; de Oliveira Junior, José Martins; Chaud, Marco Vinícius; Grotto, Denise: Biosorption of pharmaceutical products by mushroom stem waste. In: *Chemosphere* Bd. 237 (2019), 124515.

3 Mushroom Biosorbent

Varsha Meshram, Khemraj Sahu, Anjali Kosre, Deepali Koreti, Pramod Kumar Mahish, and Nagendra Kumar Chandrawanshi

3.1 INTRODUCTION

Numerous sectors produce many dangerous pollutants each year, including dyes and toxic metals. Heavy metal pollution is currently one of the most severe environmental issues due to the rapid development of industries (Liu et al., 2018). It employs numerous dyes in different processing phases in various sectors, including paper, plastic, chemical refineries, textiles, and leather (Ismail et al., 2013). About 10–15% of such dyes that cause organic and inorganic pollution are mixed into industrial sectors, which are held accountable for the passage of pollutants with toxic, carcinogenic, and genotoxicity effects to humans and microorganisms (Balakrishnan et al., 2016; Chequer et al., 2015). Environmental pollution control legislation has been introduced in numerous nations. It is essential to remove heavy metals from industrial effluent discharge effectively. Industrial waste treatment techniques vary and are based on different factors. Some techniques for treating heavy metals include chemical methods, membrane separation, electrochemistry, reduction, oxidation, and flotation (Alalwan et al., 2020). However, there are several drawbacks to the current heavy metals treatment methods, including high operating and maintenance costs, complex procedures, high chemical intake, and high levels of toxic waste production (Rizzuti et al., 2021) and Volesky (2007). Biosorption is an alternative process for treating heavy metals. A physio-chemical passive metabolite-independent method employs biosorbents derived from non-living biological materials. Biosorption is a suggested heavy metal treatment process because it is an environmentally friendly, economical, efficient, and simple technique (Javanbakht et al., 2014) for treating dyes. According to Eman et al. (2017), mechanisms for heavy metal tolerance in fungi include extracellular (chelation and cell wall binding) and intracellular (binding to substances like proteins) sequestration of heavy metals. Over the past several decades, the idea of "biosorption" has developed in various ways. A physicochemical process known as sorption allows each compound to bind to the other. A compound attaches to another through the physico-chemical process of sorption. Biological treatment techniques are used to oxidize a variety of dye solutions, but they call for a particular enzyme that catalyzes oxidation reactions (Al Prol, 2019). *Escherichia coli* and *Clostridium* sp. are examples of anaerobes that are used for the removal of dye. For dye pollutants removal, fungi and algae are also investigated, including *Aspergillus* sp., *Candida* sp., *Phanerochaete* sp., *Trametes* sp., and many more (Fu & Viraraghavan, 2001). Biosorbents from mushrooms can be prepared from mycelium or fruit bodies (live or dead) and spent mushroom substrate (SMS).

 DOI: 10.1201/9781003366058-3

The biosorption process is affected by various factors such as the presence of a microbial population, the accessibility of pollutants to these organisms, metal ion concentration, and environmental variables such as temperature, pH, and the presence of nutrients (Prakash, 2017). High accumulation potential and a shorter life span are some of the advantages of using mushrooms as biosorbents. This chapter discusses the state of mushrooms and biosorbents made from mushrooms that have been used in research to successfully remove pollutants including heavy metals and natural colors. Also describes the kinetic and isotherm models to eliminate contaminants from the environment by mushrooms based bioadsorbents.

3.2 BIOSORBENTS

"The term 'biosorption' refers to the use of biomaterials like agricultural waste, crab shells, fungi, bacteria, or composites to treat low-concentration and high-volume wastewater that contains pollutants like heavy metals, dyes, and organic compound wastes" (Nadaroglu et al., 2013). Biosorbent is a biological material (biomaterial) on which adsorption is taking place. However, the key issue of the biosorption process is identifying the most feasible biosorbent from a vast array of easily available and low-cost biosorbents. A wide range of naturally available materials can be selected as biosorbents to bind and eliminate pollutants from industrial wastewater. Biosorbents can be made from any type of biomass, including plants, animals, and microorganisms; waste from agriculture and factories; and by-products from various industries (Kurniawan et al., 2011; Volesky, 2007).

3.2.1 Kinetics of Biosorbent

Research findings on the kinetics of biosorption are crucial for sewage treatment as they provide crucial parameters to identify the biosorption process's method. To evaluate the reaction kinetics of the biosorbents, pseudo-first-order and pseudo-second-order models were usually preferred. The equation below represents the pseudo-first-order kinetic model (Mishra et al., 2020).

A simple pseudo-first-order rate equation is expressed as

$$\mathrm{Log}(q_e - q_t) = \mathrm{Log}\, q_e - \frac{k_1 t}{2.303} \tag{3.1}$$

Where (mg g^{-1}) the amount of adsorbed at equilibrium and at time *t*, respectively, and is the first-order rate constant (1 min^{-1}).

In addition, the pseudo-second-order model is also widely used. There are four types of linear pseudo-second-order kinetic models, and the most popular linear form is

$$\frac{t}{q_t} = \frac{1}{k_2 q_e^2} + \frac{t}{q_e} \tag{3.2}$$

Where, (mg g^{-1}) is the amount adsorbed at time *t*, (mg g^{-1}) is the amount of adsorbed at equilibrium, is the second-order adsorption rate constant (g mg^{-1} min^{-1}),

h(mg g^{-1} min^{-1}) is the initial sorption rate. The application of the pseudo-second-order kinetics by plotting t/qt versus t yielded the second-order rate constant (Krobba et al., 2012).

3.2.2 Isotherm of Biosorbent

The interaction between biosorbents and adsorbate is modeled using isotherms of biosorption, which is of significant importance in maximizing the utilization of adsorbents. Its absorption graphs between the amount of adsorbed data and isotherm material and concentration in a steady state of equilibrium temperature. While researchers investigate several adsorption isotherms, the Freundlich and Langmuir adsorption isotherms are the isotherm models most frequently studied. The Langmuir isotherm is a model that hypothesizes the existence of uniform active spots on the surface. Equation (3.3) provides the properties of the biosorbent and monolayer (Wei, Xu, 2020).

$$\frac{1}{q_e} = \frac{1}{q_{max}} + \frac{1}{K_L C_e q_{max}} \tag{3.3}$$

Where q_e and C_e, respectively, represented the equilibrium adsorbate concentration (L mg^{-1}) and adsorption capacity (mg g^{-1}). q_{max} represented the highest adsorption capacity, while L mg^{-1} represented the Langmuir constant. The plot of $1/C_e$ versus $1/q_e$ can be used to derive K_L and q_e values.

According to Nnadozie and Ajibade (2020), the Freundlich isotherm model predicts multilayer adsorption and heterogeneous adsorption sites on the surface of the biosorbents as

$$\ln q_e = \ln K_F + \frac{1}{n\left(\ln C_e\right)} \tag{3.4}$$

Where K_F and n, respectively, represent the Freundlich constant (mg g^{-1}) and the favorability of biosorption. K_F and n can be derived from the intersection and slope of the. $\ln q_e$ versus $\ln C_e$ plot.

3.3 MUSHROOM-BASED BIOSORBENT

Metals are naturally occurring soil components with a high adsorption capacity. Due to their multiple roles in biological systems, some metals, such as Fe^{2+}, Zn^{2+}, Mn^{2+}, and CO^{2+}, are necessary to survive all known life forms. However, some metals are toxic even at low levels and can, for instance, lead to chromosomal mutations (beryllium), slowed growth (antimony), cell lysis (silver), or enzyme deactivation (arsenic). Nickel, chromium, copper, zinc, mercury, lead, and cadmium are additional examples of metals that are thought to be toxic to humans and the environment (Ayangbenro & Babalola, 2017). Volcanic eruptions, erosion, comets, weathering of minerals, oceanic evaporation, and combustion are some natural processes that release heavy metals into the environment. The cap (pileus) of a mushroom, which has a spore-forming component (sporophore), and the stipe make up the fruiting body (stem, stalk).

TABLE 3.1
Heavy metal concentration in fruiting bodies of mushroom species

Mushroom species	Edibility	Heavy metal	Conc. in fruiting bodies (mg kg^{-1})	References
Tremella fuciformis and *Auricularia polytricha*	Edible	Pb^{2+}, Cd^{2+}, Cu^{2+}, and Zn^{2+}	97.8%, 91.1%, 78.1%, and 80.5%	Pan et al. (2010)
Agaricus arvensis	Edible	Pb, Cd, Hg, Cu, Mn, Zn, Fe, and Co	0.128, 0.842, 0.157, 34.13, 24.18, 37.85 92.74, and 0.228	Pastircakova (2004)
Amanita vaginata	Edible	Zn, Cu, Fe, Mn, Mg, Se, Cd, Cr, Ni, Sr, Pb, Co, Ti	1.49,1.49, 0.62, 0.62, 0.12, 0.51,1.22,0.5, 0.10, 015, 0.46, 0.24, 0.12	Isildak et al. (2004)
Agaricus campestris	Non-edible	Zn, Cu, Fe, Mn, Mg, Se, Cd, Cr, Ni, Sr, Pb	1.37, 0.45, 0.47, 0.33, 1.31, 0.50, 0.06, 0.03, 0.39, 0.32	Radulescu et al. (2010)
Hypholoma fasciculare	Non-edible	Zn, Cu, Fe, Mn, Mg, Se, Cd, Cr, Ni, Pb, Co, Ti	0.78, 0.47, 0.30, 0.49, 1.0, 0.25, 0.33, 0.07, 0.36, 0.38, 0.1, 0.10	Radulescu et al. (2010)
Amanita phalloides	Non-edible	Zn, Cu, Fe, Mn, Mg, Se, Cd, Cr, Ni, Sr, Pb, Co, Ti	1.65, 0.49, 0.55, 0.32, 1.34, 0.47, 0.46, 0.28, 0.32, 0.11, 0.52, 0.36, 0.17	Radulescu et al. (2010)
Lentinula edodes	Edible	Cd, As, Pb, Hg	(0.170–3.50), (0.081–1.520), (0.020–0.520), (0.020–0.079)	Yu et al. (2021)

Mushrooms take up heavy metals from a substrate with their broad mycelium. The proportion of metal contents that result from atmospheric depositions appears to be less significant due to the fruiting body's brief lifespan, typically 10–14 days. The age of the mycelium and the time between fructifications significantly influence the metal content of fruiting bodies (the formation of fruiting bodies). In general, mushrooms contain the most phosphorus, potassium, and calcium. The iron content was lowest in the fruiting bodies (Mleczek et al., 2021). Furthermore, fungi contain compounds of magnesium, iron, fluorine, copper, manganese, cobalt, titanium, and lead (Santos et al., 2022), shown in Table 3.1. Several scientists have reported that mushrooms are a potential biosorbent of heavy metals (Raj et al., 2011).

3.4 MECHANISM OF BIOSORBENT–SORBATE INTERACTION

Commonly, four mechanisms—adsorption, ion exchange, complexation, and precipitation—are responsible for binding metals, pollutants, and dyes to mushrooms and SMS. Van der Waals forces and electrostatic forces are the foundation of physical adsorption. The cation transport system occasionally transports the metal ions needed for metabolism along with other ions of the same charge and ionic radius. According to some reports, mushroom biomass creates defenses against heavy metals by secreting chelating agents that can bind with metal ions. The modifications to the metal transport system also result in fewer metal ions accumulation. The association of a metal ion with an intracellular molecule as if metallothionein is another way resistance can be developed (Ayangbenro & Babalola, 2017). The primary method of heavy metal binding is the chemical and physical ionization of metals on the substrate of the decomposing mushroom. Nevertheless, it is possible to see that ion exchange is a significant adsorption process. Sometimes, the counter ions of polysaccharides switch places with bivalent metal ions. Complexation, as was previously mentioned, is crucial to the adsorption procedure. The substrate and mushroom surface charge from wasted mycelium serves as its foundation. A negative charge is present on the surface of the mushroom mycelium thanks to chitin, a substance found in the cell wall of mushrooms. Both have negative charges due to carboxyl, amino, thiol, amide, imine, thioether, and phosphate, which allow them to form complexes during interactions with metals and ligands adsorbates and adsorbents (Javanbakht et al., 2014). A chemical interaction occurs between the metal and cell surface of the mushroom and the used mushroom substrate during precipitation, a metabolically independent process. As a result, heavy metals are deposited in solutions and on the mushroom mycelium's surface. While the adsorption of metal ions occurs passively in the dead mushroom biomass through various physicochemical methods, it depends on metabolic activity in the growing mushroom mycelium. However, a detailed understanding of metabolism-dependent activities is necessary to maximize and sustain the adsorption in the living system. The rate of respiration, the by-products of metabolism, and nutrient uptake all affect the metabolic activities of living biomass, which in turn influence adsorption, ion exchange, complication, and precipitation. Chitosan, an isolated chitin derivative, is preferable to other materials for dye adsorption. Various methods, including surface adsorption, chemisorption, pore diffusion, fluid diffusion, and chemical processes like adsorption complexation and ion exchange, are used to adsorb dyes on the chitosan surface. The amine group is the primary group involved in the adsorption of dyes, but the hydroxyl group may also play a role in the process (Javanbakht et al., 2014). It is vital to remember that the efficacy of the substrate in adsorbing the contaminants is more significant than the mechanism used in the adsorption.

3.5 ADSORBATES

Dyes are complicated organic molecules with a wide range of chemical structures and varying degrees of toxicity. Over 10,000 dyes and pigments are manufactured in industrial settings globally, and 10–15% of these products are mixed with wastewater when used. According to their ionic structure, dyes can be classified as non-ionic, anionic, or cationic (Ahmadipouya et al., 2021), some are exhibited in Figure 3.1. Due to their aromatic structure, synthetic dyes used in the dye industry have stable structures and cannot be broken down by the body. Due to the presence of dyes and other pollutants, wastewater from the textile industry is subject to hazardous

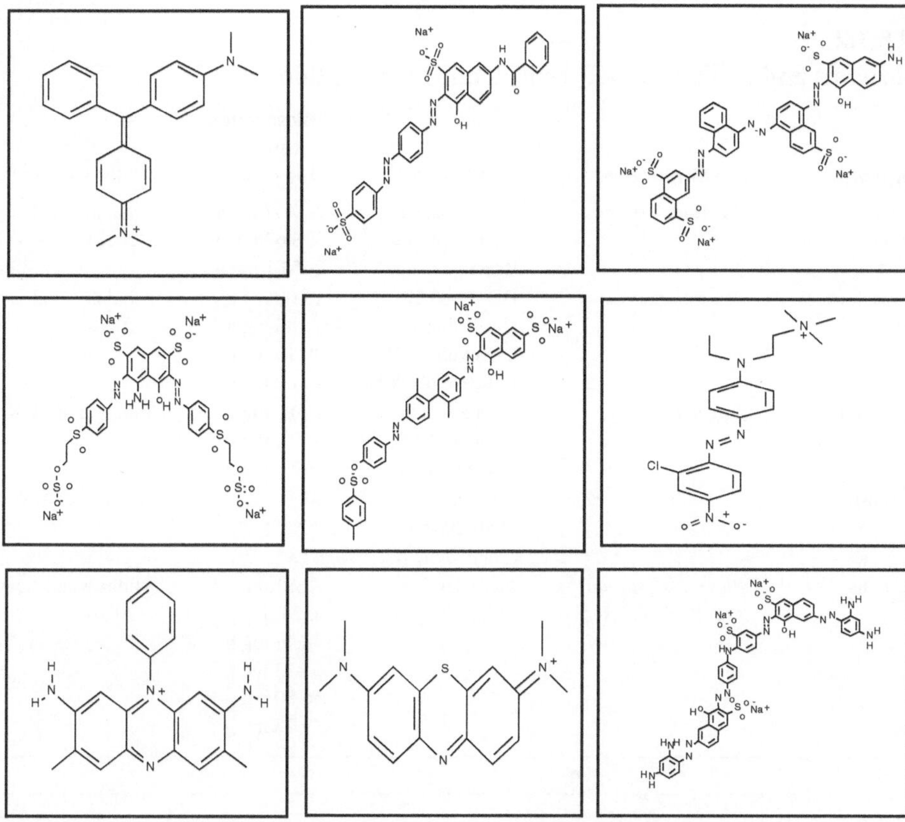

FIGURE 3.1 Chemical structure of various dyes adsorbed by mushroom and its by-products

environmental issues. As a result, before the wastewater is released into rivers, the dyes from these industries must be removed. Numerous varieties of mushrooms, waste from mushrooms, composite materials made of mushrooms, mushroom biomass, spent mushroom compost, SMS, etc., have all been used as biosorbents for dye removal in the literature. There are some specific mushroom-based biosorbents displayed in Tables 3.2 and 3.3. Because they contain different groups, such as amino, carboxyl, and thiol, which are in charge of tying dye molecules in the cell wall, they are preferred in dye biosorption. The highest data among mushroom-based materials was found to be the biosorption capacity of edible fungus substrates as porous carbon material biosorbents for Rhodamine dye removal, measured at 1497 mg g^{-1} (Chen et al., 2021). In this instance, the pore structure grown on the material's surface is essential to the adsorption procedure. The porous carbon material used as an edible fungus substrate has a large specific surface area (2,767.3 m^2 g^{-1}) and a high total pore volume (1.3936 cm^3 g^{-1}), which can offer numerous pore filling sites and a significant amount of adsorption capacity. It has proved the biosorption of dyes and mushrooms following monolayer biosorption and chemisorption.

TABLE 3.2
Mushrooms and its by-products reported for biosorption

Biosorbents	Mushrooms	Adsorbate	Biosorption capacity (mg g^{-1})/(%)	References
Modified spent mushroom waste	*Pleurotus ostreatus*	Direct red 5B Direct blue 71 Reactive black 5	249.57 mg g^{-1} 338.67 mg g^{-1} 265.01 mg g^{-1}	Alhujaily et al. (2018)
Spent mushroom waste	*Pleurotus ostreatus*	Direct red 5B Direct black 22 Direct black 71 Reactive black 5	18.00 mg g^{-1} 15.46 mg g^{-1} 20.19 mg g^{-1} 14.62 mg g^{-1}	Alhujaily et al. (2020)
Spent mushroom compost	*Agaricus bisporus*	Acid Red 111 Basic Red 18 Levafix Braun	140.9 mg g^{-1} 400.0 mg g^{-1} 169.5 mg g^{-1}	Toptas et al. (2014)
Spent substrate	*Ganodorma lucidum*	Malachite green, Safranine T Methylene blue	94% 85% 97%	Wu et al. (2018)
Mushrooms	*Pleurotus ostreatus* *Armillaria tabescens* *Morchella conica*	Methylene blue Malachite green	82.81 mg g^{-1} 43.90 mg g^{-1} 38.47 mg g^{-1} 64.13 mg g^{-1} 56.80 mg g^{-1} 39.28 mg g^{-1}	Yildirim and Acay (2020)

TABLE 3.3
Mushroom species reported for metal biosorption

Biosorbent	Mushrooms	Adsorbate	Kinetic model	Isotherm model	Biosorption capacity (mg g^{-1})/(%)	References
Spent mushroom	*Tricholomalobayense*	Pb(II)	–	Langmuir	210 mg g^{-1}	Dai et al. (2012)
Composted SMS	*Lentinus edodes*	Pb(II)	–	Langmuir and Freundlich	31.56 mg g^{-1}	Li et al. (2018)
Spent substrate	*Auricularia auricula*	Cr(IV)	Pseudo second	Langmuir and Freundlich	9.327 mg g^{-1}	Dong et al. (2018)
Pleurotus spent mushroom compost	*Pleurotus ostreatus*	Mn(II)	–	–	3.3 mg g^{-1}	Kamarudzaman et al. (2015a)
Spent mushroom compost	*Pleurotus ostreatus*	Ni(II)	Pseudo second	Langmuir	3.04 mg g^{-1}	Tay et al. (2011)
Spent mushroom compost	*Pleurotus ostreatus*	Fe(II) P	Pseudo second	Langmuir	45%	Kamarudzaman et al. (2015b)
Biochar derived from the SMS	*Pleurotus ostreatus* *Shiitake*	Pb(II)	Pseudo second	Langmuir	326 mg g^{-1} 398 mg g^{-1}	Wu et al. (2019)

3.6 ANALYTICAL TECHNIQUES FOR ELUCIDATION OF BIOSORPTION MECHANISM

Many analytical techniques have been used to study the biosorption process. These techniques may complement each other in giving insights into the mechanisms of biosorption (Park et al., 2010). They are various application based used for further study (display in Figure 3.2) such as atomic absorption spectrophotometry, ion selective electrodes, electron spin resonance spectroscopy, 18 nuclear magnetic resonance, X-ray photoelectron spectroscopy, thermogravimetric analysis, and differential scanning calorimetry, UV-Vis spectrophotometry, potentiometric titration, scanning or transmission electron microscopy coupled with energy dispersive X-ray spectroscopy, infrared (IR) spectroscopy or Fourier-transform IR spectroscopy, X-ray absorption spectroscopy, X-ray diffraction analysis (Fomina & Gadd, 2003; Michalak et al., 2013; Ngwenya, 2007; Park et al., 2010; Wang & Chen, 2006).

3.7 FACTORS AFFECTING BIOSORPTION

The factors like the presence of microbial population, the availability of contaminants to these organisms, metal ion concentration, and environmental factors like temperature, pH, and the presence of nutrients affect the biosorption process.

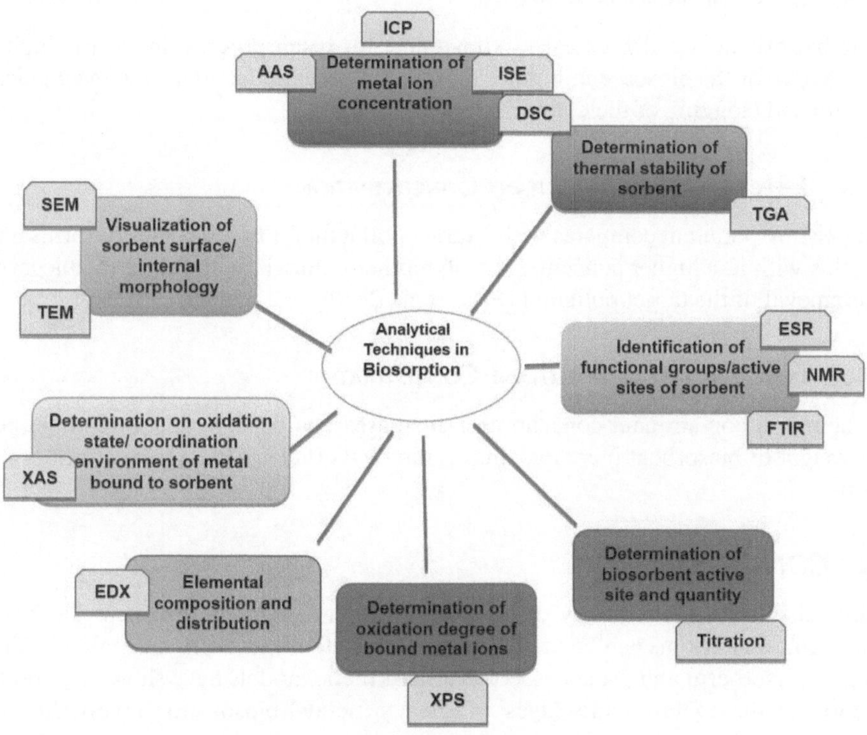

FIGURE 3.2 Analytical techniques in biosorption research

3.7.1 Effect of pH

The pH of a solution is one of the most critical parameters that control the biosorption process because it affects the activity of the functional groups in the biomass, the metal ions' competition, and the solution chemistry of the metals (Corral-Bobadilla et al., 2019). The pH of an aqueous medium is an essential factor that influences the uptake of metal ions by biosorbent in many ways. In the case of lead, as the pH increases, the biosorption also increases, reaches equilibrium at pH 5.0, and gradually decreases.

3.7.2 Effect of Temperature

A temperature rise typically improves the biosorption removal of adsorptive contaminants by raising the adsorbate's surface activity and kinetic energy, but it can also harm the biosorbent's structural properties (Park et al., 2010).

3.7.3 Effect of Ionic Strength

As ionic strength rises, it competes with the adsorbate for complex formation in the biosorbent, which decreases biosorption removal of adsorptive pollutants (Park et al., 2010).

3.7.4 Effect of biosorbent size

If the biosorbent size decreases, it is favorable for batch process due to the higher surface area of the biosorbent, but not for column process due to its low mechanical strength and clogging of the column (Park et al., 2010).

3.7.5 Effect of Other Pollutant Concentration

If coexisting pollutant competes with a target pollutant for binding sites or forms any complex with it, a higher concentration of another pollutant will reduce the biosorption removal of the target pollutant (Park et al., 2010).

3.7.6 Effect of Initial Pollutant Concentration

As the initial contaminant concentration increases, the biosorbent contaminant per unit weight of biosorbent increases, but its removal efficiency decreases (Park et al., 2010).

3.8 CONCLUSION

In this chapter, the possibility of using mushrooms as green adsorbent has been discussed. Mushrooms can be modified to enhance their adsorption capacities. The Langmuir isotherm and pseudo-second-order kinetic models have shown a good fit for most of the contaminants (dyes and heavy metals) biosorption investigations.

Temperature, pH, the nature of the biosorbents, initial pollutant concentration, other pollutant concentrations, and ionic strength are some variables that influence the biosorption rate. The removing pesticides, pharmaceutical waste, phenolic compounds, and other environmental contaminants have yet to be well addressed in recent studies on removing dyes and heavy metals from mushrooms. It is an environmentally friendly approach for adsorbing pollutants from the environment. To advance the field of study, more analyses should concentrate on examining the wastes from pharmaceutical, pesticide, and personal care products, which are regarded as emerging pollutants and are becoming increasingly toxic due to the biosorption process.

ACKNOWLEDGMENT

The authors are thankful to the Head, School of Studies in Biotechnology, Pt. Ravishankar Shukla University Raipur, for providing necessary facilities. The authors are also thankful to the Junior Research Fellowship (No. F. 82-44/2020 (SA-III, UGC-Ref. No.: 201610136180) Ministry of Education, Govt. of India, Bahadurshah Zafar Marg, New Delhi-110002, for providing funding support.

REFERENCES

Ahmadipouya, S., Haris, M. H., Ahmadijokani, F., Jarahiyan, A., Molavi, H., Moghaddam, F. M. & Arjmand, M. (2021). Magnetic Fe3O4 UiO-66 nanocomposite for rapid adsorption of organic dyes from aqueous solution. *Journal of Molecular Liquids*, *322*, 114910.

Al Prol, A. E. (2019). Study of environmental concerns of dyes and recent textile effluents treatment technology: a review. *Asian Journal of Fisheries and Aquatic Research*, *3*(2), 1–18.

Alalwan, H. A., Kadhom, M. A., & Alminshid, A. H. (2020). Removal of heavy metals from wastewater using agricultural byproducts. *Journal of Water Supply: Research and Technology-Aqua*, *69*(2), 99–112.

Alhujaily, A., Yu, H., Zhang, X., & Ma, F. (2018). Highly efficient and sustainable spent mushroom waste adsorbent based on surfactant modification for the removal of toxic dyes. *International Journal of Environmental Research and Public Health*, *15*(7), 1421.

Alhujaily, A., Yu, H., Zhang, X., & Ma, F. (2020). Adsorptive removal of anionic dyes from aqueous solutions using spent mushroom waste. *Applied Water Science*, *10*(7), 1–12.

Ayangbenro, A. S., & Babalola, O. O. (2017). A new strategy for heavy metal-polluted environments: a review of microbial biosorbents. *International Journal of Environmental Research and Public Health*, *14*(1), 94.

Balakrishnan, V. K., Shirin, S., Aman, A. M., de Solla, S. R., Mathieu-Denoncourt, J., & Langlois, V. S. (2016). Genotoxic and carcinogenic products arising from reductive transformations of the azo dye, Disperse Yellow 7. *Chemosphere*, *146*, 206–215.

Chen, S., Zhang, B., Xia, Y., Chen, H., Chen, G., & Tang, S. (2021). Influence of mixed alkali on the preparation of edible fungus substrate porous carbon material and its application for the removal of dye. *Colloids and Surfaces A: Physicochemical and Engineering Aspects*, *609*, 125675.

Chequer, F. M. D., Lizier, T. M., de Felício, R., Zanoni, M. V. B., Debonsi, H. M., Lopes, N. P., & de Oliveira, D. P. (2015). The azo dye Disperse Red 13 and its oxidation and reduction products showed mutagenic potential. *Toxicology In Vitro*, *29*(7), 1906–1915.

Corral-Bobadilla, M., González-Marcos, A., Vergara-González, E. P., & Alba-Elías, F. (2019). Bioremediation of waste water to remove heavy metals using the spent mushroom substrate of *Agaricus bisporus*. *Water*, *11*(3), 454.

Dai, J., Cen, F., Ji, J., Zhang, W., & Xu, H. (2012). Biosorption of lead (II) in aqueous solution by spent mushroom Tricholomalobayense. *Water Environment Research*, *84*(4), 291–298.

Dong, L., Liang, J., Li, Y., Hunang, S., Wei, Y., Bai, X., & Qu, J. (2018). Effect of coexisting ions on Cr (VI) adsorption onto surfactant modified Auricularia auricula spent substrate in aqueous solution. *Ecotoxicology and Environmental Safety*, *166*, 390–400.

Eman, M. F., Fatma, F. A. M., & Soad, A. E. Z. (2017). Biosorption of heavy metals onto different eco-friendly substrates. *Journal of Toxicology and Environmental Health Sciences*, *9*(5), 35–44.

Fomina, M., & Gadd, G. M. (2003). Metal sorption by biomass of melanin-producing fungi grown in clay-containing medium. *Journal of Chemical Technology & Biotechnology: International Research in Process, Environmental & Clean Technology*, *78*(1), 23–34.

Fu, Y., & Viraraghavan, T. (2001). Fungal decolorization of dye wastewaters: a review. *Bioresource Technology*, *79*(3), 251–262.

Isildak, Ö., Turkekul, I., Elmastas, M., & Tuzen, M. (2004). Analysis of heavy metals in some wild-grown edible mushrooms from the middle black sea region, Turkey. *Food Chemistry*, *86*(4), 547–552.

Ismail, B., Hussain, S. T., & Akram, S. (2013). Adsorption of methylene blue onto spinel magnesium aluminate nanoparticles: adsorption isotherms, kinetic and thermodynamic studies. *Chemical Engineering Journal*, *219*, 395–402.

Javanbakht, V., Alavi, S. A., & Zilouei, H. (2014). Mechanisms of heavy metal removal using microorganisms as biosorbent. *Water Science and Technology*, *69*(9), 1775–1787.

Kamarudzaman, A. N., Chay, T. C., Amir, A., & Abdul Talib, S. (2015a). Biosorption of Mn (II) ions from aqueous solution by Pleurotus spent mushroom compost in a fixed-bed column. *Procedia-Social and Behavioral Sciences*, *195*, 2709–2716.

Kamarudzaman, A. N., Chay, T. C., Amir, A., & Abdul Talib, S. (2015b). Fe (II) biosorption using Pleurotus spent mushroom compost as biosorbent under batch experiment. In *Applied Mechanics and Materials* (Vol. 695, pp. 314–318). Trans Tech Publications Ltd, Baech.

Krobba, A., Nibou, D., Amokrane, S., & Mekatel, H. (2012). Adsorption of copper (II) onto molecular sieves NaY. *Desalination and Water Treatment*, *37*(1–3), 31–37.

Kurniawan, A., Sisnandy, V. O. A., Trilestari, K., Sunarso, J., Indraswati, N., & Ismadji, S. (2011). Performance of durian shell waste as high capacity biosorbent for Cr (VI) removal from synthetic wastewater. *Ecological Engineering*, *37*(6), 940–947.

Li, X., Zhang, D., Sheng, F., & Qing, H. (2018). Adsorption characteristics of Copper (II), Zinc (II) and Mercury (II) by four kinds of immobilized fungi residues. *Ecotoxicology and Environmental Safety*, *147*, 357–366.

Liu, X., Bai, X., Dong, L., Liang, J., Jin, Y., Wei, Y.,& Qu, J. (2018). Composting enhances the removal of lead ions in aqueous solution by spent mushroom substrate: biosorption and precipitation. *Journal of Cleaner Production*, *200*, 1–11.

Michalak, I., Chojnacka, K., & Witek-Krowiak, A. (2013). State of the art for the biosorption process—a review. *Applied Biochemistry and Biotechnology*, *170*(6), 1389–1416.

Mishra, A., Gupta, B., Kumar, N., Singh, R., Varma, A., & Thakur, I. S. (2020). Synthesis of calcite-based bio-composite biochar for enhanced biosorption and detoxification of chromium Cr (VI) by Zhihengliuella sp. ISTPL4. *Bioresource Technology*, *307*, 123262.

Mleczek, M., Budka, A., Kalač, P., Siwulski, M., & Niedzielski, P. (2021). Family and species as determinants modulating mineral composition of selected wild-growing mushroom species. *Environmental Science and Pollution Research*, *28*(1), 389–404.

Nnadozie, E. C., & Ajibade, P. A. (2020). Adsorption, kinetic and mechanistic studies of Pb (II) and Cr (VI) ions using APTES functionalized magnetic biochar. *Microporous and Mesoporous Materials*, *309*, 110573.

Nadaroglu, H., Celebi, N., Kalkan, E., &Tozsin, G. (2013). Water purification of textile dye Acid red 37 by adsorption on laccase-modified silica fume. *Jökull Júlíusson*, *63*(5), 87.

Ngwenya, B. T. (2007). Enhanced adsorption of zinc is associated with aging and lysis of bacterial cells in batch incubations. *Chemosphere*, *67*(10), 1982–1992.

Pan, R., Cao, L., Huang, H., Zhang, R., & Mo, Y. (2010). Biosorption of Cd, Cu, Pb, and Zn from aqueous solutions by the fruiting bodies of jelly fungi (Tremella fuciformis and Auricularia polytricha). *Applied Microbiology and Biotechnology*, *88*(4), 997–1005.

Park, D., Yun, Y. S., & Park, J. M. (2010). The past, present, and future trends of biosorption. *Biotechnology and Bioprocess Engineering*, *15*(1), 86–102.

Pastircakova, K. (2004). Determination of trace metal concentrations in ashes from various biomass materials. *Energy Education Science and Technology*, *13*, 97–104.

Prakash, V. (2017). Mycoremediation of environmental pollutants. *International Journal of ChemTech Research*, *10*(3), 149–155.

Radulescu, C., Stihi, C., Busuioc, G., Popescu, I. V., Gheboianu, A. I., & Cimpoca, V. G. (2010). Evaluation of essential elements and heavy metal levels in fruiting bodies of wild mushrooms and their substrate by EDXRF spectrometry and FAA spectrometry. *Romanian Biotechnological Letters*, *15*(4), 5444–5456.

Raj, D. D., Mohan, B., & Vidya Shetty, B. M. (2011). Mushrooms in the remediation of heavy metals from soil. *International Journal of Environmental Pollution Control and Management*, *3*(1), 89–101.

Rizzuti, A. M., Winston, R. J., & Orr, I. S. (2021). Biosorption of copper from aqueous solutions utilizing agricultural wastes. *Remediation Journal*, *31*(2), 39–46.

Santos, M. P. O., Santos, M. V. N., Matos, R. S., Van Der Maas, A. S., Faria, M. C. S., Batista, B. L., & Bomfeti, C. A. (2022). Pleurotus strains with remediation potential to remove toxic metals from Doce River contaminated by Samarco dam mine. *International Journal of Environmental Science and Technology*, *19*(7), 6625–6638.

Tay, C. C., Liew, H. H., Redzwan, G., Yong, S. K., Surif, S., & Abdul-Talib, S. (2011). Pleurotusostreatus spent mushroom compost as green biosorbent for nickel (II) biosorption. *Water Science and Technology*, *64*(12), 2425–2432.

Toptas, A., Demierege, S., Mavioglu Ayan, E., & Yanik, J. (2014). Spent mushroom compost as biosorbent for dye biosorption. *CLEAN–Soil, Air, Water*, *42*(12), 1721–1728.

Volesky, B. (2007). Biosorption and me. *Water Research*, *41*(18), 4017–4029.

Wang, J., & Chen, C. (2006). Biosorption of heavy metals by Saccharomyces cerevisiae: a review. *Biotechnology Advances*, *24*(5), 427–451.

Wu, J., Zhang, T., Chen, C., Feng, L., Su, X., Zhou, L., & Wang, X. (2018). Spent substrate of Ganodorma lucidum as a new bio-adsorbent for adsorption of three typical dyes. *Bioresource Technology*, *266*, 134–138.

Wu, Q., Xian, Y., He, Z., Zhang, Q., Wu, J., Yang, G., & Long, L. (2019). Adsorption characteristics of Pb (II) using biochar derived from spent mushroom substrate. *Scientific Reports*, *9*(1), 1–11.

Yu, H., Shen, X., Chen, H., Dong, H., Zhang, L., Yuan, T., & Li, Y. (2021). Analysis of heavy metal content in Lentinula edodes and the main influencing factors. *Food Control*, *130*, 108198.

4 Algal and Plant Biosorbents

Victor Rezende Moreira, Giovanni Souza Casella, Thais Girardi Carpanez, Lucilaine Valéria de Souza Santos, Eduardo Coutinho de Paula, and Míriam Cristina Santos Amaral

4.1 BIOSORPTION PROCESSES

Algae- and plant-based biosorption processes involve a solid phase and a liquid phase that contain the dissolved species that will be adsorbed through the high affinity of the adsorbent. It is a metabolically independent and passive process in which the solute species (biosorbate) can diffuse from the liquid phase to the surface of the solid biomass (biosorbent) (Thirunavukkarasu et al., 2021). Algae and plants can be effectively utilized to translocate the contaminants or pollutants from the wastewater and concentrate them on their surface (Rangabhashiyam et al., 2018). The process is recognized as an alternative to water and wastewater treatment compared with other traditional technologies (Demey et al., 2019).

The processes involving algae and plant biosorption are driven by the concentration gradient that exists among the liquid and solid biosorbents, and the transport of solute molecules across the interphase is completely mediated by molecular diffusion (Thirunavukkarasu et al., 2021). In general, processes based on algae and plants are also characterized by low operating and manufacturing costs and high efficiency, providing purified effluents with high qualities (Elgarahy et al., 2021). The biosorbent and the adsorbate characteristics are also important as briefly introduced in Figure 4.1 for processes using algae- and plant-based materials.

Adsorbates are ions and molecules, normally pollutants or contaminants, present in different concentrations in wastewater. They may exist in different forms such as ionic species, particles, colloidal dispersions, or precipitates, each of these forms with a different relative affinity for the solid biosorbent (Gadd & Fomina, 2011; Thirunavukkarasu et al., 2021). These species can also be categorized between metals and metalloids (e.g., arsenic, cadmium, chromium, cobalt, copper, iron, lead, mercury, nickel, zinc) and organic compounds (e.g., dyes, pesticides, phenolic compounds, pharmaceutical compounds).

Biosorbents are biologically derived materials conveniently used to remove organic and inorganic contaminants from wastewater and have typical properties such as mechanical weakness, small rigidity, and low density (Tural et al., 2017). These materials can have different origins and, generally, their effectiveness increases with a decrease in the particle size. They can be derived from microorganisms, plant

 DOI: 10.1201/9781003366058-4

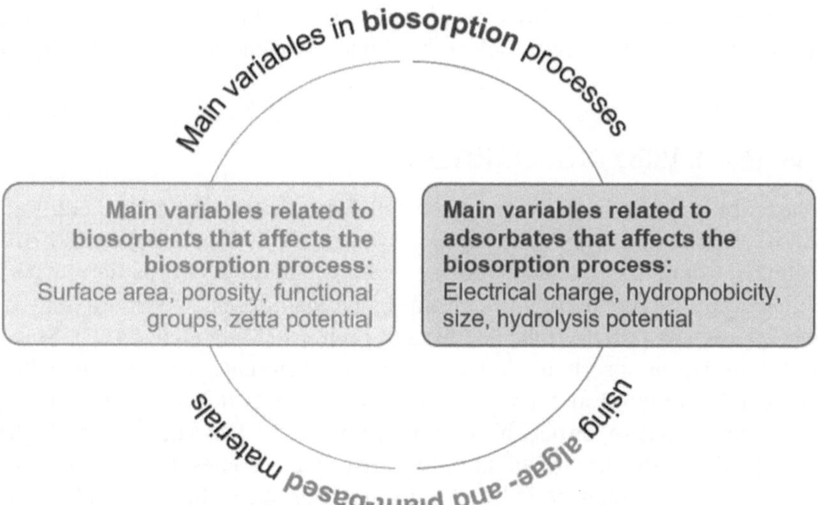

FIGURE 4.1 Main variables in biosorption processes using algae- and plant-based materials

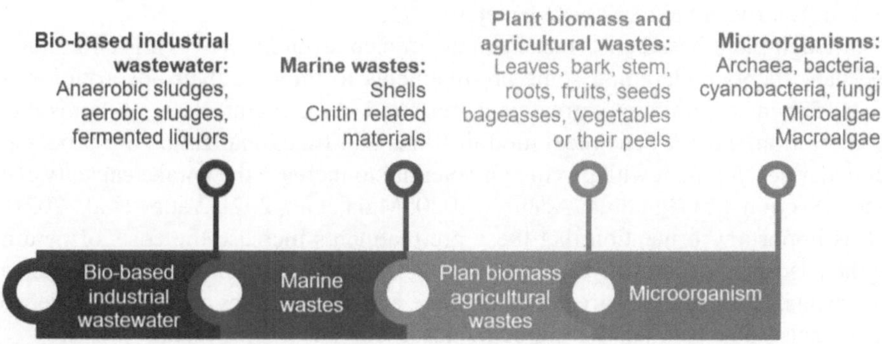

FIGURE 4.2 Materials from which biosorbents can be derived

biomass and agricultural waste derivatives, bio-based industrial effluents, and other marine wastes (Figure 4.2). In this chapter, emphasis will be given to algae-based and plant-based materials.

These materials, in general, may exhibit several functional groups on their surfaces due to their phytochemical constituents. This is an important behavior that will be presented and discussed in the following sections (Sections 4.2 and 4.3) which might provide additional surface binding regions for the biosorbate molecules (Abegunde et al., 2020; Adewuyi, 2020). Other benefits presented by these materials are their availability, accessibility, and biodegradability, reasons for which they are often chosen over other treatment techniques. Some drawbacks still limit the application of biosorption processes, a couple of examples are biosorbent swelling and in some cases the incapability to regenerate/reuse, since solid–liquid separation is more difficult (Saravanan, 2013; Tural et al., 2017). These were aspects specially discussed in Section 4.6, which disclosures about a possible future relied on biosorption processes.

4.2 PLANT-BASED BIOSORPTION

Plant-based biosorbents usually are produced from industrial and agricultural lignocellulosic biomass, mostly residues (Agarwal et al., 2020; Jain et al., 2016). The lignocellulosic biomass is considered the most abundant raw organic material on Earth. Its composition is mainly of cellulose, hemicelluloses, and lignin, and minor components such as polyphenols, pectin, and proteins (Maia et al., 2021). Since the group of plant-based adsorbents is vast, given the different types of lignocellulosic biomasses, it is common to classify the adsorbent for its origin: leave, bark, stem, root, fruit, seed, bagasse, or peel biosorbent (Jain et al., 2016; Manzoor et al., 2022).

The appeal for the use of plant-based adsorbents comes from ideas such as "Product life-cycle management" and sustainability since the feedstock for these adsorbents would mostly be considered solid waste otherwise (Manzoor et al., 2022; Yadav et al., 2021). Besides that, these adsorbents have shown high efficiency, low cost, high availability, chemical stability, and, naturally, various functional groups on their surface (highlighting hydroxyls, carbonyls, carboxyls, sulfhydryl, amines, and amides) (Agarwal et al., 2020; Yadav et al., 2021).

Although plant-based adsorbents can be efficient even *in nature*, it is common to have biosorbents undergoing some pretreatments to improve their adsorption efficiency. The most common pretreatments reported in the literature are carbonization and activation, whereas chemical modifications and functionalization are processes still under development with promising potential to increase the uptake capacity efficiency (Section 4.5) (Ighalo & Adeniyi, 2020; Maia et al., 2021; Yadav et al., 2021).

It is important to mention that these pretreatments increase the costs of obtaining the adsorbents and/or involve chemical usage and reduce the green character of biosorbents. However, they are not mandatory, and some studies are needed to decide which one(s) must be made for a specific adsorbent–adsorbate system.

However, since lignocellulosic biomasses can be degraded by some fungi and bacteria (Souza, 2013), the process of carbonization becomes extremely important for industrial biosorption. As a brief explanation, carbonization consists of thermochemical conversion of biomass, decomposing the organic materials, and turning them into fixed carbon content (Nizamuddin et al., 2017). Once this process decomposes the organic materials, it plays an important role in adsorptive continuous systems. Especially for water treatment plants, the decomposition of biomass could add undesirable substances to the treated water. However, there is no academic report about this situation, to the authors' knowledge. When released, it is necessary to specify the maximum dosage or maximum account time that will not compromise the water quality/safety. In this application, pretreatment processes such as physical washing or chemical modification processes could also be applied to plant-based materials to limit the release of undesirable substances. On the other side, when the application is the treatment of industrial effluents, if there is no input of toxicity, the control in dosages and contact time is more flexible.

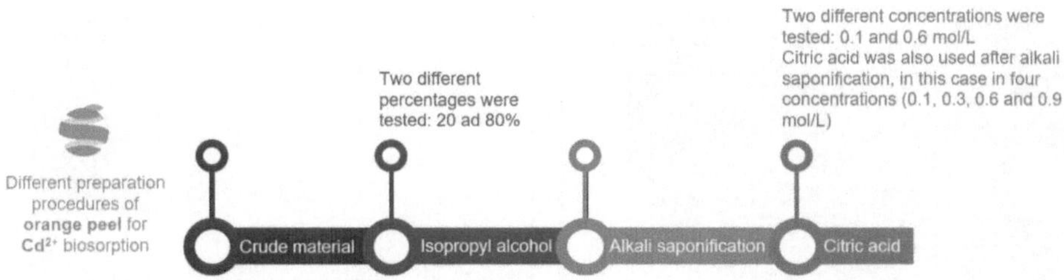

FIGURE 4.3 Summary of modification procedures applied orange peel and reported by Li et al. (2007)

As an example of a study about these pretreatments, Li et al. (2007) have studied different kinds of preparation of orange peel adsorbents for Cd^{2+} biosorption. The modifications considered are summarized in Figure 4.3. Among all these adsorbents evaluated, the one that presented the better cadmium (II) ion removal was the orange peel modified with citric acid (0.6 mol L^{-1}) after alkali saponification, which presented an adsorption capacity of 0.90 mol kg^{-1} (initial pH 6.0; initial cadmium concentration 0.001 mol L^{-1}; contact time 2 h). While this biosorbent has removed 90% of the cadmium ions, the crude orange peel has removed 45%. Other examples are also found in current literature and are summarized in Section 4.5.

4.3 ALGAE-BASED BIOSORPTION

Algae are non-vascular plants often lacking stems, vascular tissues, roots, and leaves (Ankit et al., 2022). They generally differ in size and could be categorized as macro- and microalgae, all possibly to be used as biosorbents for pollutants remediation; their use in biosorption processes was summarized in Section 4.5 (Table 4.1). These biological materials have several advantages when considered as biosorbents, including their availability in large quantities, possible to be cultivated in different climate conditions, their processing is relatively cheap, and their performance is comparable with current commercial adsorbents (Al-Homaidan et al., 2014). Bilal et al. (2018) mentioned the existence of approximately 30,000 different species of marine algae, broadly categorized between green, red, and brown. When classified as microalgae, they are found through phytoplankton in the oceanic ecosystem and can survive under high saline conditions and varied light sources. The diversity is also observed for freshwater algae, which summed up with the species from marine environments could represent at least 72,500 different species (Guiry, 2012).

Apart from their origin, algae-based materials are composed of cellulose and hemicellulose, but also inorganic compounds in smaller proportions. Images obtained from electronic microscopes and x-ray spectroscopy as presented in Figure 4.4 evidence these elements in algae surface, the same elements that compose functional groups (e.g., hydroxyls, carboxylates, amino, and phosphates (Lebron et al., 2019a,b)) that can efficiently interact with certain contaminants present in water. For algae-based materials, the biosorption process occurs either by passive uptake or by the metabolic activity of living organisms (Moreira et al., 2019). However, the use of dry biomass has some advantages over living organisms, including the possibility

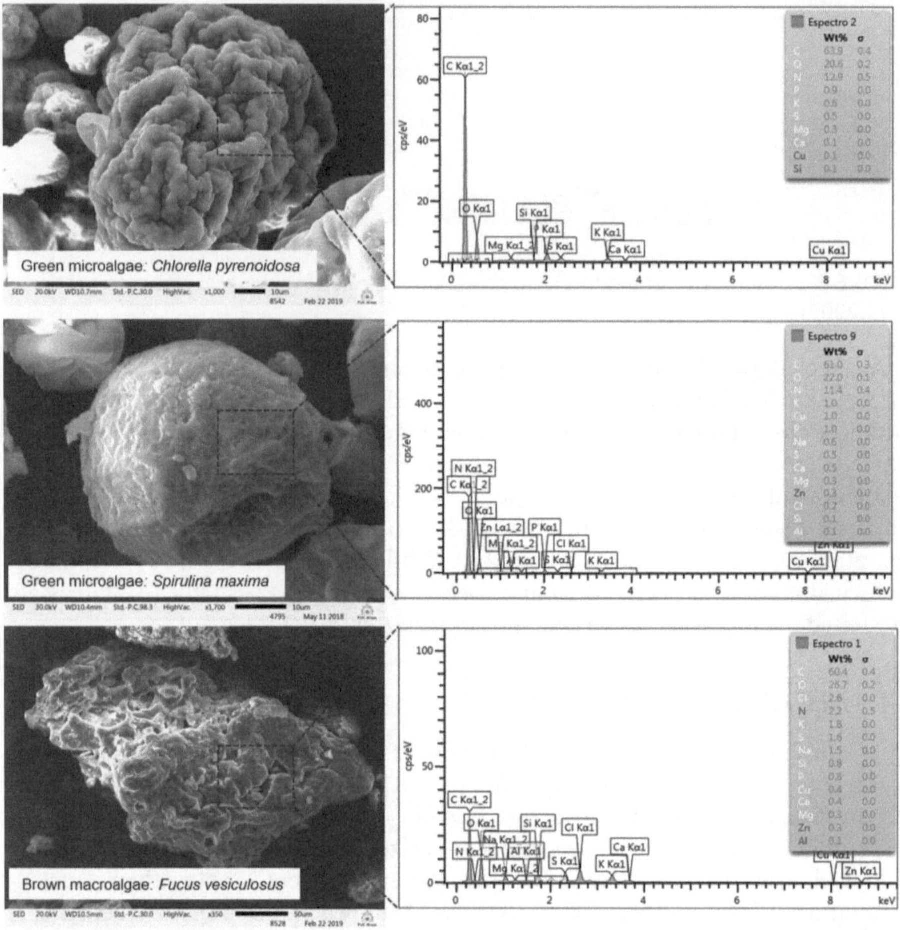

FIGURE 4.4 Images from micro- and macroalgae samples along their elemental composition

of storage at room temperature, applicability as a biosorbent for long periods without losing its biosorption capacity, inert to pollutants toxicity, and in general higher removal capacities compared to living algae (Gautam et al., 2014).

Moreira et al. demonstrated the passive uptake mechanism previously mentioned when studying the biosorption of copper ions from an aqueous solution by *Chlorella pyrenoidosa* (green microalgae) (Moreira et al., 2019). The biosorption process derived from combined electrostatic interactions and ion exchange principles, evidenced by the total amount of cations (Ca, Mg, and Na), released from the algae surface to the medium that positively correlated with the total amount of copper ions removed from the solution. The following equations (Equations 4.1–4.3) were proposed by the authors to represent the mechanisms involved, in which S represents the surface of *C. pyrenoidosa* (Moreira et al., 2019).

$$\text{S-Na} + Cu^{2+}_{(aq)} \rightarrow \text{S-Cu} + 2Na^{+} \tag{4.1}$$

$$\text{S-Ca} + Cu^{2+}_{(aq)} \rightarrow \text{S-Cu} + Ca^{2+} \tag{4.2}$$

$$\text{S-Mg} + Cu^{2+}_{(aq)} \rightarrow \text{S-Cu} + Mg^{2+} \tag{4.3}$$

The remediation process by living algae is also referred to as phytoremediation and comprises volatilization, detoxication, and transformation procedures of pollutants, all correlated in a complex metabolic pathway. The species interact with pollutants as trace metals to form complexes that are further stored in their interior in a phytoaccumulation process (Ankit et al., 2022). This capacity is intrinsically correlated with the presence of alginates and sulfated polysaccharides inside their cell walls, from which metals have high affinity (Davis et al., 2003). Some factors that interfere in the remediation capacity are moisture, pH, organic matter and nutrient available, salinity, and trace metal type (Ankit et al., 2022). Moreover, due to their capacity for bioaccumulation, algae are also widely used as metal availability bioindicator.

Jasrotia et al. (2017) reported the use of two algae (*Chlorodesmis* sp. and *Cladophora* sp.) to uptake arsenic in a contaminated water body. Parameters like pH, chemical oxygen demand, and arsenic concentration were monitored in a period of 15 days being reported that *Cladophora sp.* was found to survive up to an arsenic concentration of 6 mg L^{-1}, *Chlorodesmis sp.* could survive up to arsenic concentrations of 4 mg L^{-1}, respectively. After the monitoring period, the arsenic concentration was reduced to 0.1 mg L^{-1} and the treated water could be used for irrigation.

An important step toward the algae's widespread use as biosorbents relies on their cultivation processes. Kumar et al. (2021) summarized different cultivation methods that could be classified between raceway ponds and photobioreactors. Alterations and modifications made in these systems aim mostly to increase their productivity, especially for open systems and reduce their costs, in cases of photobioreactors. Strategies currently used to improve productivity include the supplementation with organic (e.g., glycerol, sodium acetate, and sucrose) and inorganic carbon sources (e.g., CO_2 and bicarbonate), besides improved light intensities and precise temperature and pH control.

For raceway ponds, the algae biomass productivity varies between 73 and 109,000 kg $(ha.yr)^{-1}$ (Shen et al., 2009), with the possibility to reach 127,000 kg $(ha.yr)^{-1}$ in high-rate race ponds accounted for active photon flux with data for insolation and radiation (Bharathiraja et al., 2015). When it comes to photobioreactors, the productivity rate could be 3,267,480 kg $(ha.yr)^{-1}$ (Kumar et al., 2021), improved values favored by the higher photosynthetic activity observed in these systems and the lower hydrodynamic stress on cells (Bharathiraja et al., 2015).

4.4 CONTROL VARIABLES IN PLANT- AND ALGAE-BASED BIOSORPTION

As it is a process that occurs at the solid–solute interface, the adsorption efficiency will depend on both process variables and the chemical structure of the biosorbent used and the pollutants to be removed (Ahmed & Hameed, 2018). In addition, the

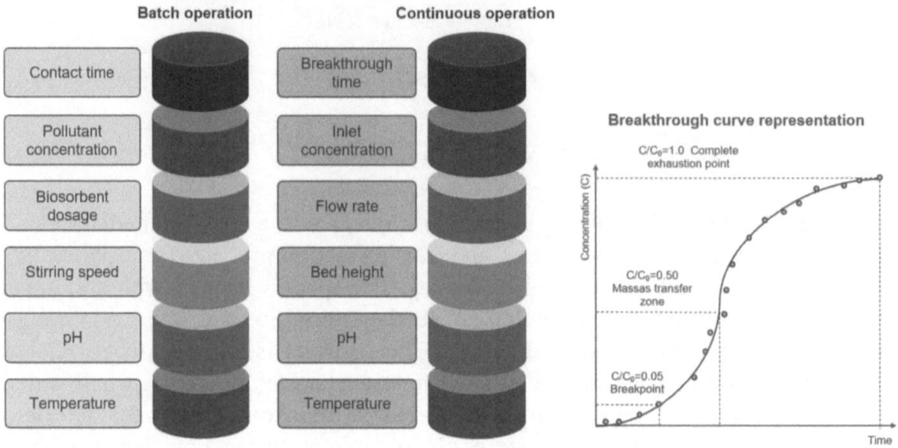

FIGURE 4.5 Control variables for batch and continuous operations, and breakthrough curve representation

operation type of adsorption that is to be performed changes the control variables of the system (Ahmed & Hameed, 2018; Gahlout et al., 2021), as shown in Figure 4.5.

As seen in Figure 4.5, there are clear differences between operating a batch operation and a continuous one. A brief explanation of the impact of each one of the variables presented in Figure 4.5 will be done ahead, starting from those belonging to batch operation.

Contact Time: The contact time between the adsorbent and the adsorbate represents an important control variable. The time must be long enough for the interaction/reaction to take place (Crini & Badot, 2008).

For batch adsorption, contact time is restricted to the batch reaction time itself. In this kind of operation, the "ideal" contact time is given by the equilibrium. The equilibrium can be determined by using one of the kinetic models already developed, highlighting pseudo-first order, pseudo-second order, intra-particle diffusion, and the Elovich model as the most common ones (Gahlout et al., 2021). Kumar et al. (2007) studied the removal of zinc from an aqueous solution by adsorption on *Ulva fasciata* sp., a marine green alga. When evaluating the contact time, it is possible to note that independently of the pollutant concentration the biosorption capacity reached a maximum at 20 minutes, after that point, there was no significant increase in the biosorption capacity over time.

Pollutant Concentration: In a simplified way, it is possible to state that the driving force of any adsorption is found in the difference in pollutant concentration between the surface of the adsorbent and the effluent (Eckenfelder et al., 2009). Thus, the higher the pollutant concentration, for the same amount of adsorbent, the greater removal is expected, until equilibrium is reached (Crini & Badot, 2008). It should be clear that in real-case applications, the pollutant concentration is not controlled even though it affects the biosorption process.

Adsorbent Dosage: It is expected an increase in the adsorption efficiency at higher adsorbent dosages since the total number of active sites also increases. However, excessive dosages may lead to a decrease in removal efficiency since the adsorbent particles start to agglomerate and overlap the active sites (Agarwal et al., 2020).

Moreira et al. (2019) have studied the biosorption capacity of copper by dried *C. pyrenoidosa* evaluating the impact of copper concentration (mg L^{-1}), pH, and biosorbent concentration (g L^{-1}). It was found that, for the data evaluated, both biosorbent concentration and copper concentration were statistically significant. The first one has shown a negative interaction, the lower the biosorbent concentration the higher the biosorption capacity, while the second one has shown a positive interaction, the higher the copper concentration the higher the biosorption capacity. Changing the biosorbent concentration from 1.6 to 0.4 g L^{-1} and the copper concentration from 4 to 16 mg L^{-1}, the biosorption capacity went from approximately 1.5 to 12 mg g^{-1}.

Solid–Liquid Ratio: The solid–liquid ratio comes from the last two variables presented. Through it, it is possible to determine the proportion of adsorbent for a given amount of pollutant to maximize the removal, without excessive adsorbent expenditure.

Li et al. (2007) studied the biosorption of Cd^{2+} by seven different orange peel adsorbents; for all of them, an increase in solid–liquid rate led to an increase in removal rate until maximum. Taking just the raw orange peel biosorbents as an example, the removal rate went from approximately 20 to 50% for a solid–liquid ratio from 1 to 5g L^{-1}.

Stirring Speed: On batch operation, the way to diminish mass transfer resistance is by increasing the stirring speed (Crini & Badot, 2008). According to Agarwal et al. (2020), agitation speed is usually kept between 120 and 200 rpm. The superior limit is determined to avoid fragmentation of the adsorbent. Nuhoğlu et al. (2021) have studied the impact of the stirring speed on the Pb (II) biosorption by *Camellia sinensis*, after its usage in a tea factory. *C. sinensis* is a plant highly grown in the Eastern Black Sea region. The study evaluated the biosorption efficiency of Pb (II) in the system at 50–300 rpm. The results showed that at the higher stirring speed, the efficiency was increased, going from 81.35 to 95.70%.

pH: The pH optimization depends on both adsorbent and adsorbate. Adsorbents are commonly characterized by their "zero-point charge" (pH_{zpc}), which is the pH range when their surface charge is equal to zero. When an adsorbent is in an environment where $pH > pH_{zpc}$, its surface will be negatively charged. Similarly, an adsorbent is in an environment where $pH < pH_{zpc}$, its surface will be positively charged (Agarwal et al., 2020; Crini & Badot, 2008).

Taking this into account and considering the electrostatic interaction between an adsorbent and a pollutant, it is preferable to work in a range of pH so that the surface of the adsorbent will have a neutral or the opposite charge of the pollutant.

Barka et al. (2011) have studied the adsorption capacity of methylene blue (MB) and eriochrome black T (ET) by *Scolymus hispanicus L.*, an easily available plant in Mediterranean countries. The pH was evaluated for this system, the adsorption of ET decreases with the increase of pH, with biosorption capacity going from approximately 160 to 70 mg g^{-1} in pH 3–9. At the same time, the MB adsorption has increased with the increase of pH, biosorption capacity going from approximately 0 to 150 mg g^{-1} in pH 2–7.

Despite being an important variable, when on an industrial scale, especially for water treatment, there is no economic interest in changing the effluent pH just to increase adsorption efficiency. It is preferable to study other adsorbents than to change the operation variable, whether by the costs of making this modification or the volume of fluid to be treated.

Temperature: It is well known that temperature is a key variable in any chemical process. At higher temperatures, chemical reactions occur faster, thus diminishing contact time, and hence making feasible the treatment of a greater quantity of effluent (Crini & Badot, 2008). Not only this, it is known that most biosorption systems are endothermic, thus higher temperature would increase global efficiency.

Some reports point out that an increase in temperature on algae systems can increase the removal percentage as described by Pahlavanzadeh et al. (2010), which evaluated the biosorption of Ni (II) by four brown algae, *Cystoseira indica, Nizamuddinia zanardini, Sargassum glaucescens, and Padina australis*. For all of them, the increase in the temperature led to an increase in the sorption capacity at equilibrium. However, when manipulating living adsorbents, such as algae, a higher temperature can be fatal. In addition, the thermal energy can change the structure of plant-based adsorbent, affecting the active site, structure, and pore size (Agarwal et al., 2020).

Finally, when on an industrial scale, it is not common to modify the temperature to achieve a greater adsorption efficiency. The great amount of energy required implies higher operational costs, resulting in an economically unfeasible process. So, the most common is to use room temperature (Agarwal et al., 2020).

Breakthrough Time: In a similar way to the equilibrium, the breakthrough time represents the moment when all the adsorbent of the column is saturated with the pollutant (Ahmed & Hameed, 2018). This variable is dependent on both the volume of the reactor (column) and flow rate. The breakthrough curve (previously illustrated in Figure 4.5) is evaluated by plotting a curve of outlet-to-inlet concentration (C/C_0) ratio against time (t) or throughput volume, and the data can be analyzed by using one of the models already developed, highlighting Adams–Bohart, Thomas, and Yoon–Nelson as the most common ones (Gahlout et al., 2021). The breakthrough time is arbitrarily defined as the moment in time that C/Co equals 0.05 (Tan & Hameed, 2017).

Inlet Concentration: As exposed in "Pollutant concentration", the pollutant concentration in the medium can be considered the driving force of any adsorption. On continuous operation, an increase in inlet concentration led to an earlier saturation of the column. In other words, an increase in the inlet will diminish the breakthrough time (Ahmed & Hameed, 2018). Cruz-Olivares et al. (2013) studied the lead (II) biosorption by allspice (*Pimenta dioica* L. Merrill) residue from the food industry. While evaluating the inlet concentration, it was shown that higher inlet concentrations diminish the expected breakthrough time.

Flow Rate: As presented in breakthrough time, the flow rate is an important variable since it determines the contact time the effluent will have with the adsorbent (Ahmed & Hameed, 2018). The flow rate must be low sufficient for reactions/interactions to occur. Calero et al. (2009) have shown that for the same inlet concentration, the flow rate of 2 mL min^{-1} presented a bigger breakthrough time rather than the tests with 4 and 6 mL min^{-1}. This study evaluated the Cr (III) biosorption by olive stone.

Bed Height: The quantity of adsorbent in the column is directly related to the bed height. A bigger column represents a longer contact time between the effluent and the adsorbent, for this reason, the bed height is related to the column performance and to a longer breakthrough time (Ahmed & Hameed, 2018). Asif and Chen (2015) studied the removal of arsenic from drinking water using rice husks. Evaluating the bed height, it was reported that a bigger bed had a lower C/Co at the same time. In addition, the effect of the diameter has also been reported, as a bigger diameter results in a greater removal efficiency, at the same time. Although the same result could virtually be achieved by changing the column radius or the column height, industrially it is easier to adjust the height than the radius of a system. For this reason, normally just the height is evaluated.

4.5 STRATEGIES TO ENHANCE BIOSORPTION CAPACITY

It is also possible to enhance the biosorption capacity of algae- and plant-based materials by thermal, physical, and chemical treatments. In most cases, it is aimed to alter their functional groups by cross-linking and functionalizing reactions (Feng et al., 2011).

Thermal treatments at high temperatures are intended to obtain materials of greater basicity as they act on reducing the number of functional groups containing oxygen. Pereira and Alves explained that the greater basicity results from the remaining ketonic groups remaining on the biosorbents surface and the delocalization of π-electrons of the carbon basal planes (Pereira & Alves, 2012). Matheickal et al. (1999) assessed the efficiency of marine algae *Durvillaea potatorum* for cadmium uptake. By a thermal treatment, the authors intended to prevent the alginic acid from leaching out the biosorbent surface and to increase the biosorbent stability, which resulted in a material with an adsorption capacity of 124.3 $mg_{Cd(II)}/g$ biosorbent.

Most physical treatments intend to reduce the biosorbents' size, therefore increasing the surface area available for interaction. Chemical modifications, in turn, could be made with different agents. Oxidizing treatments with HNO_3 and O_2, for example, can lead to biosorbents with a higher density of functional groups containing oxygen (e.g., carboxylic acids) and, therefore, greater availability of active sites for interaction with pollutants. In cases where carboxylic acids are formed, and when the medium pH is greater than their dissociation pH, there is a higher interaction rate between biosorbent–and biosorbate. On the contrary, medium pHs lower than their dissociation could favor pollutants complexation once interacted with the biosorbent (Lebron et al., 2019a,b). A different side effect that is expected for algae- and plant-based materials when chemically modified with acids is the solubilization of both cellulose and hemicelluloses via hydrolysis reactions (Wassie & Srivastava, 2016).

Another alternative is the use of salts. Zinc chloride is highly effective in separating hemicellulose from biomass and presents high selectivity in its hydrolysis. Lebron et al. (2019a,b) compared unmodified and modified samples of green and brown algae species in dye removal from surface water. The species of *Fucus vesiculosus* (brown macroalgae) were visually more porous and rougher after being treated with orthophosphoric acid. These alterations were observed in species of *C. pyrenoidosa* and *Spirulina maxima*, however, when treated with zinc chloride. The alterations visually observed through scanning electron microscope micrographs were later confirmed by alterations in infrared spectra when untreated and treated samples were compared. The main outcomes presented by Lebron et al. (2019a,b) were summarized in Table 4.1, being possible to compare the improvement in biosorption capacity of modified algae samples.

An interesting approach was also presented by Zafar et al. (2015), however for rice bran. Their results were also summarized in Table 4.1. The authors considered chemical modifications made with bases (NaOH, $CaOH_2$, and $AlOH_3$), acids (HCl, H_2SO_4, and HNO_3), organic solvents (formaldehyde, formic acid, acetic acid, methanol, acetic anhydride), and salts (Na_2CO_3 and $NaHCO_3$). Better results were found when rice bran was treated with NaOH and HCl, explained by their potential to increase the biosorbent porosity and their affinity for nickel uptake. Acid hydrolysis yields relatively pure amino sugar and d-glucosamine, which is more easily protonated and could lead to more binding sites. The base, in turn, removes the protein and lipid fractions from the rice bran surface, exposing more metal binding sites and improving the adsorption capacity of biomass. In the study presented by Zafar et al. (2015), no improvement in biosorption capacity was observed when rice bran was treated with salts, and inorganic solvents reduced the rice bran capacity for nickel removal.

Formaldehyde may cause cross-linking of adjacent hydroxyl groups, and the treatment with acetic anhydride may be responsible for the acetylation of amino and hydroxyl groups. Thus, the acetylation reaction reduces the number of positively charged sites on the biomass surface, which ultimately leads to a reduction in nickel uptake capacity. Methanol, in turn, may cause the esterification of the carboxylic acid groups which reduced the metal binding ability of carboxyl groups (Zafar et al., 2015).

It is important to distinguish processes intended for modification from processes intended for functionalization. Whereas the first intends to enhance the overall biosorption capacity, the functionalization aims to confer selectivity to the algae- and plant-based materials. Besides selectivity, functionalization can also improve the regeneration capacity of biosorbents (Teodoro et al., 2022), a positive aspect that favors its application on large scales. Teodoro et al. (2022) demonstrated that while preparing a biosorbent based on sugarcane bagasse modified with succinate pyromellitate. The authors aimed for the selective removal of copper and zinc in a binary solution, and their results showed copper uptake in the presence of nickel when the biomass was functionalized with succinate, whereas zinc selectivity was achieved when the biomass was modified with pyromellitate. Overall, the authors demonstrated that the functionalized materials have the potential to be used in water treatment processes.

4.6 A FUTURE RELIED ON ALGAE- AND PLANT-BASED BIOSORPTION PROCESSES: OPPORTUNITIES AND LIMITATIONS

There are different research and development initiatives to increase algae biomass production, and biomass derived from plants tends to increase as we increase crop production to attain a global demand. A partnership between Spain and the United States invested over 90 million dollars in research projects targeting an annual algae biomass production of 25,000 ton. Investments were made by Italy as well (Italian Energy Company Eni), which is currently testing pilot photobioreactors at 1 ha (Kumar et al., 2021).

Another initiative that could positively impact algae use in remediation processes is the opportunity to integrate their production into wastewater treatment processes (Gouveia et al., 2016). This is due to their ability to utilize the organic and inorganic content, as well as inorganic nitrogen and phosphorus in wastewater for their growth. Mohsenpour et al. (2021) presented a critical overview of the role of microalgae cultivation and mentioned that systems for microalgae cultivation incur little or no additional operational costs, but emphasize that the non-sterile environment associated with wastewater is still a major challenge that still limits the cultivation of algae in wastewater treatment plants.

In general, these are positive aspects seen as opportunities for the use of algae- and plant-based materials; however, advancements are to be made in experimental trials to validate their performance on large scales. When Table 4.1 is analyzed in detail, it is possible to observe that most studies limited their efforts in laboratory scales, in experiments conducted in batch mode, and for a synthetic aqueous solution containing in most cases only a single pollutant. Scale-up effects should be assessed in pilot trials, and continuous operations are to be considered to better assess the effectiveness of algae- and plant-based adsorbents.

These aspects, aligned with preliminary economic results, should be considered in future studies to better support the use of biosorbents in full-scale applications. By the time this chapter was written, no study was found that reported operational and capital costs related to algae- and plant-based materials in biosorption processes. That is certainly an aspect that deserves attention since most studies claim to be a low-cost process without proper data that supports it. The complexity of real wastewater or surface water should be considered as well. In these samples, not only a single component exists but other pollutants as well with the potential to compete for the active sites present in the biomass structure.

To exemplify, the scale-up of a biosorption process of MB by *F. vesiculosus* (brown macroalgae), *S. maxima* (green microalgae), and *C. pyrenoidosa* (green microalgae) was presented by Lebron et al. (2019a,b). The authors theoretically designed a single-stage batch biosorption system considering a material balance and the materials biosorption capacity, setting a removal efficiency between 65 and 95% of MB from wastewater containing 500 mg-dye L^{-1}. For removals greater than 90%, it was estimated a necessity of 0.1 g-algae L^{-} of wastewater treated. The mass requirement, however, should increase when real wastewater is considered.

In a multi-component system, Zhou et al. (2014) assessed the simultaneous removal of inorganic and organic compounds (nitrogen, phosphorus, metals, pharmaceuticals and personal care products, endocrine-disrupting chemicals, and estrogenic activity) in wastewater by different species of green microalgae (*Chlamydomonas reinhardtii*, *Scenedesmus obliquus*, *C. pyrenoidosa,* and *Chlorella vulgaris*). That is an approach that deserves to be mentioned due to its representation of real case application of biosorption processes. The treatment was monitored for seven days being reported removals between 76.7 and 92.3% for nitrogen and 67.5 and 82.2% for phosphorous.

For heavy metals, the values reported were greater than 40% for most species except for lead, nickel, and cobalt. In terms of organic compounds, removals were greater than 50%, and the estrogenic activity was reduced by 46.2–81.1%.

For a multimetallic system, V.R. et al. (2019) reported a 65–73% decrease in the biosorption capacity of cadmium, nickel, and lead when a single system (containing only one species) was compared with a ternary system (containing all three species). Despite the decrease in biosorption capacity, the biosorbent considered by the authors (*F. vesiculosus*; a brown macroalgae) maintained its efficiency when considered for trace metals removal from real water samples collected from a river contaminated by the rupture of a mining tailing dam. It was reported that removals of copper, iron, manganese, nickel, and aluminum were greater than 78.2%, and from all species, only aluminum was quantified above the threshold values. The removals also reduced the risks associated with water consumption. After the biosorption process, the risks were classified between medium and negligible, compared with the high risks associated with raw surface water.

Future studies should also focus on strategies to improve algae production, including strategies related to cultivation mode (Liu et al., 2019), system design that makes better usage of sunlight (Cervera Sardá & Gómez Pioz, 2015), and most probably a shift from closed to open cultivation systems, the first currently considered capital intensive and justified only when algae are considered for fine applications (Chaumont, 1993).

The results presented in Table 4.1 were also summarized in Figure 4.6. In general, algae-based materials might present higher biosorption capacities compared with plant-based ones, mostly explained by their composition favorable for interaction with pollutants. In the same figure is also explicit the difference in the number of studies that considered a batch and continuous mode, and it is also possible to observe that most studies focused on trace metal remediation. Trace metals and dyes are often investigated (Table 4.1 and Figure 4.6) even though they are not representative of all contaminants that could occur in water and wastewater. Emerging contaminants became a point of concern for being detected at increased concentrations and frequency (e.g., pharmaceutically active compounds [Santos et al., 2020] and phenolic compounds [Ramos et al., 2021]) in a scenario of no cost-effective technologies for their remediation.

That was a concern raised by Xiong et al. (2018) while discussing the microalgae potential for pharmaceutical contaminants removal from surface water in a phytoremediation process. The authors demonstrated that there still exist opportunities for improvement in remediation efficiency including genetic modifications to increase the algae resistance to these compounds, currently a limiting factor.

Other process characteristics are presented in Figure 4.6. The thermodynamic characteristics should be closely related to the species used, and no clear trend was observed to generalize the biosorption processes involving algae- and plant-based materials as endothermic or exothermic. Different from that, processes involving algae- and plant-based materials seem to follow, in general, a pseudo-second-order model. The model describes a reversible biosorption process as function of the adsorbate concentration in the biosorbent surface and its biosorption capacity and is generally associated with a process controlled by chemical interactions between the biosorbent and the contaminant.

TABLE 4.1
Summary of studies that assessed the use of algae-based and plant-based biosorbents for environmental pollutant remediation

Biosorbent	Classification	Scale and operation mode	Target pollutant	Biosorption capacity (q_e mg g^{-1})	Kinetics	Thermodynamics	References
C. pyrenoidosa	Algae—green microalgae	Lab scale, batch mode	Copper	11.8	Better described by a pseudo-second-order model	Endothermic process	Moreira et al. (2019)
C. pyrenoidosa	Algae—green microalgae	Lab scale, batch mode	MB	114.1	Better described by a pseudo-second-order model	Exothermic process	Lebron et al. (2018)
C. pyrenoidosa	Algae—green microalgae—chemically modified with H_3PO_4	Lab scale, batch mode	MB	124.0	n.r.	n.r.	Lebron et al. (2019a,b)
C. pyrenoidosa	Algae—green microalgae—chemically modified with $ZnCl_2$	Lab scale, batch mode	MB	212.0	Better described by a pseudo-second-order model	Exothermic process	Lebron et al. (2019a,b)
Fucus vecisulosus	Algae—brown macroalgae	Lab scale, batch mode	MB	698.4	Better described by a pseudo-second-order model	n.r.	Lebron et al. (2019a,b)
F. vecisulosus	Algae—brown macroalgae—chemically modified with H_3PO_4	Lab scale, batch mode	MB	1162.9	Better described by a pseudo-second-order model	Exothermic process	Lebron et al. (2019a,b)
F. vecisulosus	Algae—brown macroalgae—chemically modified with $ZnCl_2$	Lab scale, batch mode	MB	967.7	n.r.	n.r.	Lebron et al. (2019a,b)

(*Continued*)

TABLE 4.1 (*Continued*)
Summary of studies that assessed the use of algae-based and plant-based biosorbents for environmental pollutant remediation

Biosorbent	Classification	Scale and operation mode	Target pollutant	Biosorption capacity (q_e mg g^{-1})	Kinetics	Thermodynamics	References
S. maxima	Algae—green microalgae	Lab scale, batch mode	MB	411.0	Better described by a pseudo-second-order model	Exothermic process	Lebron et al. (2018)
S. maxima	Algae—green microalgae—chemically modified with H_3PO_4	Lab scale, batch mode	MB	291.5	n.r.	n.r.	Lebron et al. (2019a,b)
S. maxima	Algae—green microalgae—chemically modified with $ZnCl_2$	Lab scale, batch mode	MB	343.6	Better described by a pseudo-second-order model	n.r.	Lebron et al. (2019a,b)
F. vesiculosus	Algae—brown macroalgae	Lab scale, batch mode	Multimetallic system composed of nickel, cadmium, and lead	Cd: 146.2, Ni: 70.0, Pb: 516.2	Better described by a pseudo-second-order model for all three species	Exothermic process	Moreira et al. (2019)
F. vesiculosus	Algae—brown macroalgae	Lab scale, batch mode	MB and ET	MB: 698.4; ET: 24.3	Better described by a pseudo-second-order model for both species	Endothermic processes for both species	Lebron et al. (2019a,b)
C. pyrenoidosa	Algae—green microalgae	Lab scale, batch mode	Copper	44.6	n.r.	n.r.	Moreira et al. (2020)

(*Continued*)

TABLE 4.1 (*Continued*)
Summary of studies that assessed the use of algae-based and plant-based biosorbents for environmental pollutant remediation

Biosorbent	Classification	Scale and operation mode	Target pollutant	Biosorption capacity (q_e mg g^{-1})	Kinetics	Thermodynamics	References
D. potatorum	Algae—brown macro marine algae—chemically modified with $CaCl_2$ followed by thermal pretreatment	Lab scale, batch, and continuous mode	Cadmium	123.2	n.r.	n.r.	Matheickal et al. (1999)
Rice bran	Plant—chemically treated with NaOH	Lab scale, batch mode	Nickel	153.6	Better described by a pseudo-second-order model	Spontaneous process proven by a negative change in Gibbs free energy, however, no temperature effect was evaluated	Zafar et al. (2015)
Rice bran	Plant—chemically treated with $CaOH_2$	Lab scale, batch mode	Nickel	149.4	Better described by a pseudo-second-order model	Spontaneous process proven by a negative change in Gibbs free energy, however, no temperature effect was evaluated	Zafar et al. (2015)

(*Continued*)

TABLE 4.1 (*Continued*)
Summary of studies that assessed the use of algae-based and plant-based biosorbents for environmental pollutant remediation

Biosorbent	Classification	Scale and operation mode	Target pollutant	Biosorption capacity (q_e mg g^{-1})	Kinetics	Thermodynamics	References
Rice bran	Plant—chemically treated with $AlOH_3$	Lab scale, batch mode	Nickel	144.0	Better described by a pseudo-second-order model	Spontaneous process proven by a negative change in Gibbs free energy, however, no temperature effect was evaluated	Zafar et al. (2015)
Rice bran	Plant—chemically treated with HCl	Lab scale, batch mode	Nickel	140.0	Better described by a pseudo-second-order model	Spontaneous process proven by a negative change in Gibbs free energy, however, no temperature effect was evaluated	Zafar et al. (2015)
Rice bran	Plant—chemically treated with H_2SO_4	Lab scale, batch mode	Nickel	131.8	Better described by a pseudo-second-order model	Spontaneous process proven by a negative change in Gibbs free energy, however, no temperature effect was evaluated	Zafar et al. (2015)

(*Continued*)

TABLE 4.1 (*Continued*)
Summary of studies that assessed the use of algae-based and plant-based biosorbents for environmental pollutant remediation

Biosorbent	Classification	Scale and operation mode	Target pollutant	Biosorption capacity (q_e mg g^{-1})	Kinetics	Thermodynamics	References
Rice bran	Plant—chemically treated with HNO_3	Lab scale, batch mode	Nickel	129.9	Better described by a pseudo-second-order model	Spontaneous process proven by a negative change in Gibbs free energy, however, no temperature effect was evaluated	Zafar et al. (2015)
Rice bran	Plant—chemically treated with formaldehyde	Lab scale, batch mode	Nickel	118.0	n.r.	n.r.	Zafar et al. (2015)
Rice bran	Plant—chemically treated with formic acid	Lab scale, batch mode	Nickel	116.0	n.r.	n.r.	Zafar et al. (2015)
Rice bran	Plant—chemically treated with acetic acid	Lab scale, batch mode	Nickel	111.9	n.r.	n.r.	Zafar et al. (2015)
Rice bran	Plant—chemically treated with methanol	Lab scale, batch mode	Nickel	91.4	n.r.	n.r.	Zafar et al. (2015)
Rice bran	Plant—chemically treated with acetic anhydride	Lab scale, batch mode	Nickel	86.7	n.r.	n.r.	Zafar et al. (2015)
Rice bran	Plant—chemically treated with Na_2CO_3	Lab scale, batch mode	Nickel	80.2	n.r.	n.r.	Zafar et al. (2015)
Rice bran	Plant—chemically treated with $NaHCO_3$	Lab scale, batch mode	Nickel	77.0	n.r.	n.r.	Zafar et al. (2015)

(*Continued*)

TABLE 4.1 (*Continued*)
Summary of studies that assessed the use of algae-based and plant-based biosorbents for environmental pollutant remediation

Biosorbent	Classification	Scale and operation mode	Target pollutant	Biosorption capacity (q_e mg g^{-1})	Kinetics	Thermodynamics	References
Sugarcane-based	Plant—functionalized with succinate pyromellitate	Lab scale, batch, and continuous mode	Copper and zinc	Cu: 75.6 Zn: 62.1	n.r.	Endothermic processes for both species	Teodoro et al. (2022)
U. fasciata sp.	Algae—marine green microalgae	Lab scale, batch mode	Zinc	13.5	Pseudo-second-order model	n.r.	Kumar et al. (2007)
Orange peel	Plant—raw material	Lab scale, batch mode	Cadmium	47.21	Pseudo-first-order model	n.r.	Li et al. (2007)
Orange peel	Plant—washed with isopropyl alcohol	Lab scale, batch mode	Cadmium	57.32	Pseudo-first-order model	n.r.	Li et al. (2007)
Orange peel	Plant—modified with citric acid	Lab scale, batch mode	Cadmium	64.07	Pseudo-first-order model	n.r.	Li et al. (2007)
Orange peel	Plant—modified with alkali saponification	Lab scale, batch mode	Cadmium	80.93	Pseudo-first-order model	n.r.	Li et al. (2007)
Orange peel	Plant—modified with citric acid after alkali saponification	Lab scale, batch mode	Cadmium	101.16	Pseudo-first-order model	n.r.	Li et al. (2007)
C. sinensis	Plant—chemically treated with H_2SO_4	Lab scale, batch mode	Lead	22.11	Pseudo-second-order model	Spontaneous endothermic process	Nuhoğlu et al. (2021)

(*Continued*)

TABLE 4.1 (*Continued*)
Summary of studies that assessed the use of algae-based and plant-based biosorbents for environmental pollutant remediation

Biosorbent	Classification	Scale and operation mode	Target pollutant	Biosorption capacity (q_e mg g^{-1})	Kinetics	Thermodynamics	References
S. hispanicus L.	Plant—raw material	Lab scale, batch mode	MB and ET	MB: 263.92 ET: 165.77	Pseudo-second-order model for both species	Spontaneous exothermic process for both species	Barka et al. (2011)
C. indica	Algae—brown algae	Lab scale, batch mode	Nickel	47.6	Pseudo-second-order model	Spontaneous endothermic process	Pahlavanzadeh (2010)
N. zanardini	Algae—brown algae	Lab scale, batch mode	Nickel	45.4	Pseudo-second-order model	Spontaneous endothermic process	Pahlavanzadeh (2010)
S. glaucescens	Algae—brown algae	Lab scale, batch mode	Nickel	47.6	Pseudo-second-order model	Spontaneous endothermic process	Pahlavanzadeh (2010)
P. australis	Algae—brown algae	Lab scale, batch mode	Nickel	23.2	Pseudo-second-order model	Spontaneous endothermic process	Pahlavanzadeh (2010)
P. dioica L. Merrill	Plant	Lab scale, continuous mode	Lead	n.r.	n.r.	n.r.	Cruz-Olivares et al. (2013)
Olive stone	Plant	Lab scale, continuous mode	Chromium	n.r.	n.r.	n.r.	Calero et al. (2009)
Rice husk	Plant	Lab scale, continuous mode	Arsenic	n.r.	n.r.	n.r.	Asif and Chen (2015)

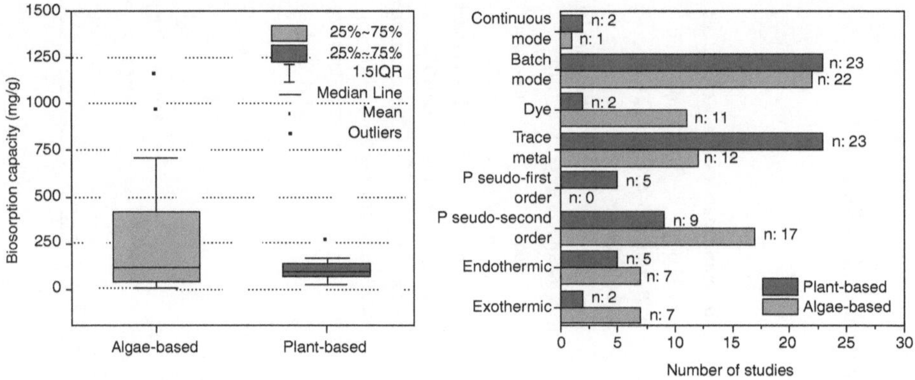

FIGURE 4.6 Summary of main outcomes related to algae- and plant-based materials used for environmental pollution remediation

A database including a higher number of studies could potentially lead to more assertive conclusions related to kinetics and thermodynamics aspects. For kinetics, in specific, special attention should be given when analyzing the experimental data and comparing it with other models as the pseudo-first-order. Simonin (2016) presented a discussion related to that and highlighted a biased interpretation of biosorption studies favoring the pseudo-second-order model. The authors presented experimental and statistical issues that would compromise the fairness in comparisons made, and new criteria are presented to enable a proper comparison, including a new kinetic model.

4.7 FINAL CONSIDERATIONS

The use of algae- and plant-based biosorption processes was demonstrated to be an effective alternative for the treatment of pollutants present in water and wastewater. Studies discussed in this chapter corroborate the potential of the technique and emphasize their performance that is comparable with current commercial adsorbents. The algae-based biosorption stands out, in addition to the performance of the process, for the expected growth of algae production and for presenting a great variety and diversity of species that could be used as biosorbents. At the same time, the plant-based process has a vast group of adsorbents, with high efficiency, low cost, high availability, and chemical stability. However, even with this high performance, these processes have some limitations like the need for experimental trials and the development of strategies to improve algae production, including strategies related to cultivation mode that should be better investigated and improved in future research, which makes the field of research broad and visionary.

ACKNOWLEDGMENTS

This writing process was supported by the Coordination of Superior Level Staff Improvement (CAPES), the National Council for Scientific and Technological Development (CNPq), and the Foundation for Research Support of the State of Minas Gerais (FAPEMIG).

REFERENCES

Abegunde, S.M., Idowu, K.S., Adejuwon, O.M., Adeyemi-Adejolu, T., 2020. A review on the influence of chemical modification on the performance of adsorbents. *Resources, Environment and Sustainability* 1, 100001. https://doi.org/10.1016/j.resenv.2020.100001

Adewuyi, A., 2020. Chemically modified biosorbents and their role in the removal of emerging pharmaceutical waste in the water system. *Water* 12, 1551. https://doi.org/10.3390/w12061551

Agarwal, A., Upadhyay, U., Sreedhar, I., Singh, S. A., Patel, C. M. 2020. A review on valorization of biomass in heavy metal removal from wastewater. *Journal of Water Process Engineering*, 38, 101602.

Ahmed, M. J., Hameed, B. H. 2018. Removal of emerging pharmaceutical contaminants by adsorption in a fixed-bed column: a review. *Ecotoxicology and Environmental Safety*, 149, 257–266.

Al-Homaidan, A.A., Al-Houri, H.J., Al-Hazzani, A.A., Elgaaly, G., Moubayed, N.M.S., 2014. Biosorption of copper ions from aqueous solutions by Spirulina platensis biomass. *Arabian Journal of Chemistry* 7, 57–62. https://doi.org/10.1016/j.arabjc.2013.05.022

Ankit, Bauddh, K., Korstad, J., 2022. Phycoremediation: Use of algae to sequester heavy metals. *Hydrobiology* 1, 288–303. https://doi.org/10.3390/hydrobiology1030021

Asif, Z., Chen, Z. 2017. Removal of arsenic from drinking water using rice husk. *Applied Water Science*, 7, 1449–1458.

Barka, N., Abdennouri, M., Makhfouk, M. E. 2011. Removal of Methylene Blue and Eriochrome Black T from aqueous solutions by biosorption on Scolymus hispanicus L.: Kinetics, equilibrium and thermodynamics. *Journal of the Taiwan Institute of Chemical Engineers*, 42(2), 320–326.

Bharathiraja, B., Chakravarthy, M., Ranjith Kumar, R., Yogendran, D., Yuvaraj, D., Jayamuthunagai, J., Praveen Kumar, R., Palani, S., 2015. Aquatic biomass (algae) as a future feed stock for bio-refineries: A review on cultivation, processing and products. *Renewable and Sustainable Energy Reviews* 47, 634–653. https://doi.org/10.1016/j.rser.2015.03.047

Bilal, M., Rasheed, T., Sosa-Hernández, J., Raza, A., Nabeel, F., Iqbal, H., 2018. Biosorption: An interplay between marine algae and potentially toxic elements—A review. *Marine Drugs* 16, 65. https://doi.org/10.3390/md16020065

Calero, M., Hernáinz, F., Blázquez, G., Tenorio, G., Martín-Lara, M. A. 2009. Study of Cr (III) biosorption in a fixed-bed column. *Journal of Hazardous Materials*, 171(1–3), 886–893.

Cervera Sardá, R., Gómez Pioz, J., 2015. Architectural bio-photo reactors: Harvesting microalgae on the surface of architecture, in: *Biotechnologies and Biomimetics for Civil Engineering*. Springer International Publishing, Cham, pp. 163–179. https://doi.org/10.1007/978-3-319-09287-4_7

Chaumont, D., 1993. Biotechnology of algal biomass production: A review of systems for outdoor mass culture. *Journal of Applied Phycology* 5, 593–604. https://doi.org/10.1007/BF02184638

Crini, G., Badot, P. M. (2008). Application of chitosan, a natural aminopolysaccharide, for dye removal from aqueous solutions by adsorption processes using batch studies: A review of recent literature. *Progress in Polymer Science* 33(4), 399–447.

Cruz-Olivares, J., Pérez-Alonso, C., Barrera-Díaz, C., Ureña-Nuñez, F., Chaparro-Mercado, M. C., Bilyeu, B. 2013. Modeling of lead (II) biosorption by residue of allspice in a fixed-bed column. *Chemical Engineering Journal*, 228, 21–27.

Davis, T.A., Volesky, B., Mucci, A., 2003. A review of the biochemistry of heavy metal biosorption by brown algae. *Water Research* 37, 4311–4330. https://doi.org/10.1016/S0043-1354(03)00293-8

Demey, Barron-Zambrano, Mhadhbi, Miloudi, Yang, Ruiz, Sastre, 2019. Boron removal from aqueous solutions by using a novel alginate-based sorbent: Comparison with Al2O3 particles. *Polymers (Basel).* 11, 1509. https://doi.org/10.3390/polym11091509

Eckenfelder, W. W., Ford, D. L., Englande, A. J. 2009. *Industrial Water Quality*. McGraw-Hill Education.

Elgarahy, A.M., Elwakeel, K.Z., Mohammad, S.H., Elshoubaky, G.A., 2021. A critical review of biosorption of dyes, heavy metals and metalloids from wastewater as an efficient and green process. *Cleaner Engineering and Technology* 4, 100209. https://doi.org/10.1016/j.clet.2021.100209

Feng, N., Guo, X., Liang, S., Zhu, Y., Liu, J., 2011. Biosorption of heavy metals from aqueous solutions by chemically modified orange peel. *Journal of Hazardous Materials* 185, 49–54. https://doi.org/10.1016/j.jhazmat.2010.08.114

Gahlout, M., Prajapati, H., Tandel, N., Patel, Y. 2021. Biosorption: An Eco-Friendly Technology for Pollutant Removal, in: Panpatte, D.G., Jhala, Y.K. (eds) *Microbial Rejuvenation of Polluted Environment*. Microorganisms for Sustainability, vol. 26. Springer, Singapore. https://doi.org/10.1007/978-981-15-7455-9_9

Gadd, G.M., Fomina, M., 2011. Uranium and fungi. *Geomicrobiology Journal* 28, 471–482. https://doi.org/10.1080/01490451.2010.508019

Gautam, R.K., Mudhoo, A., Lofrano, G., Chattopadhyaya, M.C., 2014. Biomass-derived biosorbents for metal ions sequestration: Adsorbent modification and activation methods and adsorbent regeneration. *Journal of Environmental Chemical Engineering*. https://doi.org/10.1016/j.jece.2013.12.019

Gouveia, L., Graça, S., Sousa, C., Ambrosano, L., Ribeiro, B., Botrel, E.P., Neto, P.C., Ferreira, A.F., Silva, C.M., 2016. Microalgae biomass production using wastewater: Treatment and costs. *Algal Research* 16, 167–176. https://doi.org/10.1016/j.algal.2016.03.010

Guiry, M.D., 2012. How many species of algae are there? *Journal of Phycology* 48, 1057–1063. https://doi.org/10.1111/j.1529-8817.2012.01222.x

Jain, M., Garg, V. K., Kadirvelu, K., Sillanpää, M. 2016. Adsorption of heavy metals from multi-metal aqueous solution by sunflower plant biomass-based carbons. *International Journal of Environmental Science and Technology*, 13, 493–500.

Jasrotia, S., Kansal, A., Mehra, A., 2017. Performance of aquatic plant species for phytoremediation of arsenic-contaminated water. *Applied Water Science* 7, 889–896. https://doi.org/10.1007/s13201-015-0300-4

Kumar, B.R., Mathimani, T., Sudhakar, M.P., Rajendran, K., Nizami, A.-S., Brindhadevi, K., Pugazhendhi, A., 2021. A state of the art review on the cultivation of algae for energy and other valuable products: Application, challenges, and opportunities. *Renewable and Sustainable Energy Reviews* 138, 110649. https://doi.org/10.1016/j.rser.2020.110649

Kumar, Y. P., King, P., Prasad, V. S. R. K. 2007. Adsorption of zinc from aqueous solution using marine green algae—Ulva fasciata sp. *Chemical Engineering Journal*, 129(1–3), 161–166.

Lebron, Y.A.R., Moreira, V.R., Santos, L.V.S., Jacob, R.S., 2018. Remediation of methylene blue from aqueous solution by Chlorella pyrenoidosa and Spirulina maxima biosorption: Equilibrium, kinetics, thermodynamics and optimization studies. *Journal of Environmental Chemical Engineering* 6, 6680–6690. https://doi.org/10.1016/j.jece.2018.10.025

Lebron, Y.A.R., Moreira, V.R., de Souza Santos, L.V., 2019a. Biosorption of methylene blue and eriochrome black T onto the brown macroalgae Fucus vesiculosus: Equilibrium, kinetics, thermodynamics and optimization. *Environmental Technology*, 1–59. https://doi.org/10.1080/09593330.2019.1626914

Lebron, Y.A.R., Moreira, V.R., de Souza Santos, L.V., 2019b. Studies on dye biosorption enhancement by chemically modified Fucus vesiculosus, Spirulina maxima and Chlorella pyrenoidosa algae. *Journal of Cleaner Production* 240, 118197. https://doi.org/10.1016/j.jclepro.2019.118197

Li, X., Tang, Y., Xuan, Z., Liu, Y., & Luo, F. 2007. Study on the preparation of orange peel cellulose adsorbents and biosorption of Cd2+ from aqueous solution. *Separation and Purification Technology*, 55(1), 69–75.

Liu, X., Hong, Y., Liu, P., Zhan, J., Yan, R., 2019. Effects of cultivation strategies on the cultivation of Chlorella sp. HQ in photoreactors. *Frontiers of Environmental Science and Engineering* 13, 78. https://doi.org/10.1007/s11783-019-1162-z

Maia, L. C., Soares, L. C., Gurgel, L. V. A., 2021. A review on the use of lignocellulosic materials for arsenic adsorption. *Journal of Environmental Management*, 288, 112397.

Manzoor, K., Batool, M., Naz, F., Nazar, M. F., Hameed, B. H., Zafar, M. N., 2022. A comprehensive review on application of plant-based bioadsorbents for Congo red removal. *Biomass Conversion and Biorefinery*, 1–27.

Matheickal, J.T., Yu, Q., Woodburn, G.M., 1999. Biosorption of cadmium(II) from aqueous solutions by pre-treated biomass of marine alga DurvillAea potatorum. *Water Research* 33, 335–342. https://doi.org/10.1016/S0043-1354(98)00237-1

Mohsenpour, S.F., Hennige, S., Willoughby, N., Adeloye, A., Gutierrez, T., 2021. Integrating micro-algae into wastewater treatment: A review. *Science of the Total Environment* 752, 142168. https://doi.org/10.1016/j.scitotenv.2020.142168

Moreira, V.R., Lebron, Y.A.R., Freire, S.J., de Souza Santos, L.V., Palladino, F., Jacob, R.S., 2019. Biosorption of copper ions from aqueous solution using Chlorella pyrenoidosa: Optimization, equilibrium and kinetics studies. *Microchemical Journal* 145, 119–129. https://doi.org/10.1016/j.microc.2018.10.027

Moreira, V.R., Lebron, Y.A.R., de Souza Santos, L.V., 2020. Predicting the biosorption capacity of copper by dried Chlorella pyrenoidosa through response surface methodology and artificial neural network models. *Chemical Engineering Journal Advances* 4, 100041. https://doi.org/10.1016/j.ceja.2020.100041

Nizamuddin, S., Baloch, H. A., Griffin, G. J., Mubarak, N. M., Bhutto, A. W., Abro, R., ... Ali, B. S. 2017. An overview of effect of process parameters on hydrothermal carbonization of biomass. *Renewable and Sustainable Energy Reviews*, 73, 1289–1299.

Nuhoğlu, Y., Ekmekyapar Kul, Z., Kul, S., Nuhoğlu, Ç., Ekmekyapar Torun, F. 2021. Pb (II) biosorption from the aqueous solutions by raw and modified tea factory waste (TFW). *International Journal of Environmental Science and Technology*, 1–12.

Pahlavanzadeh, H., Keshtkar, A. R., Safdari, J., Abadi, Z. 2010. Biosorption of nickel (II) from aqueous solution by brown algae: Equilibrium, dynamic and thermodynamic studies. *Journal of Hazardous Materials*, 175(1–3), 304–310.

Pereira, L., Alves, M., 2012. Dyes—Environmental Impact and Remediation, in: Malik, A., Grohmann, E. (eds) *Environmental Protection Strategies for Sustainable Development.* Springer Netherlands, Dordrecht, pp. 111–162. https://doi.org/10.1007/978-94-007-1591-2_4

Ramos, R.L., Moreira, V.R., Lebron, Y.A.R., Santos, A. V., Santos, L.V.S., Amaral, M.C.S., 2021. Phenolic compounds seasonal occurrence and risk assessment in surface and treated waters in Minas Gerais—Brazil. *Environmental Pollution* 268, 115782. https://doi.org/10.1016/j.envpol.2020.115782

Rangabhashiyam, S., Lata, Sujata, Balasubramanian, P., 2018. Biosorption characteristics of methylene blue and malachite green from simulated wastewater onto Carica papaya wood biosorbent. *Surfaces and Interfaces* 10, 197–215. https://doi.org/10.1016/j.surfin.2017.09.011

Santos, A.V., Couto, C.F., Lebron, Y.A., Moreira, V.R., Foureaux, A.F.S., Reis, E.O., de Souza Santos, L.V., de Andrade, L.H., Amaral, M.C.S., Lange, L.C., 2020. Occurrence and risk assessment of pharmaceutically active compounds in water supply systems in Brazil. *Science of the Total Environment* 141011. https://doi.org/10.1016/j.scitotenv.2020.141011

Saravanan, 2013. Biosorption of textile dye using immobilized bacterial (Pseudomonas Aeruginosa) and fungal (Phanerochate Chrysosporium) cells. *American Journal of Environmental Sciences* 9, 377–387. https://doi.org/10.3844/ajessp.2013.377.387

Shen, Y., Yuan, W., Pei, Z.J., Wu, Q., Mao, E., 2009. Microalgae mass production methods. *Transactions of the ASABE* 52, 1275–1287. https://doi.org/10.13031/2013.27771

Simonin, J.-P., 2016. On the comparison of pseudo-first order and pseudo-second order rate laws in the modeling of adsorption kinetics. *Chemical Engineering Journal* 300, 254–263. https://doi.org/10.1016/j.cej.2016.04.079

Tan, K. L., Hameed, B. H. 2017. Insight into the adsorption kinetics models for the removal of contaminants from aqueous solutions. *Journal of the Taiwan Institute of Chemical Engineers*, 74, 25–48.

Teodoro, F.S., Soares, L.C., Filgueiras, J.G., de Azevedo, E.R., Patiño-Agudelo, Á.J., Adarme, O.F.H., da Silva, L.H.M., Gurgel, L.V.A., 2022. Batch and continuous adsorption of Cu(II) and Zn(II) ions from aqueous solution on bi-functionalized sugarcane-based biosorbent. *Environmental Science and Pollution Research* 29, 26425–26448. https://doi.org/10.1007/s11356-021-17549-5

Thirunavukkarasu, A., Nithya, R., Sivashankar, R., 2021. Continuous fixed-bed biosorption process: A review. *Chemical Engineering Journal Advances* 8, 100188. https://doi.org/10.1016/j.ceja.2021.100188

Tural, B., Ertaş, E., Enez, B., Fincan, S.A., Tural, S., 2017. Preparation and characterization of a novel magnetic biosorbent functionalized with biomass of Bacillus Subtilis: Kinetic and isotherm studies of biosorption processes in the removal of Methylene Blue. *Journal of Environmental Chemical Engineering* 5, 4795–4802. https://doi.org/10.1016/j.jece.2017.09.019

V.R., M., Y.A.R., L., Lange, L.C., L.V.S., S., 2019. Simultaneous biosorption of Cd(II), Ni(II) and Pb(II) onto a brown macroalgae fucus vesiculosus: Mono- and multi-component isotherms, kinetics and thermodynamics. *Journal of Environmental Management* 251, 109587. https://doi.org/10.1016/j.jenvman.2019.109587

Wassie, A.B., Srivastava, V.C., 2016. Chemical treatment of teff straw by sodium hydroxide, phosphoric acid and zinc chloride: Adsorptive removal of chromium. *International Journal of Environmental Science and Technology* 13, 2415–2426. https://doi.org/10.1007/s13762-016-1080-6

Xiong, J.-Q., Kurade, M.B., Jeon, B.-H., 2018. Can microalgae remove pharmaceutical contaminants from water? *Trends Biotechnology* 36, 30–44. https://doi.org/10.1016/j.tibtech.2017.09.003

Yadav, S., Yadav, A., Bagotia, N., Sharma, A. K., Kumar, S. 2021. Adsorptive potential of modified plant-based adsorbents for sequestration of dyes and heavy metals from wastewater-A review. *Journal of Water Process Engineering*, 42, 102148.

Zafar, M.N., Aslam, I., Nadeem, R., Munir, S., Rana, U.A., Khan, S.U.-D., 2015. Characterization of chemically modified biosorbents from rice bran for biosorption of Ni(II). *Journal of the Taiwan Institute of Chemical Engineers* 46, 82–88. https://doi.org/10.1016/j.jtice.2014.08.034

Zhou, G.-J., Ying, G.-G., Liu, S., Zhou, L.-J., Chen, Z.-F., Peng, F.-Q., 2014. Simultaneous removal of inorganic and organic compounds in wastewater by freshwater green microalgae. *Environmental Science Process Impacts* 16, 2018. https://doi.org/10.1039/C4EM00094C

5 Animal-Based Biosorbents

Johnson O. Oladele, Meichen Wang, and Timothy D. Phillips

5.1 INTRODUCTION

There is an upsurge in the rate at which contaminants are being released into the biosphere as a result of the increase in man activities and industrial processes. Scientists around the world have been faced with the discovery of biomaterials that can successfully remove these contaminants without contributing to their deleterious effects. For the intended pollutant removal, a wide range of biomaterials found in nature have been used as biosorbent. Biomass of every kind, including those produced by microorganisms, plants, and animals, as well as the products that are derived from it, has attracted a lot of attention [1–3]. However, in recent years, research has focused on polysaccharides, industrial waste biomaterials, and agricultural waste materials [4–7]. Due to its high amount of amino and hydroxyl functional groups, chitosan, a naturally occurring amino polysaccharide, has drawn significant attention as one of these biomaterials for the treatment of numerous aquatic contaminants.

In addition, a wide range of biological substances, particularly algae (including seaweeds, macroalgae, and microalgae), lichens, fungi, yeasts, bacteria, and cyanobacteria, have attracted significant attention for the removal and recovery of heavy metal ions due to their successful use, low cost, and widespread availability. These biological materials have a higher affinity for metal ions due to the abundance of chelating functional groups [8]. The following biomaterials have garnered the most attention in the literature, outside the abovementioned natural biosorbents: peat moss [9], sugarcane bagasse [10], sawdust [11, 12], leaves [13, 14], plant barks [15, 16], coconut shell [17], and rice husk [18]. From the previously stated biomaterials, particular focus was placed on the utilization of fly ash, which is produced during coal burning, as a practical sorbent [19–21]. At higher pH levels, fly ash has a strongly alkaline surface that is negatively charged. Thus, it is reasonable to assume that electrostatic attraction, precipitation, and ion exchange are all effective ways to remove metal ions from aqueous solutions [20, 21]. The use of biological ashes (wood and bone ash) will be a promising substitute for traditional adsorbents used in wastewater treatment, according to a study by Chojnacka and Michalak [22]. Figure 5.1 presents a broad diagram of the many biosorbent types employed in the biosorption process.

Biosorbents are often created by inactivating waste biomass that is naturally plentiful and then pre-treating by washing with acid or base before final drying [1]. To create particles with the necessary mechanical properties, certain types of biomass

DOI: 10.1201/9781003366058-5

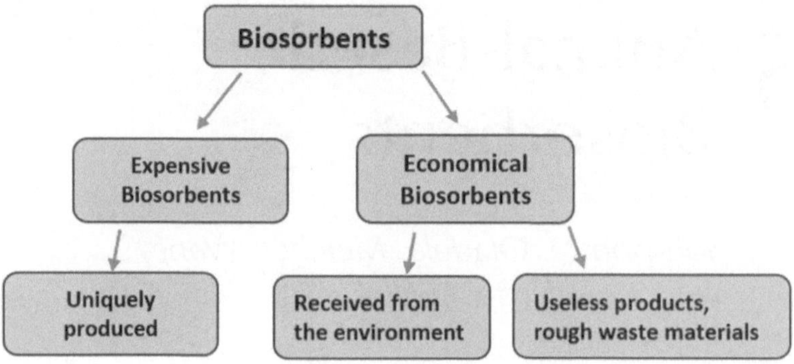

FIGURE 5.1 Chart of kinds of biosorbents employed in biosorption applications

must either be immobilized by a synthetic polymer matrix [23] or grafted onto an inorganic support material like silica [24]. Additionally, dry biomass can be simply chopped or ground to produce stable biosorbent particles of a suitable size [25]. Overall, dead biomass was used as the majority of the biosorbent; this has some distinct benefits over the use of living microorganisms such as easy storage of dead cells or usage for longer periods of time, nutrient supply is not necessary, dead biomass is not constrained by metal toxicity, and metal ion-loaded biosorbents can be easily desorbed and reused [26, 27]. The use of non-living biomass in powdered form does have some drawbacks, though, including small particle size, low mechanical strength, mass loss after regeneration, and inability to easily remove the biomass from the reaction system, which makes it challenging to use in batch and continuous systems [28]. However, by employing an appropriate immobilization technique, these issues can be resolved.

5.2 BIOSORPTION

5.2.1 History Overview of Biosorption

Even though research on the capacity of living microorganisms to absorb metals from aqueous solutions dates back to the 18th and 19th centuries [29], it has only been in the last three decades that living or non-living microorganisms have been used as adsorbents for the removal and recovery of materials from aqueous solutions. Treatment of sewage and wastewater was one of the earliest technological applications of biosorption techniques [30]. The possibility of using it to upgrade wastewater produced by the chemical sector was also explored [31]. The Ames Crosta Mills & Company Ltd. filed the first patent application for a biosorption device used for biological wastewater treatment in 1973 [32].

Investigators in the life sciences mainly concentrated on the harmful effects and increase in concentration of microorganisms, while environmental engineers and investigators used this ability of microorganisms as a technique for evaluating heavy metal pollution as well as for removal/recovery of metals from metal-bearing wastewaters [32]. The earliest quantitative study of metal biosorption, according to various review papers, was conducted by L. Hecke in 1902 on the copper uptake by the fungus spores tritici and crameri [33]. Two co-researchers published similar tests in 1922 that assessed how well corn smut absorbed Ag, Cu, Ce, and Hg [32].

Volesky and Tsezos received the first patent on the application of biosorption technology in 1982 [34], while Ruchloft was the first to report that activated sludge was capable of successfully extracting radioactive elements like plutonium-239 from contaminated residential wastewater in 1949 [33]. The practical application of biosorption technique for detecting minute amounts of heavy metals in the environment was reported by Goodman and Roberts in 1971 [35]. In their 1975 study on the kinetics of biosorption by activated sludge, Neufeld and Hermann [36] found that Cd, Hg, and Zn were rapidly absorbed in the first few minutes, followed by a sluggish uptake over the following three hours.

For a metal-binding investigation, Friedmann and Dugan (1968) employed a pure culture of *Zoogloea* [37]. The ability of microorganisms to acquire uranium ions was found to be in the order of fungus<yeast< bacteria<actinomycetes by Nakajima and his coworkers (1978) after extensive screening of microorganisms for metal uptake [38, 39]. When Gould and Genetelli (1984) looked at the rivalry of metal ions for anaerobic sludge's binding sites, they found that Ni<Zn<Cd<Cu [39, 40] was the order of binding affinities. While Tsezos and Volesky (1981) concentrated on the biosorptive removal of thorium and uranium by dead fungal biomass of *Rhizopus arrhizus* [41], Chiu and his collaborators (1976) examined the biological sorption of uranium on mycelia of Penicillium C-1 [42]. Steen and Karickhoff (1981) observed that mixed microbial populations ingested hydrophobic organic contaminants [43]. Given the current status of the biomass employed in this study, it should be emphasized that there was some ambiguity surrounding the words "bioaccumulation" and "biosorption" in the literature at this point. Volesky and Tsezos, pioneers in this field, offer their perspectives on the early development of biosorption research in their special review papers [41].

5.2.2 Scope and Description of Biosorption

The process of removing heavy metals from an aqueous solution by their passive adsorption to non-living biomass is known as biosorption. This suggests that the removal process is not regulated by metabolism. Contrarily, the term "bioaccumulation" refers to an active process in which the metabolism of a living creature is necessary for the elimination of metals. Since biomass can be used to recover precious metals from processing solutions or sequester heavy metals from industrial effluents (such as those from the mining or electroplating industries), research on the mechanisms of biosorption has increased in recent years [45].

The use of biological technology to remove pollutants has three main benefits. The first is that biological processes can be carried out on-site at the contaminated location; second, bioprocess technologies are typically cost-effective and environmentally friendly (no secondary pollutants). Bioaccumulation and biosorption have been shown to have the best potential to replace current procedures for the removal of dyes and metals among the many biological techniques. The phenomena of living cells are thus classified as bioaccumulation in this context, whereas biosorption methods are based on the usage of dead biomass. Toxic substances being taken up by living cells is the precise definition of bioaccumulation. The toxicant can enter the cell, build up intracellularly, cross the cell membrane, and go through the metabolic process of the cell. In contrast, biosorption is the passive uptake of toxins by inert/dead biological materials or by substances produced from living organisms. The methods for pollutant uptake vary depending on the kind of biomass, and biosorption is the result of a number of metabolism-independent processes that primarily occur in the cell wall. There are several intrinsic benefits of biosorption over bioaccumulation methods. As a result of these positive traits, biosorption has, unsurprisingly, attracted a lot of attention recently [46].

5.2.3 Merits of Biosorption

i. Since cells involved are not living things, processing conditions are not limited to those that promote cell growth. In other words, it is feasible to operate in a larger range of conditions, including metal concentration, temperature, and pH. This procedure doesn't require aseptic conditions. If the value and amount of metal recovered are significant, metal can be easily desorbed and then recovered. If the biomass is abundant, metal-loaded biomass can be burned, obviating the need for additional treatment.
ii. The process is extremely quick and lasts only a few minutes to a few hours since non-living biomass acts as an ion exchanger. Biomass frequently has a very high metal loading, which makes metal uptake particularly effective.
iii. Since biomass is essentially a waste product, it may be obtained easily.
iv. The physiological restraint of living microbial cells is not what controls the process.
v. The cytotoxic restriction of cells is not applicable to non-living, growth-independent biomass. Because expensive nutrients are not necessary for cell growth in feed solutions, there are no issues with disposing of extra nutrients or metabolic waste.

5.2.4 Demerits of Biosorption

i. Because cells do not metabolize, the possibility of improving biological processes (such as through genetic engineering of cells is constrained. Because the adsorptive agent is produced during pre-growth, the properties of the biosorbent are not under biological control. If waste biomass from a fermentation unit is used, this will be especially true.
ii. There is no chance that the metal valency status may be changed naturally, for the breakdown of organometallic complexes, for instance, or even in less soluble forms [47].
iii. Early saturation can be a concern; regardless of the metal value, metal desorption is required prior to future usage when metal interaction sites are occupied.

5.3 MECHANISMS INVOLVED IN BIOSORBENT-BASED POLLUTANT REMOVAL

Numerous kinds of biosorbents, such as those made from algae, yeasts, fungi, and bacteria, are produced from different types of raw material. The complexity of raw biomass suggests that there are numerous, but as-yet-unknown, methods by which these biosorbents remove various contaminants. For instance, microbes' extracellular polymeric substances and their functions in metal biosorption are related in terms of their structure and functionality. Accordingly, a variety of functional and chemical groups can draw out and ensnare contaminants depending on the biosorbent selected. Phosphodiester, phosphate, phenolic, thioether, sulfhydryl, sulfonate, imidazole, imine, hydroxyl, carboxyl, carbonyl, amine, and amidate groups can all be found in these compounds [44].

Although some functional groups may increase the likelihood of successful biosorption of contaminants, other obstacles such as conformational, steric, or other ones might also exist. The significance of any given group for the biosorption of a given pollutant by a given biomass depends on a number of variables, such as the chemical state of the sites (i.e., availability), accessibility of the sites, quantity of reactive sites in the biosorbent, and affinity between the sites and the specific pollutant of interest (i.e., binding strength). Biosorption of metals or dyes occurs primarily through interactions like precipitation, entrapment in inner spaces, adsorption by physical forces, complexation, and ion exchange [44].

5.3.1 Complexation

The number of protons released into solution decreased in the following order: $Mg^{2+} < Mn^{2+} < Ni^{2+} < Co^{2+} < Ca^{2+} < Sr^{2+} < Ba^{2+} < Cd^{2+} < Cu^{2+} < Pb^{2+}$, according to metal–ion binding to alginic acid extracted from *Laminaria digitata*. The relative capacity of the binding metal to compete with protons for organic binding sites was used to interprete these findings. Alginate's metal–ion binding affinity sequence was taken from *L. digitata* with similar patterns observed: $Co^{2+} < Ca^{2+} < Ba^{2+} < Cu^{2+}$. It was discovered that alkaline earth metals lose their ability to attach to polymannuronate and polyguluronate in the following order: $Mg^{2+} < Ca^{2+} < Sr^{2+} < Ba^{2+}$. The preferential binding of heavier ions was attributed by Haug and Smidsrod to stereochemical effects because larger ions may fit a binding site with two distant functional groups better [45].

5.3.2 Ion Exchange

Ion exchange is a crucial idea in biosorption since it provides an explanation for many of the findings from studies on the absorption of heavy metals. Moreover, given that it has been demonstrated that ion exchange occurs between metals when binding to alginate, it is a logical extension to the idea that alginate plays a significant part in biosorption by brown algae. When the alga, *Ascophyllum nodosum*, reacted with a cobalt-containing aqueous solution as opposed to a cobalt-free solution, Kuyucak and Volesky reported an elevated release of ions (Na^+, Mg^{2+}, K^+, and Ca^{2+}). Ca^{2+} release and Co^{2+} absorption were shown to have a 2:3 stoichiometric connection when the alga was pre-treated with $CaCl_2$ and HCl. The ratio would have been closer to one, however, according to Schiewer and Volesky [48], if protons were taken into account when balancing the charge. Ion exchange was found to be the predominant process. The majority of untreated biomass contains light metal ions like Mg^{2+}, Ca^{2+}, Na^+, and K^+. These were originally absorbed from seawater and bonded to the algal acid functional groups [45].

5.3.3 Coordination or Complex Formation

Coordination, sometimes referred to as complex formation, is the joining of cations with molecules or anions that have free electron pairs (bases). Covalent or electrostatic complex formation, or coordination, can take either of these two forms. When a heavy metal cation is bonded, it is frequently referred to as the central atom and separated from the anions or molecules that make up the coordination complex, or ligand(s). The ligand atom is the one that gives the ligand its basic or nucleophilic properties when the ligand is made up of several atoms. A base in a multidentate complex, which contains more than one ligand atom, may occupy more than one coordination position. Chelation is the process of forming complexes from multidentate ligands; complexes are chelates. Additionally, the coordination number refers to the number of ligand atoms surrounding the central atom. The most typical coordination numbers for metal cations are four and six. Due to steric effects, these values may be lower in the case of polymers. In contrast to the greater coordination numbers observed in metal complexes, a proton complex has a coordination number of one [45]. This outline is meant to serve as a foundation for the terminology, even if it is normally used for aqueous complexation with tiny ligands. This is because the phrases are frequently used in the literature when discussing more complex compounds. The words "inner-sphere complex" and "outer-sphere complex" are used to distinguish between binding that is mostly electrostatic or predominantly covalent in nature. In the first scenario, the metal cation and the interacting ligand are situated right next to one another. In the second instance, ions with opposing charges are drawn to one another and come close enough to one another to effectively form an ion pair. When an outer-sphere complex is formed, the metal ion, the ligand, or both often retain its coordinated water. In other words, one or more water molecules are often present between the metal ion and the ligand [45].

5.4 BATCH BIOSORPTION

Biosorption procedures are usually carried out using batch and continuous modes of operation, while a biosorption process can be carried out in a variety of ways. Despite the fact that the majority of industrial applications favor a continuous mode of operation, batch experiments must be utilized to assess the necessary fundamental data, such as the potential for biomass regeneration, biosorption rate, ideal experimental conditions, and biosorbent efficiency.

5.4.1 Features That Affect Batch Biosorption

Agitation rate, solute concentration, biosorbent size, biosorbent dosage, ionic strength, temperature, and solution pH are crucial factors that affect batch biosorption. Among them, the pH of the solution typically plays a significant role in biosorption and appears to have an impact on the chemistry of metals and dyes in solution as well as the reaction of the functional groups in biomass.

i. Influence of Agitation Speed
As agitation speed increases, adsorptive pollutant removal rates are increased by reducing mass transfer resistance, but the physical structure of the biosorbent may be harmed [44].

ii. Influence Concentration of Other Pollutant
If a target pollutant and a coexisting pollutant compete for binding sites or create a complex, a larger concentration of the other pollutant will inhibit the target pollutant's biosorptive clearance [44].

iii. Influence of Biosorbent Size
If biosorbent size decreases, batch processes benefit from the biosorbent's increased surface area, but column processes do not because of the biosorbent's poor mechanical strength and potential column clogging [44].

iv. Influence of Biosorbent Dose
A biosorbent's dosage significantly affects how much biosorption occurs. Higher uptakes are produced in higher dosages. A rise in biosorption capacity because the biosorbent's surface area has been increased, which also raises the number of binding sites. On the other hand, as biosorbent dosage is increased, less solute is biosorbed per unit weight of biosorbent, which may be the result of the intricate interactions among a number of different parameters. A crucial aspect in limited solute uptake at high sorbent doses is that the available solute is insufficient to entirely cover the available exchangeable sites on the biosorbent. Additionally, as noted by, it is impossible to ignore the interference between binding sites brought on by higher biosorbent doses because this would result in a low specific uptake [49].

v. Influence of Initial Pollutant Concentration
If the starting pollutant concentration is increased, it will have a negative impact on the removal efficiency of biosorbent but enhance the quantity of biosorbed pollutant per unit weight.

vi. Influence of Ionic Strength
If ionic strength increases, it decreases the amount of biosorption removal of adsorbate that can bind to the biosorbent's binding sites.

vii. Influence of Temperature
An increase in temperature tends to accelerate the biosorptive removal activity of the adsorptive pollutant and kinetic energy of the adsorbate, but it can also harm the biosorbent's physical structure [44].

viii. Influence of pH
It has been demonstrated that the pH of the solution has a significant impact on the affinity of cationic species for the functional groups found on

cellular surfaces. pH until it reaches an ideal pH M (OH). Due to competition between pH rises and more ligands, including imidazole, phosphate, carboxyl, and amino groups, cell wall ligands are strongly connected with hydronium ions of mn^+ at low pH values, would be exposed and carry negative charges which attract mn^+ and absorb it on the cell surface [50].

5.5 ANIMAL-BASED SORBENTS

The production and consumption of meat have increased as a result of the growing world population. Due to this, millions of tonnes of animal manure are being produced annually by cattle farms and other enterprises involved in the food chain. Utilizing the disposal is necessary to protect the environment and, by recycling garbage, to promote the circular economic system. Valorizing animal manure for use as an environmentally friendly sorbent is one excellent way to achieve these goals [51–53]. Animal bone is useful as a sorbent material, because it houses carbon and other components that are good for binding pollutants, like apatite calcium phosphate, which has ionic exchange properties to attract in dispersed pollutants [54].

5.5.1 Biosorbent from Pig Bone

It would be interesting to explore using pig bone as alternative sorbent material. According to a previous research, pig bone-based sorbent that had first undergone calcination and mineralization was successful in adsorbing Pb(II) metal ions [55]. The adsorption capacity could rise from 96.1 mg g^{-1} to 312.5 mg g^{-1} as a result of activation procedures. Furthermore, successful research was done to create a hydroxyapatite nanocrystal sorbent from pig bone to trap dangerous radioisotope pollutants [56]. This sorbent's surface is negatively charged as a result of chemical impregnation with NaHCO3, which increases its sorption capacity. In this instance, the presence of CO3 two sites may play a crucial role in improving the ability of the sorbent to adsorb various metal ions, such as Pb(II), Cd(II), and Cu (II).

5.5.2 Biosorbent from Bovine Bone

The use of sorbent based on bovine bone has been documented. To start, it was reported that bovine bone-based sorbent was successful in separating chromium ions from raw chrome electroplating effluent after being first activated using NaOH impregnation [57]. Additionally, a study used activated carbon that was generated from this bone. Sulfamethazine and chloramphenicol, two antibiotic pollutants, were effectively eliminated by the sorbent by approximately 1,194 mg g^{-1} and 1240 mg g^{-1}, respectively, due to its outstanding surface area of 3231.8 m^2 [58]. As a result, the removal efficiency of the brushite sorbent made from bovine bone, which was employed to separate Cr(III) metal ions from aqueous solutions, may reach up to 95% [59]. All of the investigations concurred that using bovine bone as a different choice for sorbent could be more environmentally friendly. In a subsequent study, different bovine bone parts—including vertebrae, scapulae, ribs, and legs—were separately tested for performance under identical operating conditions. The maximum sorption capacity was then reported for biochar made from cow ribs, followed by biochar made from scapulae, vertebrae, and legs [60].

5.5.3 Biosorbent from Camel Bone

Due to rising demand from the meat business, tourism, and conventional transportation, camel populations have expanded, particularly in Mediterranean nations. The increasing number of camel carcasses could then be a problem. Fortunately, using them as sorbent material could contribute to environmentally friendly solutions. According to several studies, this inexpensive material is capable of adsorbing various contaminants, particularly heavy metal ions. For example, camel bone-based sorbent for the removal of Cd(II) and Pb(II) ions has been reported [61]. The study determined that 800°C was the ideal pyrolysis temperature since it resulted in the elimination of Cd(II) and Pb(II) by around 99.4% and 99.89%, respectively. Other pyrolysis temperatures of 500, 600, 800, and 900°C were also used. By applying a chemical coating of Fe_3O_4 to the sorbent surface, other research created magnetic composite sorbent derived from camel bones [62]. It was discovered that the presence of P, Ca, and C on the modified sorbent had an impact on its sorption capacity, which was 300.8 mg g^{-1} for Cd(II), 315 mg g^{-1} for Pb(II), and 271.7 mg/g for Co(II), respectively. Furthermore, it was noted that camel bone activated by carbonization and sodium hydroxide impregnation significantly removed Cu(II) metal ions from solution in an industrial semi-pilot column mode [63].

5.5.4 Biosorbent from Fish Bone

Fish bone can be used since it is a plentiful source of animal waste and because it is as effective as other organic materials. According to a study, bone char from pleco fish (*Pterygoplichthys* spp.) performed well in terms of removing fluoride and Cd(II) metal ions [64]. Bone char from fish performed better at higher pH for Cd(II) removal and lower pH for fluoride removal with a surface area of approximately 110 m^2 g^{-1}. Furthermore, a more thorough characterization examination revealed that physisorption, ion exchange, and electrostatic attractions occurred throughout the adsorption process.

Due to the complexity of the solution in nature, there may be some changes in how this sorbent is used to treat raw and synthetic aquatic media. One of those intricate mixtures is raw wastewater from butcheries. In a study on this subject, fish bone chitoprotein was used as a sorbent and coagulant agent to lower particle burden [65]. Expected results were observed since both the biochemical oxygen demand (BOD) concentration and the majority of the particle load could be decreased. Thus, by successfully removing around 87% of Congo red dyes and 77% of crystal violet within 75 minutes in addition to photo catalytic treatment, the bone formation of other fish species, specifically Catla fish, was also reliably informed [54]. The elimination of tetracycline from an aqueous solution was then successfully accomplished by carbonizing bone debris made from Tilapia fish bones [66]. Based on the experimental findings, it is possible to assume that tetracycline may exhibit anionic properties at pH values close to neutral when the sorbent surface is positively charged, resulting in better adsorption capacity.

5.5.5 Biosorbent from Chicken Bone

As a natural sorbent, valorized chicken bone has been valued for a variety of pollutant types and activation modalities. First, it was discovered that biochar made from chicken bone that had been carbonized at a lower temperature was advantageous for adsorbing metal Cr(VI) ions because of its polar functional group [67]. The ability to adsorb metal lead ions was further investigated by pyrolyzing various chicken bone fragments [68]. Comparative investigation revealed that the biochar made from chicken tibia bones performed better than that made from humerus, vertebrae, clavicle, ribs, and pelvis bones. Later, it was reported that the significant rate of regeneration was confirmed by the fact that most metal ions trapped on the sorbent surface could desorb using acid solution. Another study claimed that cadmium metal ions present in solution could be mainly decreased by a uniquely produced chicken bone with double coating employing iron and magnesium hydroxides [69]. In separate tests, the dye pollutant basic fuchsine could also be eliminated from aqueous medium using activated carbon and biochar made from chicken bone. According to this study, activated carbon is more reliable than biochar [70].

5.5.6 Biosorbent from Cow Bone

The most significant sector of the world meat business, the cow farm, produces enormous amounts of cow bone waste. Therefore, there are plenty of cow bones available for making organic sorbent. A subsequent study used a ball-milling approach to create nitrogenous biochar from cow bone, and it reported that this innovation was successful in achieving sufficient heavy metals reduction [71]. Additionally, another study found that cow bone-based sorbent would be the best choice for cleaning raw cassava wastewater because it has a high capacity for adsorbing many heavy metals at a dosage of at least 0.02 g of sorbent [72]. The aforementioned result was extremely competitive when compared to any commercial sorbents that were readily available.

5.5.7 Biosorbent from Bird's Bone

The outstanding ability of ostrich bone-based sorbent to collect pollutants has been reported. Only five minutes after contact time with heavy metal such as lead ions, they were removed from an aqueous solution [73]. The properties of ostrich bone were subsequently improved by additional research, resulting in a unique composite ash/nanoscale zerovalent iron that is capable of adsorbing both Hg(II) and Pb(II) metal ions in a fixed bed column mode [74]. The elimination efficiency of Hg(II) and Pb(II) is higher than that of Ni(II) and Cd(II); therefore, the effectiveness of this modified ostrich bone was not equivalent to all types of metal ions. Due to the relative stability of sorbent performance even after several adsorption stages, both fresh and used sorbent were considered economical in terms of their reusability [75].

5.5.8 Other Biosorbent of Animal Origin

Animal wastes also include various body parts such as fur, hooves, beaks, feathers, and excretion like feces in addition to the aforementioned bones. The growing number of studies looking at diverse animal remains for sorbent production is a result of the growing interest in examining the possibilities of those wastes. To begin with, by utilizing chicken feathers, hazardous Cr(VI) metal ions have been efficiently eliminated from solutions [76]. This sorbent may have a high monolayer capacity at very acidic pH. The adsorption capability of Cr(VI) ions in fixed bed mode was reported to be more adequate when the feathers were acidified with nitric acid [77]. Other forms of metal, such as Mn(II), Cu(II), and Co(II) ions, could not, according to this data, be efficiently adsorbed. This phenomenon has to be investigated further since it may be related to how some metal characteristics behave differently from sorbent characteristics found in nature.

Additionally, chemically activated chicken feathers can be used in batch mode to perform the defluorination procedure [78]. The experimental results showed that increasing the sorbent dosage by eight times resulted in a twofold increase in fluoride removal. Another study compared the development of sorbent from waste chicken beak, waste chicken femur, and waste fishbone to adsorb Cd(II) metal ions from raw shipbuilding wastewater at a carbonization temperature of 900°C [79]. It claimed that sorbent made from chicken femur functioned better than sorbents made from fishbone and chicken beak. Furthermore, pH near neutral point was discovered to be ideal to eliminate targeted metal ions from the solution utilized in the investigation, in contrast to earlier works that showed acidic pH to be advantageous.

To adsorb Pb(II) metal ions, a sorbent derived from chondroitin extraction waste found in animal bone was developed [80]. It became clear that metal removal was influenced by the initial pH, with pH 3 being the optimum. Ultimately, other research has discussed the use of dried animal feces for making alternate sorbents. For instance, the effectiveness of activated carbon made from cow dung was demonstrated by the material's high active surface area and good removal percentage [81]. A different study evaluated the effectiveness of bivalve mollusk-based sorbent that had previously undergone pyrolysis treatment to remove the antibiotic rifampicin from water [82]. The sufficient sorption capacity of the sorbent, which might reach 7 mg g^{-1}, can be linked to residual $CaCO_3$ in the form of aragonite and calcite together with its porous structure on heterogeneous sorbent surface, according to the characterization analysis.

Based on the findings of the study that compared the analysis of animal and plant wastes, it was possible to use the hydrothermal liquefaction technique to individually activate rice husk and cow dung under the same operating circumstances for the purpose of removing dye. According to the study, employing rice husk and cow dung-based sorbents, respectively, the amount of Congo red dye left in raw textile effluent may be reduced by up to 96.9% and 98.8% [83]. This makes it clear that plant and animal wastes are equally as trustworthy as other natural and man-made sorbents for removing different contaminants from a variety of aqueous mediums.

5.6 CONCLUSION

Global population growth and the emergence of new industries have a direct and indirect impact on the growth of aqueous medium pollution. To stop additional damage to the environment and to human health, this issue must be adequately addressed. Therefore, the first step in controlling the wastewater stream and solid waste should prioritize the use of green technologies. There are many water and wastewater treatment technologies, such as chemical oxidation, biological processes, and membrane separation, which can be used to remove pollutants and present as options for performing an integrated adsorption system. This chapter has presented the current use of organic solid waste, specifically animal remains as sorbent materials to clean polluted aquatic media. The properties and performance of the sorbents have been compared and contrasted in light of pertinent publications. In addition, waste materials from animals including bone and hair have been observed to work effectively as sorbents to adsorb different contaminants present in synthetic and untreated wastewaters. In order to increase the active surface area and functional group on the sorbent surface, several modification procedures, such as physical activation, chemical impregnation, and thermal activation process, were also proven to be beneficial. Furthermore, it can be assumed that utilizing an animal-based sorbent generally produces better results, notably for the removal of heavy metals. Further study is required for other kinds of pollutants. In terms of sorbent performance and adsorption capacity, these two properties of animal-baseds biosorbent can be influenced by a variety of factors, including species of those animals, operating conditions, the modification processes applied to the raw sorbents, the characteristics and concentration of targeted pollutants, the properties of synthetic and raw aqueous media, and the parts of biomass used as sorbents, such as nail, feather, bone, or fur. In conclusion, it is anticipated that increased interest in using the circular economy system and growing awareness of maintaining environmental quality will result in an increase in the exploration and use of biomass as sorbent material, especially animal-based biosorbents in the near future.

REFERENCES

1. Volesky, B. (1990). Biosorption and biosorbents. In B. Volesky (Ed.), *Biosorption of heavy meatals* (pp. 3–5). Boca Raton, FL: CRC Press.
2. Volesky, B. (2003). Potential of biosorption. In B. Volesky (Ed.), *Sorption and biosorption* (pp. 5–12). Montreal: BV Sorbex, Inc.
3. Al-Masri, M. S., Amin, Y., Al-Akel, B., & Al-Naama, T. (2010). Biosorption of cadmium, lead, and uranium by powder of poplar leaves and branches. *Applied Biochemistry and Biotechnology*. 160: 976–987.
4. Witek-Krowiak, A., & Reddy, D. H. K. (2013). Removal of microelemental Cr(III) and Cu(II) by using soybean meal waste--unusual isotherms and insights of binding mechanism. *Bioresour. Technol.* 127: 350–357.
5. Witek-Krowiak, A., Szafran, R. G., & Modelski, S. (2011). Biosorption of heavy metals from aqueous solutions onto peanut shell as a low-cost biosorbent. *Desalination*. 265: 126–134.
6. Reddy, D. H. K., Lee, S.-M., & Seshaiah, K. (2012). Biosorption of toxic heavy metal ions from water environment using honeycomb biomass—an industrial waste material. *Water Air Soil Pollut.* 223: 5967–5982.
7. Blázquez, G., Martín-Lara, M. A., Tenorio, G., & Calero, M. (2011). Batch biosorption of lead(II) from aqueous solutions by olive tree pruning waste: Equilibrium, kinetics and thermodynamic study. *Chem. Eng. J.* 168: 170–177.
8. Volesky, B. (2007). Biosorption and me. *Water Res.* 41: 4017–4029.
9. Ma, W., & Tobin, J. M. (2003). Development of multimetal binding model and application to binary metal biosorption onto peat biomass. *Water Res.* 37: 3967–3977.

10. Khoramzadeh, E., Nasernejad, B., & Halladj, R. (2013). Mercury biosorption from aqueous solutions by Sugarcane Bagasse. *J. Taiwan Inst. Chem. Eng.* 44: 266–269.
11. Witek-Krowiak, A. (2011). Analysis of influence of process conditions on kinetics of malachite green biosorption onto beech sawdust. *Chem. Eng. J.* 171: 976–985.
12. Witek-Krowiak, A. (2013). Application of beech sawdust for removal of heavy metals from water: biosorption and desorption studies. *Eur. J. Wood Wood Products.* 71: 227–236.
13. Reddy, D. H. K., Harinath, Y., Seshaiah, K., & Reddy, A. V. R. (2010). Biosorption of Pb(II) from aqueous solutions using chemically modified *Moringa oleifera* tree leaves. *Chem. Eng. J.* 162: 626–634.
14. Reddy, D. H. K., Seshaiah, K., Reddy, A. V. R., & Lee, S. M. (2012). Optimization of Cd(II), Cu(II) and Ni(II) biosorption by chemically modified Moringa oleifera leaves powder. *Carbohydr. Polym.* 88: 1077–1086.
15. Reddy, D. H. K., Seshaiah, K., Reddy, A. V. R., Rao, M. M., & Wang, M. C. (2010). Biosorption of Pb2+ from aqueous solutions by Moringa oleifera bark: equilibrium and kinetic studies. *J. Hazard. Materials.* 174: 831–838.
16. Reddy, D. H. K., Ramana, D. K. V., Seshaiah, K., & Reddy, A. V. R. (2011). Biosorption of Ni(II) from aqueous phase by *Moringa oleifera* bark, a low cost biosorbent. *Desalination.* 268: 150–157.
17. Acheampong, M. A., Pakshirajan, K., Annachhatre, A. P., & Lens, P. N. L. (2013). Removal of Cu(II) by biosorption onto coconut shell in fixed-bed column systems. *J. Industr. Eng. Chem.* 19: 841–848.
18. Manique, M. C., Faccini, C. S., Onorevoli, B., Benvenutti, E. V., & Caramão, E. B. (2012). Rice husk ash as an adsorbent for purifying biodiesel from waste frying oil. *Fuel.* 92: 56–61.
19. Alinnor, I. J. (2007). Adsorption of heavy metal ions from aqueous solution by fly ash. *Fuel.* 86: 853–857.
20. Cho, H., Oh, D., & Kim, K. (2005). A study on removal characteristics of heavy metals from aqueous solution by fly ash. *J. Hazard. Mater. B.* 127: 187–195.
21. Erol, M., Küçükbayrak, S., Ersoy-Meriçboyu, A., & Ulubaş, T. (2005). Removal of Cu^{2+} and Pb^{2+} in aqueous solutions by fly ash. *Energy Convers. Manage.* 46: 1319–1331.
22. Chojnacka, K., & Michalak, I. (2009). State of the art for the biosorption process—a review. *Global Nest: Int. J.* 11: 205–217.
23. Jeffers, T. H., & Corwin, R. R. (1993). In A. E. Torma, M. L. Apel, & C. L. Brierley (Eds.), *Biohydrometallurgical technologies, proceedings of the international biohydrometallurgy symposium, wastewater remediation using immobilized biological extractants* (pp. 1–14). Warrendale, PA: The Minerals, Metals and Materials Society.
24. Mahan, C. A., & Holcombe, J. A. (1992). Immobilization of algae cells on silica gel and their characterization for trace metal preconcentration. *Analyt. Chem.* 64: 1933–1939.
25. Fourest, E., & Roux, J. C. (1994). Improvement of heavy metal biosorption by mycelial dead biomasses (*Rhizopus arrhizus, Mucor miehei* and *Penicillium chrysogenum*): pH control and cationic activation. *FEMS Microbiol. Rev.* 14: 325–332.
26. Baysal, Z., Cinar, E., Bulut, Y., Alkan, H., & Dogru, M. (2009). Equilibrium and thermodynamic studies on biosorption of Pb(II) onto Candida albicans biomass. *J. Hazard. Mater. B.* 161: 62–67.
27. Selatnia, A., Madani, A., Bakhti, M. Z., Kertous, L., Mansouri, Y., & Yous, R. (2004). Biosorption of Ni^{2+} from aqueous solution by a NaOH-treated bacterial dead *Streptomyces rimosus* biomass. *Miner. Eng.* 17: 903–911.
28. Arica, M. Y., Bayramoglu, G., Yilmaz, M., Genc, O., & Bektas, S. (2004). Biosorption of Hg^{2+}, Cd^{2+}, and Zn^{2+} by Ca-alginate and immobilized wood-rotting fungus *Funalia trogii*. *J. Hazard. Mater. B.* 109: 191–199.

29. Modak, J., & Natarajan, K. A. (1995). Biosorption of metals using nonliving biomass—A review *Miner. Metall. Proc.* 12: 189.
30. Ullrich, A. H., & Smith, M. W. (1951). The Biosorption process of sewage and waste treatment. *Ind. Wastes.* 2 3: 1248.
31. Stasiak, M. (1969). Application of biosorption process for renovation of waste waters at chemical industry.*Przemysl Chemiczny.* 48: 426.
32. Mills, A.C. (1973). *Company Ltd. and J. R.Sanderson*, Great Britain Patent GB. 1324358.
33. Ruchoft, C. (1949). A critical review of the literature of 1949 on sewage and waste treatment and water pollution. *Sewage Works J.* 21: 877.
34. Volesky, B., & Tsezos, M. (1982). Separation of uranium by biosorption. US Patent US. 1: 4320093.
35. Goodman, G., & Roberts, T. M. (1971). Plants and soils as indicators of metals in the air. *Nature.* 231: 287.
36. Neufeld, R. D., & Hermann, E. R. (1975). Heavy metal removal by acclimated activated sludge. *J. Water Pollut. Control Fed.* 47: 310.
37. Friedman, B. A., & Dugan P. R. (1968). Concentration and accumulation of metallic ions by the bacterium *Zoogloea. Dev. Ind. Microbiol.* 9: 381.
38. Nakajima, A., Horikoshi, T., & Sakaguchi, T. (1982). Recovery of uranium by immobilized microorganisms. *J. Appl. Microbiol.* 16: 88.
39. Sakaguchi, T., Nakajima, A., & Horikoshi, T. (1978). Uptake of uranium from sea water by microalgae. *J. Ferment. Technol.* 56: 561.
40. Gould, M. S., & Genetelli, E. J. (1984). Effects of competition on heavy metal binding by anaerobically digested sludges. *Water Res.* 18: 123.
41. Tsezos, M., & Volesky, B. (1981). Biosorption of uranium and thorium. *Biotechnol. Bioeng.* 23: 583.
42. Chiu, Y., Asce, M., & Zajic, J.E. (1976). Biosorption isotherm for uranium recovery. *J. Environ. Eng. ASCE.* 102: 1109.
43. Steen, W. C., & Karickhoff, S. W. (1981). Biosorption of hydrophobic organic pollutants by mixed microbial populations *Chemosphere.* 10: 27.
44. Davis, T. A., Volesky, B., & Mucci, A. (2003). A review of the biochemistry of heavy metal biosorption by brown algae. *Water Res.* 37: 4311–4330.
45. Park, D., Yun, Y. S., & Moon Park, J. (2010). The past, present, and future trends of biosorption. Biotechnology and Bioprocess Engineering 15: 86–102.
46. Vijayaraghavan K., & Yun, Y. S. (2008). Bacterial biosorbents and biosorption. *Biotechnol. Adv.* 26: 266.
47. Singh Ahluwali, S. T., & Goy, D. (2007). *Bioresour. Technol.* 98, 2243.
48. Schiewer S., Volesky, B. (2000). Biosorption by marine algae. In J.J. Valdes (Ed.), *Remediation*, Dordrecht: Kluwer Academic Publishers, pp. 139–169.
49. Brenner D.J., Krieg N.R., Staley J.T. and Garrity G.M., (Eds.) (2005) *Bergey's Manual of Systematic Bacteriology*, 2nd Edition, Vol. 2 (The Proteobacteria), part C (The Alpha-, Beta-, Delta-, and Epsilonproteobacteria), Springer, New York.
50. Joo, J., & Hassan, S. H. A. (2010). Comparative study of biosorption of Zn2+ by Pseudomonas aeruginosa and Bacillus cereus. *Int. Biodeterior. Biodegrad.* 64: 734.
51. Prakash, M. O., Raghavendra, G., Ojha, S., & Panchal, M. (2020). Characterization of porous activated carbon prepared from Arhar Stalks by single step chemical activation method. *Mater Today: Proc.* 39(4): 1476–1481. https://doi.org/10.1016/j.matpr.2020.05.370.
52. Tang, Y., Lin, T., Jiang, C., Zhao, Y., & Ai, S. (2021). Renewable adsorbents from carboxylate-modified agro-forestry residues for efficient removal of methylene blue dye. *J. Phys. Chem. Solids.* 149: 109811. https://doi.org/10.1016/j.jpcs.2020.109811

53. Karimi, H., Heidari, M. A., Emrooz, H. B. M., & Shokouhimehr, M. (2020). Carbonization temperature effects on adsorption performance of metal-organic framework derived nanoporous carbon for removal of methylene blue from wastewater; experimental and spectrometry study. *Diamond Relat. Mater.* 108: 107999.
54. Sathiyavimal, S., Vasantharaj, S., Shanmugavel, M., Manikandan, E., Nguyen-Tri, P., Brindhadevi, K. & Pugazhendhi, A. (2020). Facile synthesis and characterization of hydroxyapatite from fish bones: Photocatalytic degradation of industrial dyes (crystal violet and congo red). *Progress Organ Coat.* 148: 105890.
55. Zhou, Y., Chang, D., & Chang, J. (2017). Preparation of nano-structured pig bone hydroxyapatite for high-efficiency adsorption of Pb2+ from aqueous solution. *Int. J. Appl. Ceramic Technol.* 14(6): 1125–1133. https://doi.org/10.1111/ijac.12749.
56. Sekine, Y., Nankawa, T., Yamada, T., Matsumura, D., Nemoto, Y., Takeguchi, M., Sugita, T., Shimoyama, I., Kozai, N., & Morooka, S. (2021). Carbonated nanohydroxyapatite from bone waste and its potential as a super adsorbent for removal of toxic ions. *J. Environ. Chem. Eng.* 9(2): 105114. https://doi.org/10.1016/j.jece.2021.105114.
57. Simpen, I. N., Negara, I. M. S., & Jayanto, S. D. (2020). Activated adsorbent prepared from bovine bone waste: Physico-chemical characteristics, isoterm and thermodynamics adsorption of chromium ions in wastewater. *Int. J. Eng. Technol. Manage. Res.* 7(3): 1–11. https://doi.org/10.29121/ijetmr.v7.i3.2020.534.
58. Dai, J. Qin, L. Zhang, R. Xie, A. Chang, Z. Tian, S. Li, C., & Yan, Y. (2018). Sustainable bovine bone-derived hierarchically porous carbons with excellent adsorption of antibiotics: Equilibrium, kinetic and thermodynamic investigation. *Powder Technol.* 331: 162–170. https://doi.org/10.1016/j.powtec.2018.03.005.
59. Maldonado, H. J. A., Torres García, F. A., Salazar Hernández, M. M., & Hernández Soto, R. (2017). Removal of chromium from contaminated liquid effluents using natural brushite obtained from bovine bone. *Desalinat Water Treatm.* 95: 262–273. https://doi.org/10.5004/dwt.2017.21480.
60. Wang, M., Liu, Y., Yao, Y., Han, L., & Liu, X. (2020). Comparative evaluation of bone chars derived from bovine parts: Physicochemical properties and copper sorption behavior. *Sci. Total Environ.* 700: 134470.
61. Rashed, M. N., Gad, N. M., & Fathy, A. A. -E. (2019). Adsorption of Cd (II) and Pb (II) using physically pretreated camel bone biochar. *Adv. J. Chem. Sect. A.* 2(4): 347–364. https://doi.org/10.33945/SAMI/AJCA.2019.4.8.
62. Alqadami, A. A., Khan, M. A., Otero, M., Siddiqui, M. R., Jeon, B.-H., & Batoo, K. M. (2018). A magnetic nanocomposite produced from camel bones for an efficient adsorption of toxic metals from Water. *J. Clean. Product.* 178: 293–304. https://doi.org/10.1016/j.jclepro.2018.01.023.
63. Abd-Rabboh, H. S., Fawy, K. F., & Awwad, N. S. (2019). Removal of copper (II) from aqueous samples using natural activated hydroxyapatite sorbent produced from camel bones. *Desalin Water Treatm.* 164: 300–309.
64. Medellín-Castillo, N. A., Cruz-Briano, S. A., Leyva-Ramos, R., Moreno-Piraján, J. C., Torres-Dosal, A., Giraldo-Gutiérrez, L., Labrada-Delgado, G. J., Pérez, R. O., Rodriguez-Estupiñan, S. Y., Reyes Lopez, M. S., & Berber Mendoza, J. P. (2020). Use of bone char prepared from an invasive species, Pleco Fish (*Pterygoplichthys* spp.), to remove fluoride and cadmium(II) in water. *J. Environ. Manage.* 256: 109956. https://doi.org/10.1016/j.jenvman.2019.109956.
65. C. F. Okey-Onyesolu, O. D. Onukwuli, M. I. Ejimofor, & C. C. Okoye. Kinetics and mechanistic analysis of particles decontamination from abattoir wastewater (ABW) using novel fish bone chito-protein (FBC). *Heliyon.* 6(8): e04468. https://doi.org/10.1016/j.heliyon.2020.e04468.

66. Módenes, A. N., Bazarin, G., Borba, C. E., Locatelli, P. P. P., Borsato, F. P., Pagno, V., Pedrini, R., Trigueros, D. E. G., Espinoza-Quiñones, F. R., & Scheufele, F. B. (2021). Tetracycline adsorption by tilapia fish bone-based biochar: Mass transfer assessment and fixed-bed data prediction by hybrid statistical-phenomenological modeling. *J. Clean. Product.* 279: 123775. https://doi.org/10.1016/j.jclepro.2020.123775.
67. Yang, T., Han, C., Tang, J., & Luo, Y. (2020). Removal performance and mechanisms of Cr (VI) by an in-situ selfimprovement of mesoporous biochar derived from chicken bone. *Environ. Sci. Pollut. Res.* 27: 5018–5029.
68. Park, J. H., Wang, J. J., Kim, S. H., Kang, S. W., Cho, J. S., Delaune, R. D., Ok, Y. S., & Seo, D. C. Lead sorption characteristics of various chicken bone part-derived chars. *Environ. Geochem. Health.* 41(4): 1675–1685. https://doi.org/10.1007/s10653-017-0067-7.
69. Alquzweeni, S. S., & Alkizwini, R. S. (2020). Removal of cadmium from contaminated water using coated chicken bones with double-layer hydroxide (Mg/Fe-LDH). *Water.* 12(8): 2303. https://doi.org/10.3390/w12082303.
70. Côrtes, L. N., Druzian, S. P., Streit, A. F. M., Junior, T. R. S. a. C., Collazzo, G. C., & Dotto, G. L. (2019). Preparation of carbonaceous materials from pyrolysis of chicken bones and its application for fuchsine adsorption. *Environ. Sci. Pollut. Res.* 26(28): 28574–28583. https://doi.org/ 10.1007/s11356-018-3679-2.
71. Xiao, J., Hu, R., & Chen, G. (2020). Micro-nano-engineered nitrogenous bone biochar developed with a ball-milling technique for high-efficiency removal of aquatic Cd(II), Cu(II) and Pb(II). *J. Hazard. Mater.* 387: 121980.
72. Olaoye, R. A., Afolayan, O. D., Adeyemi, K. A., Ajisope, L. O. & Adekunle, O. S. (2020). Adsorption of selected metals from cassava processing wastewater using cow-bone ash. *Sci. Afr.* 10: e00653. https://doi.org/10.1016/j.sciaf.2020.e00653.
73. Malla, K. P., Adhikari, R., JeewanYadav, R., Nepal, A., & Neupane, B. P. (2018). Removal of lead (II) ions from aqueous solution by hydroxyapatite biosorbent extracted from ostrich bone. *J. Health Allied Sci.* 7(1): 27–33. https://doi.org/10.37107/jhas.19.
74. Amiri, M. J., Abedi-Koupai, J., & Eslamian, S. (2017). Adsorption of Hg (II) and Pb (II) ions by nanoscale zero valent iron supported on ostrich bone ash in a fixed-bed column system. *Water Sci. Technol.* 76(3): 671–682. https://doi.org/10.2166/wst.2017.252.
75. Gil, A., Amiri, M. J., Abedi-Koupai, J., & Eslamian, S. (2018). Adsorption/reduction of Hg (II) and Pb (II) from aqueous solutions by using bone ash/nZVI composite: Effects of aging time, Fe loading quantity and co-existing ions. *Environ. Sci. Pollut. Res.* 25(3): 2814–2829. https://doi.org/10.1007/s11356-017-0508-y.
76. Chakraborty, R., Asthana, A., Singh, A. K., Verma, R., Sankarasubramanian, S., Yadav, S., Carabineiro, S. A. C., & Susan, M. A. B. H. (2020). Chicken feathers derived materials for the removal of chromium from aqueous solutions: Kinetics, isotherms, thermodynamics and regeneration studies. *J. Dispers. Sci. Technol.* 1–15. https://doi.org/ 10.1080/01932691.2020.1842760.
77. Sun, P., Zhu, G., Li, T., Li, X., Shi, Q., Xue, M., & Li, B. (2020). Acidification chicken feather as sorbent for selectively achicken feather as sorbent for selectively adsorbing of Cr(VI) ions in aqueous solution. *Mater. Today Commun.* 24: 101358. https://doi.org/10.1016/j.mtcomm.2020.101358.
78. Nasiebanda, R., Wambu, E. W. & Lusweti, K. (2021). Water defluoridation by adsorption using aluminium modified chicken feathers. *Afr. J. Educat. Sci. Technol.* 6(2): 278–291. http://ajest.info/index.php/ajest/article/view/518.
79. Foroutan, R., Peighambardoust, S. J., Hosseini, S. S., Akbari, A., & Ramavandi, B. (2021). Hydroxyapatite biomaterial production from chicken (Femur and Beak) and fishbone waste through a chemical less method for Cd2+ removal from shipbuilding wastewater. *J. Hazard. Mater.* 413: 125428. https://doi.org/10.1016/j.jhazmat.2021.125428.

80. Wang, H., Lv, Z., Wang, Y.-n., Sun, Y. & Tsang, Y. F. (2021). Recycling of biogenic hydroxyapatite (HAP) for cleaning of lead from wastewater: Performance and mechanism. *Environ. Sci. Pollut. Res.* 28: 29509–29520. https://doi.org/10.1007/s11356-020-10855-4.
81. Saraswat, S. K., Demir, M., & Gosu, V. (2020). Adsorptive removal of heavy metals from industrial effluents using cow dung as the biosorbent: Kinetic and isotherm modeling. *Environ. Qual. Manage.* 30(1): 51–60. https://doi.org/10.1002/tqem.21703.
82. Henrique, D. C., Quintela, D. U., Ide, A. H., Erto, A., Duarte, J. L. d. S., & Meili, L. (2020). Calcined mytella falcata shells as alternative adsorbent for efficient removal of rifampicin antibiotic from aqueous solutions. *J. Environ. Chem. Eng.* 8(3): 103782. https://doi.org/10.1016/j.jece.2020.103782.
83. Khan, N., Chowdhary, P., Ahmad, A., Shekher Giri, B., & Chaturvedi, P. (2020). Hydrothermal liquefaction of rice husk and cow dung in mixed-bed-rotating pyrolyzer and application of biochar for dye removal. *Bioresour. Technol.* 309: 123294. https://doi.org/10.1016/j.biortech.2020.123294.

6 Biosorbents Mechanism of Action

Akanksha Jain, Sonia Bajaj, Swati Tanti, and Suchitra Panigrahy

6.1 INTRODUCTION

The high rate of economic development is making it difficult to preserve a pristine and stable ecosystem from one generation to the next. Ecological harmony in all three realms—land, sea, and sky—is under attack due to careless use and abuse of natural resources. Businesses pollute the environment by carelessly dumping their garbage in landfills and waterways. Heavy metals, metalloids, organic compounds, dyes, etc. make up the bulk of industrial waste, all of which are potentially harmful to humans and the environment. Because of their persistence and inability to be reduced to a non-toxic form, heavy metals and metalloids are the most prevalent pollutants. Because of their slow rate of breakdown and bioaccumulation, heavy metals offer serious health hazards (deFreitas et al., 2020). According to Nagajyoti et al. (2010), the maximum allowable amounts of Cd, Cu, Ni, Pb, and Zn in soil in India are 3–6, 135–270, 75–150, 250–500, and 300–600 mg kg^{-1}, respectively. According to Indian regulations, the following concentrations of heavy metals are considered safe for human consumption: Fe 1 mg l^{-1}, Mn 0.3 mg l^{-1}, F 1.5 mg l^{-1}, Cu 1.5 mg l^{-1}, Zn 15 mg l^{-1}, As 0.05 mg l^{-1}, Hg 0.001 mg l^{-1}, Cr 0.05 mg l^{-1} (CWC Report, 2010). However, effects on plant and human health can be seen at concentrations well below these thresholds, demanding environmentally responsible approaches to eradicate the problem. Biomass, both live and decaying, is utilized in the biosorption process to bind metal ion pollutants. The use of biosorption as a cheap and environmentally benign method of reducing heavy metal pollution has recently emerged (Gupta et al., 2015). In an effort to help readers better understand the process, the various types of biosorbents, the current condition, and the future importance of this topic, this review attempts to outline recent findings made over the previous decade.

Heavy Metals and Toxicity—II. The toxicity of heavy metals is conditional on a number of variables, including the amount eaten, the route of exposure, the kind of metal, the age, gender, genetics, and diet of the exposed individual. Polluted soil and water allow heavy metals to bioaccumulate in organisms and then be ingested by animals and humans. Enzyme systems inside the organism

 DOI: 10.1201/9781003366058-6

are impacted, depriving the individual of essential macromolecules due to non-specific or competitive binding (Gupta, 2013). Tables 6.1 and 6.2 describe the potential origins, effects, and conventional treatments for certain significant toxicants.

TABLE 6.1
Sources, remediation strategies of some heavy metals

Heavy metal, symbol	Source	Treatment strategies	References
Iron, Fe	Natural leaching from cast-iron pipes in municipal water systems	Oxidizing filter, green-sand mechanical filter	Aroraetal (2002)
Manganese, Mn	Garbage dumps, rock and soil dumps, etc.	Ion exchange, chlorination, oxidizing filter, green-sand mechanical filter	Wu (1994) O'Neal and Zheng (2015)
Copper, Cu	Algae treatment and the leaching of copper from water pipelines and tubing Debris from factories and mines, chemicals used to treat wood raw materials	Ion exchange, reverse osmosis, distillation	Pichhode (2015)
Chromium, Cr	Systems for treating sewage, industrial wastewater, and mining	Ion exchange, reverse osmosis, distillation	
Mercury, Hg	Battery fungicides and fungicides Industries such as mining, electricity, plants, papers, and vinyl chloride Raw materials	Reverse osmosis, distillation	Messer et al. (2005) Sutton and Tchounwou (2007)
Lead, Pb	Paint, diesel fuel combustion Various types of pipes and solder, used batteries, paint, and leaded gas Natural deposits	Ion exchange, activated carbon, reverse osmosis, distillation	Mancuso et al. (2018); Yedjouet al. (2008)
Arsenic, As	Rocks that were once mined for use in insecticides (orchards), improperly disposed of, or utilized in the production of items that break glass or electronics	Activated alumina filtration, reverse osmosis, distillation, chemical precipitation, ion exchange, lime softening	Coxetal (1996) Lin et al. (2013); Yedjou and Tchounwou (2007)

(Continued)

TABLE 6.1 (*Continued*)
Sources, remediation strategies of some heavy metals

Heavy metal, symbol	Source	Treatment strategies	References
Flourine, F	Aluminum, glass, and fertilizer factories, as well as geological and	Activated alumina, distillation, reverse osmosis, coagulation, followed by precipitation, ion exchange, and adsorption	Hong et al. (2016) Swathy et al. (2017)
Cadmium, Cd	Processing and refining of metals; production of fertilizers; production of paint pigments, pesticides, plastics, and polyvinyl chloride; and the melting and refining of copper.	Sediment washing, electrochemical process, thermal treatment	Akcil et al. (2015); Guo et al. (2008); Hiroaki et al. (2014); Mohanpuria et al. (2007)
Zinc, Zn	Metals from galvanized pipes and fittings, as well as paints and dyes, can leach into the environment. Raw materials	Ion exchange, water softeners, reverse osmosis, distillation	Choi et al. (1996); Plum et al. (2010)
Nickel, Ni	Industries that use steam turbines to create electricity from non-ferrous metals, minerals, paints, and plating	Sediment washing, electrochemical process	Akcil et al. (2015); Chibuike et al. (2014); Fashola et al. (2016)
Cobalt, Co	Waste products from the glass and ink industries	Chemical–thermal process	Akcil et al. (2015); Leyssens et al. (2017); Nagajyoti et al. (2010)
Thallium, Tl	Activities such as burning fossil fuels, metal working, oil refining, cement manufacturing, and bread baking all contribute to the global carbon footprint.		Babula et al. (2008) Blais et al. (2008)

TABLE 6.2
Impact of some heavy metals

Heavy metal, symbol	Effects on plants	Effects on humans	References
Iron, Fe	Decreased photosynthesis and yield, elevated oxidative stress and ascorbate peroxidase activity, disrupted cellular structure, and permanent damage to membranes, DNA, and proteins are all consequences.	Aging muscle atrophy, viral replication, rosacea, pulmonary alveolar proteinosis, etc.	Arora et al. (2002)
Manganese, Mn	Crinkled leaves, chlorosis, and necrotic dark spots on leaves, petioles, and stems are all symptoms of this disease.	Neurotoxicity, cardiovascular toxicity, infant mortality	Wu (1994) O'Neal and Zheng (2015)
Copper, Cu	Leaf chlorosis, plant growth inhibition, and other symptoms of oxidative stress and reactive oxygen species production in numerous tree species.	Anemia, digestive disturbances, liver and kidney amage, gastrointestinal irritations, bitter or metallic taste	Pichhode (2015)
Chromium, Cr	Photosynthetic pigment breakdown, antioxidant enzyme induction and activation, and growth retardation	Skin irritation, skin and nasal ulcers, lung tumors, gastrointestinal effects, damage to the nervous system and circulatory system, accumulates in the spleen, bones, kidney, and liver	
Mercury, Hg	Plant cell membrane lipids and water transport are disturbed, and plant metabolism is slowed down as a result of this physical obstruction.	Loss of vision and hearing, intellectual deterioration, kidney and nervous system disorders, death at high levels	Messer et al. (2005); Sutton and Tchounwou (2007)
Lead, Pb	Negative effects on development, growth, and photosynthesis, inability to germinate seeds, and increased oxidative stress	Mental retardation, interference with kidney and neurological functions, hearing loss, blood disorders, hypertension, death at high levels	Yedjou et al. (2008) Mancuso et al. (2018)
Arsenic, As	Fruit production drops, leaf fresh weight drops, growth slows down, chlorosis sets in, and plants wither	Weight loss, depression, lack of energy, skin and nervous system toxicity, genotoxicity	Cox et al. (1996); Lin et al. (2013); Yedjou and Tchounwou (2007)

(Continued)

TABLE 6.2 (*Continued*)
Impact of some heavy metals

Heavy metal, symbol	Effects on plants	Effects on humans	References
Flourine, F	Causes damage to leaves and a reduction in growth, which severely reduces photosynthesis and other functions.	Brownish discoloration of teeth, bone damage, damage to the brain and kidneys, and effects on the heart	Hong et.al (2016) Swathy et al. (2017)
Cadmium, Cd	Chlorosis, growth inhibition, browning of root tip, and ultimately death, interference with uptake, transport, and usage of numerous elements (Ca, Mg, P, and K) and water by plants, altering plasma membrane permeability	It binds with respiratory enzymes causing oxidative stress and cancer, bone and kidney failure	Mohanpuria et al. (2007); Akcil et al. (2015); Guo et al. (2008); Hiroaki et al. (2014)
Zinc, Zn	Impedes plant growth and produces senescence; causes leaves to turn a purple–red color; causes manganese (Mn) and copper (Cu) deficits in plant shoots; all of these effects are result of inhibition of numerous metabolic functions in the plant.	Respiratory symptoms, metal fume fever, nausea and vomiting, epigastric pain, abdominal cramps, and diarrhea, acute brain injury	Choi et al. (1996); Plum et al. (2010)
Nickel, Ni	Decrease in water intake, chlorosis and necrosis, disruption of nutritional balance and cell membrane function	Chronic bronchitis, lung cancer	Akcil et al. (2015); Fashola et al. (2016); Gajewska et al. (2006); Chibuike et al. (2014)
Cobalt, Co	Impacts development and metabolism	Neurological (e.g., hearing and visual impairment), cardiovascular and endocrine deficits	Leyssens et al. (2017); Nagajyoti et al. (2010); Akcil et al. (2015)
Thallium, Tl	Causes growth suppression by inhibiting enzymes	Diseases such as alopecia, ataxia, burning foot syndrome, coma, convulsions, delirium, exhaustion, gastroenteritis, hair loss, and hallucinations have been linked to the virus.	Thallium, Tl

6.2 BIOSORPTION

Biosorption is increasingly viewed as a multifaceted, effective process that has developed over the past few years. It's a commendable alternative to traditional ways of wastewater treatment. The primary definition of sorption is the physical–chemical process by which molecules of sorbate adsorb (or stick) to the surface of another substance. This produces high-quality effluents that have been purified. The "bio" prefix indicates thata biological entity is involved, but the definition of the term "biosorption" is straightforward nonetheless. In terms of sorption mechanism, both bioabsorption and biosorption play a role. To absorb is to incorporate something of one state into something of another state. It also includes liquid absorption by a solid and gas absorption by water. Physical attachment occurs during adsorption when sorbate reacts with a sorbent to form an interface between the two. Biosorption is a general term for the interaction between any sorbate and a biological matrix, and it is a non-active, metabolically-independent process. It's an essential component of many processes observed in various scientific fields.

Conventional methods for cleaning polluted areas with heavy metal ions include chemical precipitation, lime coagulation, solvent extraction, membrane filtration, reverse osmosis, ion exchange, and adsorption (Bilal et al., 2017; Khalid et al., 2016; Murtaza et al., 2014). However, at high metal concentrations, these procedures lose credibility due to time requirements, rising energy and chemical requirements, final disposal of leftover metals, and sludge formation post-treatment (Abbas et al., 2014; Salman et al., 2015). However, biosorption is a more effective, inexpensive, and eco-friendly technology that can aid in the management of large quantities of contaminated land and water. Biosorption is typically thought of as a purely physical and chemical process that employs non-living, inert biological material (Fomina & Gadd, 2014). Active and passive processes that involve living or dead organisms and their parts are now included in the broader definition (Fomina & Gadd, 2014). Quickly and easily, metallic ions bind to the biomaterial's surface functional groups. Biosorption takes place when metal species are distributed in both a liquid phase (often an aqueous solvent) and a solid phase (living/dead/modified biomass). An ion exchanger is made from plant matter (Abbas et al., 2014a,b; Salman et al., 2014). The steps in the process are as follows: (a) the solute ion migrates to the biosorbent, (b) the solute ion adsorbs to the biosorbent surface, and (c) the metal migrates within the biosorbent (Barakat, 2011). Biosorption can be utilized to remove metal ions present in ppb quantities over a wide pH (pH 3–9) and temperature (pH 4–9) range. With the lowest possible production of harmful sludge, the procedure is also economically viable. Furthermore, the biosorbents employed are regenerated following metal recovery and are made from low-cost agricultural and industrial waste (Abdi & Kazemi, 2015; Adewuyi, 2020).

Biosorbents made from naturally occurring microbes (bacteria, fungi, yeast, or algae) are highly effective. These biosorbents are effective even when exposed to trace amounts of metals. Functioning groups like amide, amine, carbonyl, carboxyl aid in the elimination of heavy metals.

A biosorbent, or material formed from living organisms, is used in the biosorption process, together with a liquid phase solvent (often water) holding the sorbed component. Physiologically, biosorption includes processes like absorption, adsorption, ion exchange, surface complexation, and precipitation. Low cost, high efficiency, minimization of chemical and biological sludge, no additional nutrition requirement, regeneration of biosorbent, and the possibility of high metal recovery are just some of the advantages of the biological approach for adsorption of heavy metals over conventional physicochemical treatment methods. Despite its cell wall characteristics, ease of growth and manipulation, natural availability, and eco-friendly nature, biosorbing heavy metals and other pollutants using fungal biomass offers little advantage over other biosorbents. Heavy metal absorption capacity is seen in both living and decaying fungal biomasses. Some pretreatment methods can increase the biosorption of heavy metals by fungi by introducing functional groups into the fungal cell wall, making it more porous to metal ions. Chemical pretreatment entails subjecting biomass to acids, alkalis, detergents, solvents, enzymes, sodium hydroxide, formaldehyde, etc., while physical pretreatment encompasses heating, freezing, drying, and lyophilization. While some of these treatments thinned the fungal cells' cell wall, others increased the cells' capacity to absorb by adding functional groups according to Jadhav and Mahish (2020).

6.2.1 Fungal–Metal Interactions

Salts, oxides, sulfates, and nanoparticles are all different forms of metals. Dissociation of these compounds frees the metal ions, allowing the fungi to take them up and use them without a problem. Copper sulfate (CuSO) dissolves in water to form copper (II) sulfate pentahydrate ($CuSO_4$ $5H_2O$), which then decomposes into copper (II) ions and sulfur dioxide (SO42). Fungal proteins can then take up Cu2+ after its dissociation. Recently, nanoparticles of metal have been gaining popularity due to their potential as antifungals. Because of this, manufacturing of nanoparticles has increased dramatically. Nanoparticles are defined as small particles (100 nm) with diverse shapes, physicochemical, optical, and biological properties. Ions released from nanoparticles do so at a slower rate, but they are still available to interact with homeostatic mechanisms.

One type of metal-sensing regulatory transcription factor, known as a metalloregulatory protein, controls the expression of a subset of genes in bacteria in response to metal ion availability and excess. Metal-regulatory proteins are a class of multimeric DNA-binding proteins that undergo an allosteric transition in response to metal binding.

LamB protein was genetically modified with two different metal-binding peptide sequences, HP and CP, and expressed in *Escherichia coli*. Cd2+:HP and Cd2+:CP stoichiometries in peptides were 1:1 and 3:1, respectively. It has been demonstrated that hybrid LamB proteins in the outer membrane of *E. coli* are properly folded. With the addition of new metal-binding peptides, the capacity of isolated *E. coli* cell envelops to bind Cd2+ was enhanced by as much as 1.8-fold. Bioaccumulation of Cd2+, Cu2+, and Zn2+ by *E. coli* was examined. Exhibiting CP on the surface of *E. coli* improved its ability to remove Cd2+ from its growth medium by a factor of four. The accumulation of copper and zinc ions was unaffected by the presentation of HP peptide. With such a high affinity for HP, Cu2+ is probably no longer adding to Cd2+ accumulation. Thus, the rational design of peptide sequences possessing affinity for a particular metal requires consideration of the relative affinities of metal-binding peptide and, for instance, the cell wall to metal ion (Kotrba 2011), proteins and metabolic byproducts of microorganisms abound in the cell membrane, and they can use these to form intricate structures (chelation) with metal ions. The extracellular sequestration can be defined as the complexation of metal ions as insoluble compounds or the accumulation of metal ions by components of the cell in the periplasm. Copper-resistant *Pseudomonas syringae* strains express the copper-inducible proteins, CopA, CopB (periplasmic proteins), and CopC (outer membrane protein), which bind microbial colonies to copper ions. A zinc ion efflux mechanism is present in the Synechocystis PCC 6803 strain, allowing zinc ions to be transported out of the cytoplasm and into the periplasm (Thelwell et al., 1998). *Desulfuromonas* spp. and *Geobacter* spp. are examples of bacteria that may break down iron and sulfur, converting potentially hazardous elements into less harmful ones. Metallireducens may convert the hazardous forms of uranium (U) and manganese (Mn) into the safer forms (II) and (IV), respectively. G is the required anaerobe (Gavrilescu, 2004). In intracellular sequestration, several compounds combine metal ions in the cytoplasm. Due to their interaction with surface ligands and subsequent delayed transport into the cell, metals may accumulate to high quantities within the interiors of microorganism cells. The exceptional capacity of bacterial cells to accumulate metals intracellularly has found significant application, particularly in the field of waste treatment. The presence of low molecular weight proteins rich in cysteine allowed a cadmium-tolerant *Pseudomonas putida* strain to sequester copper, cadmium, and zinc ions within the cell (Higham et al., 1986). Rhizobium leguminosarum glutathione was found to play a role in the intracellular sequestration of cadmium ions (Lima et al., 2006). Fungi are composed of a hard cell wall that includes lipids, chitin, mineral ions, nitrogen-containing polysaccharide, polyphosphates, and proteins. Because of their propensity to deposit metals into their spores and mycelium, many fungi are capable of decontaminating metal ions through energy intake, intracellular and extracellular precipitation, and valence exchange. Efflux mechanisms can regulate the amount of HM inside the cell by removing the ions from the cell (Remenar et al., 2018). Many bacteria, including those isolated from metal-contaminated environments, have been shown to contain efflux mechanisms. In order to allow HM ion efflux, many different types of metal-exporting proteins are scattered throughout the cell membrane. Proton–cation antiporters, cation diffusion facilitators, ATPases of the P-type, and ABC transporters are all examples. In order to get rid of copper (II), cadmium (II), and zinc, gram-positive bacteria need an enzyme called P-type efflux ATPase (II). A membrane-bound, ATP-dependent export protein controls the outflow of arsenite from the cell (Soto et al., 2019; Yang et al., 2012). ABC transporters (or traffic ATPases) mediate membrane transfer of HM ions, aiding microorganisms in their resistance to HM stress (Al-Gheethi et al., 2015; Lerebours et al., 2016; Zammit et al., 2016).

6.2.2 Types of Metalloregulatory Systems

Three types of regulators are involved in metal sensing: riboswitches that directly bind metal, proteins that bind a metal-dependent cofactor, and proteins that directly bind metal. One example of a direct metal sensor that modifies transcription is Zn (II) binding to Zur. (B) To evaluate the intracellular metal levels, product-sensing metalloregulators measure the concentration of a metal-dependent metabolite. In *B. japonicum* Irr, heme is replaced for Fe (II). The enzyme ferrochelatase is responsible for catalyzing the insertion of iron into protoporphyrin, which leads to the synthesis of heme. This enzyme has a direct binding site for Irr. When there is sufficient Fe (II) in the cell, ferrochelatase synthesizes heme. After being connected to, Irr is deteriorated.

Irr has a direct binding site on ferrochelatase, the enzyme responsible for catalyzing the insertion of iron into protoporphyrin, so producing heme. Heme is synthesized by ferrochelatase when the cell has a enough supply of Fe (II). After Irr is bound to heme, it is degraded. The transcriptional repressor apo-Irr is only activated and released under Fe (II)-limiting conditions (Akcil et al., 2015). The yybP-ykoYMn (II)-sensing riboswitch is an example of a metal-sensing riboswitch that can regulate transcription and translation (Ayangbenro et al., 2017; Babula et al., 2008). An RNA shape that prevents the formation of an intrinsic transcription termination hairpin is favored by Mn (II) binding in *Bacillus subtilis.*

6.3 BIOSORBENTS

A wide range of biomaterials, including those derived from algae, bacteria, fungi, and even rubbish, have been shown to be effective at removing heavy metals from polluted environments (Adewuyi, 2020; Jadhav and Mahish 2020; Vijayaraghavan & Yun, 2008). The biosorption ability is enhanced by surface features such as cell wall composition, surface-to-volume ratio, surface groups, and affinity for contaminating metal species (Abbas et al., 2014; Michalak et al., 2013). Bioaccumulation is the process by which metals are taken up by living cells, while biosorption is the process by which they are taken up by dead cells (Abbas et al., 2014; Salman et al., 2014; Vijayaraghavan & Yun, 2008). The bioaccumulation methods are limited in their metal sequestration efficiency because they necessitate keeping living organisms alive on a growth-sustaining nutrient substrate in a bioreactor, which is subject to varying concentrations of toxicants and environmental fluctuations. Using dead biomass, on the other hand, does not restrict the process's potential applications. The concentration of metals, the viability of their recovery, the effectiveness of the biosorbent, and the cost of the biosorbent all play a role in determining whether or not dead biomass may be employed as a biosorbent. It is highly desirable to create and use biosorbent made from industrial waste since it is inexpensive and can be used to reduce pollution risks (Torres, 2020; Zhang et al., 2020). The numerous organisms and agro-industrial waste used to create biosorbents are summarized in Table 6.3.

6.4 MECHANISM OF BIOSORPTION

Different biomaterials have different physical and chemical mechanisms at play during biosorption, as seen by the variety of biomaterials that show promise as biosorbents. It is likely that a variety of elements are involved, although it is not known how they all operate together (Abbas et al., 2014; Michalak et al., 2013). Mechanisms considered to drive biosorption include the metal's physicochemical qualities, the biosorbent's chemistry, and the process's settings. The bacteria or

TABLE 6.3
Various types of biosorbents used for heavy metal biosorption

S no.	Types of biosorbents		Examples	References
1.	Natural	Algal	*Schizosaccharomyces sps. Filamentous algae* of *pithophora* sp. *Sargassum* sp. *Scenedesmus spinosus, Chlorella protothecoides, Nannochloropsis oculata, Chlorella vulgaris, Spirogyra hyaline, F. algae*	Ahmad et al. (2018), Hiremath and Theodore (2017), Kumar and Oommen (2012), Liu et al. (2014), Mart´ınez-Ju´arez et al. (2012), Rani et al. (2010) and Sulaymon et al. (2012)
		Fungal	*Rhizopus oryzae, Aspergillus niger, Aspergillus flavus, Aspergillus fumigatus* I-II *Rhizopus nigricans*	Dusengemungu et al. (2020), Haq et al. (2015), Hassan and El-Kassas (2012), and Shoaib et al. (2012)
		Plant	*Acacia leucocephala* bark, sugar beet pectins, water plants, leaves, seeds and roots of *Nymphaea lotus, Echorniacrassipes, Canna indica,* sandal wood (*Santalum album)* leaf powder *Anogeissus dhofarica,* leaf powder from neem trees, *Utricularia aurea,* wheat-based biosorbents, biomass of *Momordica charantia*, tur pod (*Cajanus cajan*), mango and neem leaves, lignocellulosic coconut fiber, banana stem	Bakar et al. (2016), Bhatnagar and Minocha (2010), Dixit et al. (2015), Farooq et al. (2010), Fu et al. (2012), Galadima et al. (2015), Johari et al. (2014), Khound and Bharali (2018), Shahin et al. (2017) and Swathy et al. (2017) and Tseveendorj (2017)
		Bacterial	*Micrococcus luteus* sulfate-reducing bacteria *Enterobacter cloacae, B. subtilis, Shewanella oneidensis* MR-1, *Geobacter sulfurreducens, Bacillus cereus cells*	Ayangbenro et al. (2017), Sinha et al. (2012) and Subhashini et al. (2011)
		Agro-waste	Rice husk, *Punica granatum* peels waste peanut husk charcoal, rice polish	Ahmad et al. (2017), Hegazi (2013) and Shen et al. (2017)
		Industrial waste	Fly ash, biochar produced from wheat straw pellets and rice husk, lettuce roots and sugarcane bagasse, date palm biochar, wetland plant derived biochar, NaOH-treated and immobilized peanut husk and dewmelon peel biochar, banana peduncle biochar: Chinese coal fly ashes	Ahmadi et al. (2016), Cui et al. (2016), Hegazi (2013), Karim et al. (2015), Milani et al. (2018), Zhong et.al. (2014), Usman et al. (2016) and Song et al. (2014)
2.	Modified		Carboxyl functionalized *Cinnamomum camphora,* jackfruit wood sawdust, activated carbon, silica and silica-activated carbon composite, lemon peel (*Citrus limonum*) modified with citric acid, compost, modified fly ash	Adewuyi (2020), Karnib et al. (2014), Kim et al. (2015), Mullassery et al. (2014), Tovar et al. (2018) and Wang and Wang (2018)

plants used to create the biosorbents store the metals in their appendages and the gaps in their cell walls and membranes until the concentration is high enough to be useful (Gupta et al., 2015). Biosorption relies heavily on the biosorbent's surface functional groups attracting the metal ion of interest (e.g., hydroxyl, carboxyl, sulfhydryl, ketones, amines, imines, and immidazols) (Salman et al., 2014; Wang & Chen, 2009). The presence of these functional groups releases the metal(loid)s from their usual positions. Consequently, by altering their oxidation state, metal ions become less poisonous (Chaturvedi et al., 2015). Biosorbent preparation is affected by structural conformations, the pH of the contaminated media, and biomass transformations. The biosorbent capacity of various biosorbent surface functional groups, for instance, is affected by the surrounding pH. For this reason, a biosorbent with a negative charge is preferable in areas contaminated by cationic metallic species. The optimal pH level is found to be between 7.0 and 8.0. Reduced pH causes hydrogen ion binding to compete with metal ion binding. The hydroxides of metals tend to precipitate out at higher pH levels. Chrome, arsenic, and molybdenum are examples of anionic metallic species that thrive in acidic conditions (pH 4.0–6.0). Those extra positive charges have a favorable effect on these anions (Zhang et al., 2020). A significant factor in biosorption is temperature, since it affects both the surface activity of the biosorbent and the kinetic energy of the metal to be biosorbed (Sedlakova-Kadukova et al., 2019). It is possible that biosorption occurs as a result of electrostatic interactions between the cell walls of microorganisms that have died and metallic ions. Mushrooms can actively absorb metal ions through their chitinous cell walls and into their mycelium and spores, which increases their tolerance to metal toxicity. Both external and intracellular precipitation or valence transformation contributes to the detoxification procedure (Gupta et al., 2015). Extracellular polymeric molecules or microbial metal-binding metabolites may also bind to the insoluble heavy metal residues (Ayangbenro et al., 2017).

Microbial biosorption of hazardous metals and associated elements has been studied for decades, and the fundamental mechanisms have been elucidated (Gadd, 2009; Kotrba, 2011). The mechanisms of biosorption, which can be broken down into adsorption, ion exchange, and complexation/coordination for the sake of a more precise characterization, are responsible for the biosorbent's fast and reversible nature, as well as its similarity to traditional ion exchange resins (Gadd, 2009). However, many different functional groups (such as carboxyl, phosphate, hydroxyl, amino, thiol) in biomass are able to interact with metal species in varied degrees, with these interactions controlled by physicochemical characteristics. Depending on the environment and the conditions provided, biosorption can actually be a somewhat mechanistically complex process (Gadd, 2009). Precipitation or crystallization is another mechanism that can occur and make sorption or desorption more challenging. There may be a trade-off between absorption and desorption, with the latter perhaps suffering. Actinides like uranium and thorium tend to precipitate out of the environment in non-living fungal biomass, making it a good example (Gadd & White, 1992).

It's possible that various interrelated biosorption processes will be at play. Understanding ligand preferences in metal complex formation is critical. It has been hypothesized that a crucial stage in the surface complex synthesis of cations like

Cu2+ is the coordination of metal ions with oxygen donor atoms, followed by proton release and the creation of bidentate surface complexes (Gadd, 2009). An outer-sphere complex forms when a chemical (i.e., largely covalent) bond forms between the metal and the electron-donating oxygen ion, while an inner-sphere complex forms when a cation approaches the surface negative groups to a critical distance but the cation and base are separated by at least one water molecule. Anion biosorption mechanisms have received little attention in the scientific literature. This seems to be affected by a number of chemical factors, most notably pH. Anionic species including TcO_4, $PtCl_{4,3}$, $CrO_{4,2}$, SeO_2, and $Au(CN)_2$ have higher biosorption rates at acidic pH values (Gadd, 2009).

Since organic pollutants come in a wide range of chemical structures, the efficacy of biosorption depends on a number of variables, including the nature of the biosorbent used and the chemical make-up of the wastewater. When hydrophobic chemicals come into contact with biomass in biosorption systems, hydrophobic sorption is unmistakably the result (Stumm & Morgan, 1996). It is possible for hydrophobic chemicals to enter the organic matrix because they are lipophilic and can diffuse through membranes. The process of biosorption of organic contaminants may involve absorption to some degree. Several mechanisms contribute to the biosorption of dye onto chitosan; these include surface adsorption, chemisorption, diffusion, and adsorption complexation; however, film diffusion, pore diffusion, and chemical reactions such ion exchange and complexation are the most important (Crini & Badot, 2008). Intermolecular interactions between dye molecules in chitosan-dye systems are characterized by amine sites as the predominant reactive groups and maybe contributing hydroxyl groups (Crini & Badot, 2008) (Figures 6.1 and 6.2).

Transport Across Cell Membrane—Bacteria are frequently studied for their ability to transport heavy metal ions across cell membranes. First, metal ions attach to binding sites on the microbe's cell wall (a process known as independent binding metabolism); then, the metal ions are transported over the cell membrane and into the cell (a process known as metabolism-dependent intracellular uptake) (Perpetuo et al., 2011). The specific method by which heavy metal ions with the same ionic radius and charge as the essential metal ions trick cellular metal transport systems into thinking they are transferring the vital metal ions is not well known at this time (Perpetuo et al., 2011).

6.4.1 Physical Adsorption

Physical adsorption occurs at the biosorbent's surface, sometimes with the cell walls of microorganisms. This process is aided by electrostatic interactions like Van der Waals forces. The physical adsorption mechanism is affected by the solution's pH and, in some cases, the biosorbent's surface area (Bashir et al., 2019; Chojnacka, 2006). The carboxylate and hydroxylate groups in the biomass increased lead biosorption by hami melon peels in basic conditions (Bashir et al., 2019). Grass and wheat straw were proposed to biosorb chromium(III) ions via a physical adsorption mechanism restricted in a monolayer form based on predictions of their maximum biosorption capacity and knowledge of the ionic radius of the metal (Chojnacka, 2006).

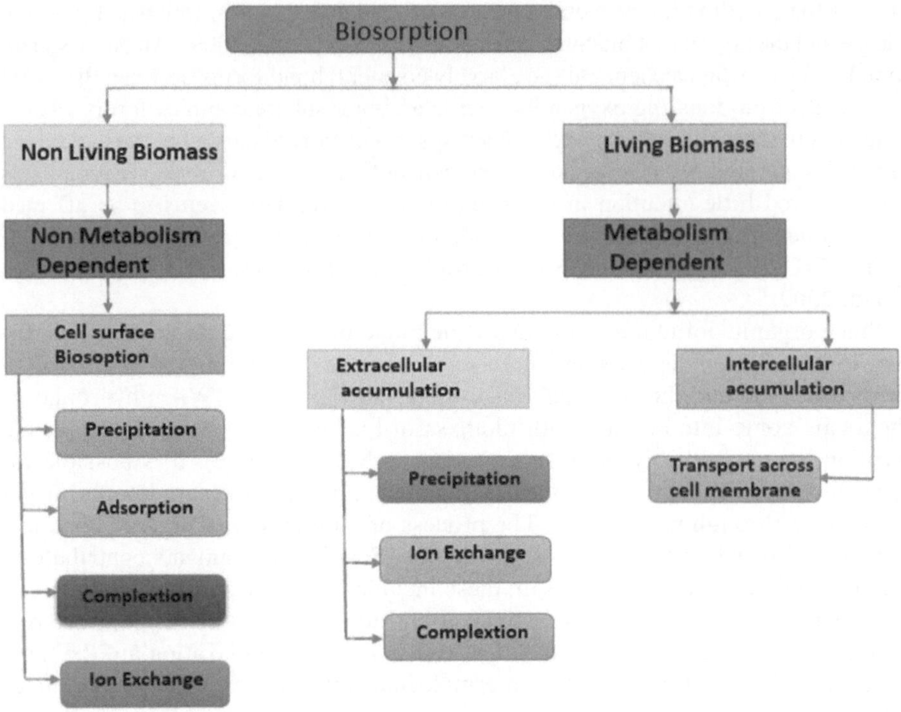

FIGURE 6.1 Classification of biosorption mechanism (Shaikh et al., 2022)

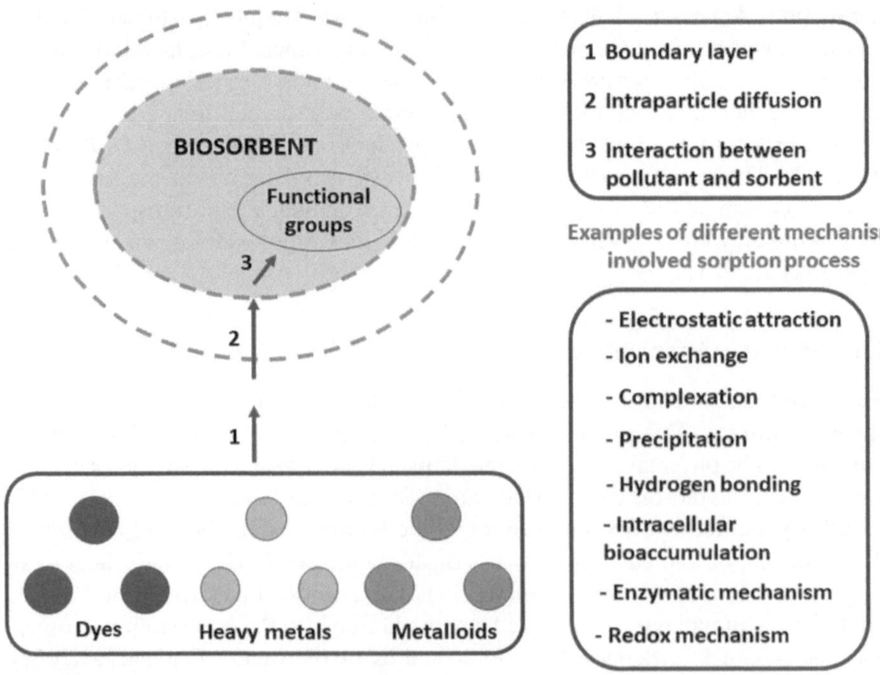

FIGURE 6.2 Mechanism involved in biosorbent process (Elgarahy et al., 2021)

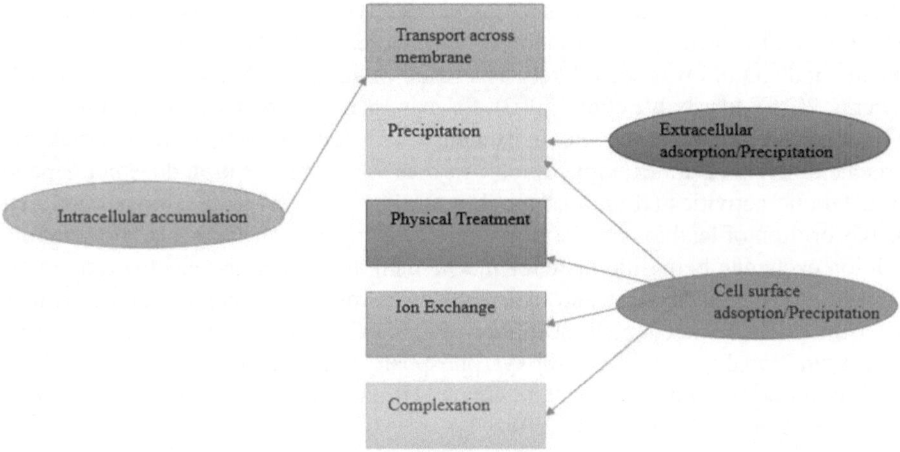

FIGURE 6.3 Biosorption mechanisms according to location where the metals removed are found (Salman et al., 2014)

Biosorption of copper ions by other bacteria, like *Zoogloea ramigera*, and by other algae, including *Chiarella vulgaris*, has been demonstrated via physical adsorption (Perpetuo et al., 2011). Biosorption mechanism acoording to location depicted in (Figure 6.3.)

Chemical Precipitation: Colloid agents such as alum, lime, iron salts, and organic polymers can be used to precipitate metals. The biggest drawback is the vast quantity of sludge containing hazardous compounds that is created throughout the process.

Hydroxide Precipitation: Heavy metals are commonly precipitated chemically using lime or sodium hydroxide as their hydroxides. Lime is the material of choice for precipitation operations due to its low cost, versatility (it may be used to adjust pH between 8.0 and 10.0), and ability to be utilized as an adsorbent for the removal of metal ions in excess. The effectiveness of the process is affected by a number of factors, including the metal ion's ease of hydrolysis, the nature of the oxidation state, the pH, the presence of complex forming ions, the standing time, the degree of agitation, the settling and filtering procedures, and the characteristics of the precipitate. It can be difficult to treat effluents containing mixed metal ions because the ideal pH for hydroxide formation varies between metals. One disadvantage is that the solubility of metal hydroxides might change over time at a given pH (Gadd, 2009; Perpetuo et al., 2011).

6.4.2 Ion Exchange

During an ion exchange process, one type of ion in solution is switched out for another type of ion that is bonded to the solid phase (Gadd, 2009). Biosorption of copper ions by *Ganoderma lucidum* occurs by an ion exchange process, which is typical among microorganisms because to the presence of cell wall polysaccharides that are capable of exchanging ions with their counter ions (Perpetuo et al., 2011).

After binding heavy metal ions such as copper(II), cadmium(III), and zinc(II) to the alginate of brown algae, an increase in the release of calcium, potassium, magnesium, and sodium was seen, likely as a result of an ion exchange process (Abdi & Kazemi, 2015). In Ahalya et al. (2003), for example, rice straws have been reported to biosorb cadmium(II) ions via ion exchange with sodium, potassium, magnesium, and calcium. This is an example of a biosorbent whose biosorption does not depend on metabolic activities (Kanamarlapudi et al., 2018). It was also demonstrated that the biosorption of lead ions by hami melon peels depends on the pH of the solution, with ion exchange being the primary mechanism at low pH and electrostatic interactions between carboxylate and hydroxylate groups on the biomass and lead ions being the primary mechanism at higher pH (Bashir et al., 2019). Ion exchange by certain functional groups like carboxyl, phosphate, and hydroxyl may be responsible for chromium(III), cadmium(II), and copper(II) biosorption in the *Cyanobacterium Spirulina* (Kanamarlapudi et al., 2018).

6.4.3 Complexation

Chelation is a more complicated form of complexation in which the organic ligand forms coordinative connections with the metal ion from many positions simultaneously to form a complex with increased stability via electrostatic attraction or covalent bonding (Tsezos et al., 2014). The HSAB theory classifies elements as either hard acids (mainly metals) or soft bases (primarily non-metals), with hard acids having a greater affinity to hard bases and weak acids and bases having a lesser affinity to weak acids and bases (Ayers, 2005). Lead, a soft acid, would rather bind covalently to organic ligands of the biosorbent containing nitrogen or sulfur, two soft bases; manganese, zinc, cadmium, and copper ions, intermediate between hard and soft ions, also have a high affinity to form complexes with nitrogen or sulfur containing organic compounds (Tsezos et al., 2014). However, the complexation process is influenced by how easily accessible the binding site is that contains the donor atom (base). Desorption experiments with SEM, TEM, and EDX can verify if the biosorption mechanism involves complexation (Han et al., 2006). These techniques confirmed that *Chlorella miniata*'s (a green microalgae) biosorption mechanism for chromium(III) is complexation by carboxyl, phosphonate, and amine ligands (Han et al., 2006).

6.4.4 Precipitation

Precipitation, or the formation of an insoluble metal precipitate, is one of the few mechanisms involved in metabolism-dependent biosorption. Precipitation, however, can also take place within the context of biosorption that is metabolically unrelated. In the metabolically dependent biosorption process, the presence of harmful metal ions triggers precipitation as an active defense mechanism in the microorganisms (Perpetuo et al., 2011). The biosorbent's functional groups react with the metal ion in the cell wall during metabolism-independent biosorption, leading to precipitation. In some cases, oxidation–reduction reactions may be involved in these procedures.

6.4.5 Biosorption Modeling

The formula for determining how many metal ions can be biosorbed by one gram of biomass is as follows:

$$Q = Ci - Cf / M \times V$$

where M is the mass of biomass added to the reaction mixture, V is the volume of the mixture in liters, Ci is the concentration of metal ions at the beginning of the reaction (in milligrams per liter), and Cf is the concentration of metal ions at the end of the reaction (in milligrams per liter).

The equilibrium of biosorption can be described by several different models. Values of the isotherm parameters express the surface properties and affinity of biosorbent for different heavy metal ions. For his Cu(II) and Co(II) kinetics and modeling study, Masoumeh Akbari analyzed equilibrium data using the Langmuir, Freundlich, and Temkin isotherm models (II).

The Langmuir model can be stated as the following equation under the assumption of a surface with an infinite number of identical sites with homogeneous biosorption energy.

$$qe = \frac{qmbLCe}{1 + bLCe}$$

The affinity between the biosorbent and the metal ion is represented by the constant bL (mg g^{-1} mg^{-1} L^{-1}), hence the maximum biosorption capacity is the biosorption capacity when the surface is completely coated with metal ions (qm).

Taking into account binding site heterogeneity and multilayer biosorption on the biosorbent's surface, the Freundlich isotherm is an empirical expression.

$$qe = kFCe\frac{1}{n}$$

where kF is the Freundlich constant reflecting biosorbent capacity and n is the Freundlich exponent expressing biosorbent intensity. The Freundlich isotherm model parameters can be determined from a plot of log(ln)qe versus log(ln)Ce.

The following Temkin isotherm takes into account the effect of the biosorption heat, which decreases with increasing biosorbent coverage, and the interaction between the biosorbate and the biosorbent:

$$qe = \frac{RT \ln atCe}{bt}$$

T is the absolute temperature, while R is the gas constant (8.314 J mol K^{-1}) (K). To determine at and bt, plot qe against Ce (Karim et al., 2015).

The work "The numerical model of biosorption of Zn2+ and its application to the bio-electro tower reactor (BETR)" by Tau Wang explored the 2D numerical kinetic model based on flow velocity and adsorption. The old pseudo-first-order model and the pseudo-second-order model both treat qe as a constant, but the new model is preferred since it considers the adsorbed quantity at equilibrium to be a transient variable (Karnib et al., 2014).

Hall et al. developed the dimensionless equilibrium parameter RL from the Langmuir isotherm coefficient b, and it can be used to identify effective adsorbents and adsorption procedures.

$$RL = \frac{1}{1+bC0}$$

Hall et al. and Chen and Wang used the following equation to predict whether a sorption process would be beneficial ($0 < RL < 1$) or unfavorable ($RL < 1$): the adsorbent's affinity for the adsorbed species increases as RL lowers. Since the RL values for Pb2+ ions are lower (0.061–0.537) under similar environmental circumstances, yeast biomass has a stronger affinity for Pb2+ ions than Cu2+ ions (0.155–0.768). When the biomass dose in the biosorption tank was increased from 0.75 to 2.5 g l^{-1}, the RL values for Cu2+ ions changed only slightly.

Metal biosorption mechanisms are discovered to be influenced by a number of factors:

- life or death for biomass;
- biomaterials, categorize;
- chemical characteristics of metal solutions; and
- pH, temperature, and other environmental factors.

Because of this ambiguity, it is often challenging to characterize the mechanisms involved in biosorption, with perhaps the exception of the simplest laboratory systems. The complexity of biological matter means that many mechanisms may be at play under any given set of circumstances, not all of which are well understood. The mechanisms of biosorption can be broken down into the following categories.

Here, biosorption mechanisms are split into two groups, those that rely on metabolism and those that don't;

- In accordance with the site at which biosorption occurs, the mechanisms involved in biosorption are classified as either extracellular accumulation/precipitation, cell surface sorption/precipitation, or intracellular accumulation.

Physical adsorption, ion exchange, complexation, precipitation, and intracellular trapping are all mechanisms by which metals are bioabsorbed (Lawrence et al., 2010).

Chemically processed biomass, along with living or decomposing biological cells, can all be used in the biosorption process. Uptake of metal ions by living cells and non-living materials uses two different mechanisms. Bioaccumulation, the process by which living things absorb something, and biosorption, by which dead things get rid of something, are two completely separate processes.

Surface binding of metal ions to cell walls and extracellular material constitutes the first form of absorption, which is unrelated to metabolic activity within the cell. Biosorption, often known as "passive uptake," describes this process.

Intracellular uptake, active uptake, or bioaccumulation is the second way metals that are taken up by cells; it is dependent on cellular metabolism. Live cells are required for intracellular uptake of toxic metal ions via metabolism. Surface sorption of metal ions can occur with functional groups such as carboxylate, hydroxyl, sulfate, phosphate, and amino acids.

6.4.5.1 Metal Biosorption Process Using Living Cells

For metals to be taken up by living cells, a two-stage process is required:

- First, the metal ions adsorb to the surface of the cells by interacting with functional groups on the cell wall.
- Second, metal ions can enter cells through the cell membrane, thanks to active biosorption. Activation state is associated with metal transport and deposition, which is in turn dependent on the metabolic state.

6.4.5.2 Metal Biosorption Using Dead Cells

Using metal-binding mechanisms such as complexation, ion exchange, physical adsorption, inert materials can remove metals in a "passive mode" that is metabolism-independent and rapid. Adsorption and desorption are in a state of dynamic equilibrium where they can occur simultaneously. Energy is not required.

There are a few positives to working with cadaverous materials:

i. The need for expansion is unnecessary;
ii. There is no need for a growth medium; and

Wastes and byproducts can be used to make these substances.

Ion exchange, complexation, and physical adsorption are the mechanisms responsible for these interactions. Dead plant and animal matter appear to be the preferable option because:

i. There are no toxicity constraints, no nutrient requirements in the feed solution, and regenerated biomass can be used again;
ii. Fungal and yeast cells, both living and non-living, can extract toxic metals from wastewater;

As is commonly known, heavy metals can severely stunt a mushroom's development

pH: The solubility of the metal ions, the concentration of the counter ions on the functional groups of the adsoption/biosorption (i.e., the surface charge of adsoption/biosorption), and the degree of ionization of the adsorbate during reaction are all influenced by pH, making it an essential factor in the adsoption/biosorption of metal ions from aqueous solution. Therefore, it affects not only the adsorbate's solution chemistry but also the biosorbent's or adsorbent's functional group activity and the ion competition between adsorbate and sorbate ions. pH has a role because hydrogen ion is a strong competitive adsorbate (Puranik et al., 2005). Table 6.4 displays the acidity constants for the functional groups in biosorbents that are responsible for sorbent binding. The surface charge is mostly negative between the pH values of 3.0 and 10.0.

TABLE 6.4
Major binding groups for biosorption

Binding group	Structural formula	pKa	Ligand atom	Occurrence in selected biomolecules
Hydroxyl	—OH	9.5–13.0	O	PS, UA, SPS, AA
Carbonyl (ketone)	C = O	–	O	Peptide bond
Carboxyl	—COOH	1.7–4.7	O	UA, AA
Sulfhydryl(thiol)	—SH	8.3–10.8	S	AA
Sulfonate	—SO_3	1.3	O	SPS
Thioether	—S	–	S	AA
Amine	—NH_2	8.0–11.0	N	Cto, AA
Secondary amine	—NH	13.0	N	Cti, PG, AA
Amide	—C = O NH	–	N	AA
Imine	=NH	11.6–2.6	N	AA
Imidazole	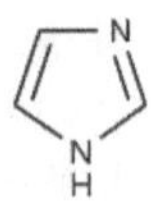	6.0	N	AA
Phosphonate	OH | —P = O | OH	0.9–2.1 6.1–6.8	O	PL
Phosphodiester	P = O | OH	1.5	O	TA, LPS

PS: polysaccharides, UA: uronicacids, SPS: sulfated PS, AA: aminoacids, Cto: chitosan, Cti: chitin, PG: peptidoglycan, PL: phospholipids, TA: teichoic acid, LPS: lipo, PS.

Under conditions with a pH less than 3.0, the net surface charge of microbial cells becomes positive.

At high solution pH, metal solubility decreases to the point that precipitation is likely, which might make the sorption process more challenging (Volesky, 2003). At pH levels between three and six, for instance, bacterial biomass is optimal for the biosorption of metal ions. This is because changing the pH might affect how active binding sites are (Nagajyoti et al., 2010). Solution pH decreases as a result of the release of H+ ions during biosorption of metals by protonated bacterial biomass. Since most of the reaction takes place early on, the pH shifts are dramatic at first, but gradually level off as the system settles into equilibrium. Maintaining a constant pH level throughout the contact time is essential for achieving equilibrium (Ayangbenro, 2017).

Heavy metal biosorption into *Pseudomonas aeruginosa* was impacted by pH value during cation uptake on the biomass surface. *P. aeruginosa*, an industrial byproduct, has also been shown to be effective in removing heavy metals at pH levels between 3 and 5 (Babula, 2008).

TABLE 6.5
Metal hydroxide precipitations

Metal	pH value
Lead	6.3
Chromium	6.5–7.3
Cadmium	9.7
Copper	7.1–7.3
Nickel	9.2–9.4
Zinc	8.3–8.5
Iron	4.3
Aluminum	5.2
Tin	1.0–4.5

In batch biosorption studies, the influence of $H_2O_2/MgCl_2$ treatment on the uptake of Pb^{2+}, Cd^{2+}, Cu^{2+}, and Zn^{2+} ions by *Azolla filiculoides* was analyzed. Biosorption levels were demonstrated to decrease as pH was reduced, with the findings owing to competition between protons and metal ions for binding sites (Taghi et al., 2005).

The effectiveness of decomposing mycelial biomass in cadmium removal was studied in batch experiments over the pH range of 3–8. The electrostatic binding of ions to their respective functional groups is affected by pH, making the adsorption of metal ions pH-dependent. In most cases, a pH of 4 was found to be the best for metal adsorption. The quantity of metal ions adsorbing to a surface is greatly reduced at high pH levels due to the development of metal hydroxide and other metal complexes. The pH values for precipitation of metal hydroxides are listed in Table 6.5 (Taghi et al., 2005).

Temperature: Although high temperatures, such as 50°C, may boost biosorption in some situations, such temperatures may permanently harm microbial living cells and therefore decrease metal uptake, whereas the efficiency of the bioaccumulation process stays unaltered within the range 20–35°C (Salman, 2015). In most cases, the rate of adsorption rises as the temperature drops, and the amount adsorbed is almost always exothermic. *S. cerevisiae*'s biosorption ability for Ni and Pb was found to be greatest at 25°C and to decrease at 40°C (Murtaza, 2014).

6.4.6 Present Status of Application

Multiple factors obscure the potential of biosorption in the business world. Granular and powdered biosorbent preparations have a problem with cell separation after biosorption and mass recovery after regeneration because of their tiny size (deFreitas et al., 2020). Immobilization of the biomass can enhance the material characteristics (Vijayarhagvan & Yun, 2008). Utilizing biochars or activated carbons that have high adsorption capacity, large volume microporosity, metal selectivity, and adaptability is another option (deFreitas et al., 2020; Xiang et al., 2019). Subsequently, it is imperative to investigate the problems associated with biosorbents that get saturated too quickly, require frequent replacement, and lack correlation between laboratory studies and the real site (Qin et al., 2020; Redha, 2020). To make matters worse, companies are reluctant to implement novel approaches since there is insufficient data to substantiate their effectiveness when compared to more established strategies (Redha, 2020).

To be truly sustainable, the development process must incorporate cutting-edge innovations from a variety of new disciplines. Using nanotechnology to create nano-products with enhanced physical properties is a good example. As adsorbents and stabilizers, these have commercial potential for reducing heavy metal pollution (Qin et al., 2020). Biocomposite sorbents are made by mixing biopolymers such as cellulose, chitosan, starch, alginate with other substances like clay, graphene, or carbon. The sorbents are safe to use, come in plenty, and feature high surface area and high cation exchange capacity (Adewuyi, 2020; Karnib et al., 2014). For a method to be economically viable, it is also important to consider the price tag associated with biosorbent development and biosorption. The cost of remediating a polluted area depends on many variables, including the type of pollution, the size of the area, whether the process is batch or continuous, where the biosorbent is sourced, and the local economy. There may be promising business opportunities in these sectors that should be explored.

6.5 CONCLUSION

Heavy metals are notoriously difficult to dissolve or remove from impacted areas due to their complex chemical composition. The toxic effects of heavy metals on human and plant physiology have been known for a long time. Therefore, strategies were developed to control and lessen the impact of heavy metals. Large amounts of sludge are created by conventional methods such as ion exchange, distillation, filtration, thermal treatment which cancel out any savings. In addition, time and energy constraints reduce the effectiveness of such approaches. Heavy metals can be safely removed and recovered by biosorption, a recently developed technique. Finding and developing novel biosorbents from higher plant species, modified agro-waste, and industrial waste can solve the problem of waste disposal and pollution management. If developed and widely adopted, this technology could pave the way to a sustainable future in a variety of ways.

REFERENCES

Abbas SH, Ismail IM, Mostafa TM, Sulaymon AH (2014a). Biosorption of heavy metals: A review. *J Chem. Sci. Technol.,* 3 (4); 74–102.

Abbas SH, Ismail IM, Mostafa TM, Sulaymon AH (2014b). Sulaymon, biosorption of heavy metals: A review. *J. Chem. Sci. Technol. (JCST)*, 3 (4), 74–102.

Abdi O, Kazemi M (2015). A review study of biosorption of heavy metals and comparison between different biosorbents. *J. Mater. Environ. Sci.,* 6 (5), 1386–1399.

Adewuyi A (2020). Chemically modified biosorbents and their role in the removal of emerging pharmaceutical waste in the water system. *Water*, 12, 1551. https//doi.org/10.3390/w12061551

Ahalya N, Ramachandra TV, Kanamadi RD (2003). Removal and recovery of heavy metals by biosorption. *Res. J. Chem. Environ.*, 7(4), 71–79. https//doi.org/10.1111/j.1742-4658

Ahmad A, Bhat AH, Buang A (2018). Biosorption of transition metals by freely suspended and Ca-alginate immobilised with Chlorella vulgaris: Kinetic and equilibrium modeling. *J. Clean. Prod.*, 171, 1361–1375.

Ahmad I, Akhtar MJ, Jadoon IBK, Imran M, Imran M, Ali S (2017). Equilibrium modeling of cadmium biosorption from aqueous solution by compost. *Environ. Sci. Pollut. Res.*, 24, 5277–5284. https//doi.org/10.1007/s11356-016-8280-y

Ahmadi M, Kouhgardi E, Ramavandi B, (2016). Physico-chemical study of dew melon peel biochar for chromium attenuation from silumated and actual wastewaters. *Korean J. Chem. Eng.*, 33 (9), 2589–2601.

Akcil A, Erust C, Ozdemiroglu S, Fonti V, Beolchini F (2015). A review of approaches and techniques used in aquatic contaminated sediments: Metal removal and stabilization by chemical and biotechnological processes. *J. Clean. Product.*, 86, 24e36.

Al-Gheethi AAS, Lalung J, Noman EA, Bala JD, Norli I (2015). Removal of heavy metals and antibiotics from treated sewage effluent by bacteria. Clean Technol. Environ. Policy 17, 2101–2123. Doi:10.1007/s10098-015-0968-z

Arora A, Sairam RK, Srivastava GC (2002). Oxidative stress and antioxidative system in plants. *Curr. Sci.*, 82, 1227–1338.

Ayangbenro AS, Babalola OO (2017). New strategy for heavy metal polluted environments: A review of microbial biosorbents. *Int. J. Environ. Res. Public Health*, 14, 94. https//doi.org/10.3390/ijerph14010094

Ayers PW (2005). An elementary derivation of the hard/ soft-acid/base principle. *J. Chem. Phys.*, 122 (14). https//doi.org/10.1063/1.1897374.2005.04698.x

Babula P, Adam V, Opatrilova R, Zehnalek J, Havel L, Kizek R (2008). Uncommon heavy metals, metalloids and their plant toxicity: A review. *Environ. Chem. Lett.*, 6, 189–213.

Bakar AABA, Ali KABM, Tarmizi NABA, Japeri AZUBM, Tammy NJB (2016). Potential of using bladderwort as a biosorbent to remove zinc in wastewater. *AIP Conf. Proc.*, 1774, 030023. https//doi.org/10.1063/1.4965079

Barakat MA (2011). New trends in removing heavy metals from industrial wastewater. *Arab. J. Chem.*, 4, 361–377.

Bashir A, Malik LA, Ahad S, Manzoor T, Bhat MA, Dar GN, & Pandith AH (2019). Removal of heavy metal ions from aqueous system by ion-exchange and biosorption methods. *Environ. Chem. Lett.*, 17 (2), 729–754. https//doi.org/10.1007/s10311-018-00828-y

Bhatnagar A, Minocha AK (2010). Biosorption optimization of nickel removal from water using Punica granatum peels waste. *Coll. Surf. B Biointerf.*, 76, 544–548.

Bilal M, Rasheed T, Iqbal HMN, Hu H, Wang W, Zhang X (2017). Novel characteristics of horseradish peroxidase immobilized onto the polyvinyl alcohol-alginate beads and its methyl orange degradation potential. *Int. J. Biol. Macromol.*, 105, 328–335.

Blais J, Djedidi Z, Cheikh RB, Tyagi R, Mercier G (2008). Metals precipitation from effluents: Review. *Pract. Period. Hazard. Toxic Radioact. Waste Manag.*, 12, 135–149.

Chaturvedi AD, Pal D, Penta S, Kumar A (2015). Ecotoxic heavy metals transformation by bacteria and fungi in aquatic ecosystem. *World J. Microbiol. Biotechnol.*, 31, 1595–1603.

Chibuike G, Obiora S (2014). Heavy metal polluted soils: Effect on plants and bioremediation methods. *Appl. Environ. Soil Sci.*, 2014, 1–12.

Choi JM, Pak CH, Lee CW (1996). Micronutrient toxicity in French marigold. *J. Plant Nutri.*, 19, 901–916.

Chojnacka K (2006). Biosorption of Cr(III) ions by wheat straw and grass: A systematic characteristics of new biosorbents. *Polish J. Environ. Stud.*, 15(6), 845–852.

Cox MS, Bell PF, Kovar JL (1996) Differential tolerance of canola to arsenic when grown hydroponically or in soil. *J. Plant Nutr.*, 19 (12), 1599–1610.

Cui X, Hao H, Zhang C, He Z, Yang X (2016). Capacity and mechanism of ammonium and cadmium sorption on different wetland plant derived biochar. *Sci. Total Environ.*, 539, 566–575.

CWC Report, (2010), *Central Water Comission, Ministry of Water Resource*, Govt. of India, P. 141.

Crini G, Badot P (2008). Application of chitosan, a natural aminopolysaccharide, for dye removal from aqueous solutions by adsorption processes using batch studies: A review of recent literature. *Prog. Polym Sci.* 33, 399–447. doi: 10.1016/j.progpolymsci.2007.11.001.

Dixit A, Dixit S, Goswami CS (2015). Eco-friendly alternatives for the removal of heavy metal using dry biomass of weeds and study the mechanism involved. *J. Bioremed. Biodeg.*, 6, 290. https//doi.org/10.4172/2155–6199.1000290.

Dusengemungu L, Kasali G, Gwanama C, Ouma KO (2020). Recent advances in biosorption of copper and cobalt by filamentous fungi. *Front. Microbiol.* 11, 582016.

Elgarahy AM, Elwakeel KZ, Mohammad SH, Elshoubaky GA (2021). A critical review of biosorption of dyes, heavy metals and metalloids from wastewater as an efficient and green process. *Clean. Eng. Technol.*, 4, 100209.

Farooq U, Kozinski JA, Khan MA, Athar M (2010). Biosorption of heavy metal ions using wheat based biosorbents – A review of the recent literature. *Bioresour. Technol.*, 101, 5043–5053.

Fashola M, Ngole-Jeme V, Babalola O (2016). Heavy metal pollution from gold mines: Environmental effects and bacterial strategies for resistance. *Int. J. Environ. Res. Public Health*, 13, 1047.

Fomina M, Gadd GM, (2014). Biosorption: Current perspectives on concept, definition and application. *Bioresour. Technol.,* 160, 3e14.

Fu YQ, Li S, Zhu HY, Jiang R, Yin LF (2012). Biosorption of copper (II) from aqueous solution by mycelia pellets of Rhizopus oryzae. *Afr. J. Biotechnol.*, 11, 1403–1411.

Gadd GM (2009). Biosorption: Critical review of scientific rationale, environmental importance and significance for pollution treatment. *J. Chem. Technol. Biotechnol.*, 84 (1), 13–28. https//doi.org/10.1002/jctb.1999

Gadd GM, White C 1992. Removal of thorium from simulated acid process streams by fungal biomass: potential for thorium desorption and reuse of biomass and desorbent. *J. Chem. Technol. Biotechnol.* 55, 39–44, 30.

Gajewska E, Sklodowska M, Slaba M, Mazur J (2006) Effect of nickel on antioxidative enzymes activities, proline and chlorophyll contents in wheat shoots. *Biol. Planta*, 50, 653–659.

Galadima LG, Wasagu RSU, Lawal M, Aliero AA, Magaji UF, Suleman H (2015). Biosorption activity of Nymphaea lotus (water lily). *Int. J. Eng. Sci. (IJES)*, 4 (3), 66–70. ISSN (e): 2319-1813 ISSN (p): 2319-1805.

Garnham G.W., 1997. The use of algae as metal biosorbents, in: Wase J., Forster C. (Eds.*), Biosorbents for Metal Ions*, Taylor&Francis, London, pp. 11–37.

Gavrilescu M. (2004). Removal of heavy metals from the environment by biosorption. Eng. Life Sci. 4 219–232. 10.1002/elsc.200420026

Guo J, Dai X, Xu W, Ma M (2008). Over expressing GSHI and AsPCSI simultaneously increases the tolerance and accumulation of cadmium and arsenic in Arabidopsis thaliana. *Chemo-Sphere*, 72, 1020–1026.

Gupta V (2013). Mammalian feces as bio-indicator of heavy metal contamination in Bikaner Zoological Garden, Rajasthan, India. *Res. J. Anim. Vet. Fish. Sci.*, 1 (5), 10–15.

Gupta VK, Nayak A, Agarwall S (2015). Bioadsorbents for remediation of heavy metals: Current status and their future prospects. *Environ. Eng. Res.*, 20 (1), 001–018. pISSN 1226-1025.

Han X, Wong YS, Tam NFY (2006). Surface complexation mechanism and modeling in Cr(III) biosorption by a microalgal isolate, Chlorella miniata. *J. Coll. Interf. Sci.*, 303 (2), 365–371. https//doi.org/10.1016/j.jcis.2006.08.028

Haq F, Butt M, Ali H, Chaudhary HJ (2015). Biosorption of cadmium and chromium from water by endophytic Kocuria rhizophila: Equilibrium and kinetic studies. *Desalin. Water Treat*, 57 (42), 1–13.

Hassan SW, El-Kassas HY (2012). Biosorption of cadmium from aqueous solutions using A local fungus Aspergillus cristatus (Glaucus Group). *Afr. J. Biotechnol.*, 11 (9), 2276–2286.

Hegazi HA (2013). Removal of heavy metals from wastewater using agricultural and industrial wastes as adsorbents. *HBRC J.*, 9, 276–282.

Higham D. P., Sadler P. J., Scawen M. D. (1986). Cadmium-binding proteins in Pseudomonas putida: pseudothioneins. Environ. Health Perspect. 65 5–11. 10.1289/ehp.86655

Hiremath PG, Theodore T (2017). Modelling of fluoride sorption from aqueous solution using green algae impregnated with zirconium by response surface methodology. *Adsorp. Sci. Technol.*, 35 (1–2), 194– 217.

Hiroaki I, Motoki I, Norie S, Ribeka T, Yoshio K, Shiro Y (2014) Dietary cadmium intake and breast cancer risk in Japanese women: A case–control study. *J. Hazard. Mater.*, 217, 70–77.

Hong BD, Joo RN, Lee KS, Lee DS, Rhie JH, Min SW, Song SG, Chung DY (2016). Fluoride in soil and plant. *Korean J. Agric. Sci.*, 43, 522–536.

Jadhav SK, Mahish PK (2020). Biosorption of Lead from Aqueous Solution using a New Fungal strain, Cunninghamella elegans TUFC20022. *Res. J. Biotechnol.*, 15 (12), 35–42.

Johari K, Saman N, Song ST, Heng JYY, Mat H (2014). Study of Hg (II) removal from aqueous solution using lignocellulosic coconut fiber biosorbents: Equilibrium and kinetic evaluation. *Chem. Eng. Commun.*, 201 (9), 1198–1220.

Kanamarlapudi SLRK, Chintalpudi VK, Muddada S (2018). Application of biosorption for removal of heavy metals from waste water. *Sch. Environ. Sci.*, 18 (69), 69–116.

Karim AA, Kumar M, Mohapatra S, Panda CR, Singh A (2015). Banana peduncle biochar: Characteristics and adsorption of hexavalent chromium from aqueous solution. *Int. Res. J. Pure. Appl. Chem.*, 7 (1), 1–10.

Karnib M, Kabbani A, Holail H, Olama Z (2014). Heavy metals removal using activated carbon, silica and silica activated carbon composite. *Energy Proc.*, 50, 113–120.

Khalid S, Shahid M, Niazi NK, Behzad Murtaza, Bibi I, et al., (2016). A comparison of technologies for remediation of heavy metal contaminated soils. *J. Geochem. Explor.*, 182, 247–268. https//doi.org/10.1016/j.gexplo.2016.11.021>.<hal-01577861v2>

Khound NJ, Bharali RK (2018). Biosorption of fluoride from aqueous medium by Indian sandalwood (Santalum Album) leaf powder. *J. Environ. Chem. Eng.*, 6 (2), pp. 1726–1735.

Kim IH, Choi J-H, Joo JO, Kim Y-K, Choi J-W, Oh B-K (2015). Development of a microbe-zeolite carrier for the effective elimination of heavy metals from seawater. *J. Microbiol. Biotechnol.*, 25, 1542–1546.

Kotrba P (2011). Microbial biosorption of metals—general introduction, in: Kotrba P., Mackova M., Macek T. (Eds.), *Microbial Biosorption of Metals*, Springer Netherlands, Dordrecht, pp. 1–6.

Kumar IN and Oommen C (2012). Removal of heavy metals by biosorption using freshwater alga Spirogyra hyaline. *J. Environ. Biol.*, 33, 27–31.

Lawrence K, Wang JT, Stephen TT, Yung-Tse H (2010). *Handbook of environmental engineering, environmental bioengineering*. Springer, New York Dordrecht Heidelberg London.

Leyssens L, Vinck B, Van Der Straeten C, Wuyts F, Maes L (2017). Cobalt toxicity in humans—A review of the potential sources and systemic health effects. *Toxicology*, 15 (387), 43–56. https//doi.org/10.1016/j.tox.2017.05.015. Epub 2017 May 29.

Lima AIG, Corticeiro SC, Figueira EM, de AP (2006). Glutathione-mediated cadmium sequestration in Rhizobium leguminosarum. Enzyme Microb. Technol. 39, 763–769. 10.1016/j.enzmictec.2005.12.009

Lerebours A, To VV, Bourdineaud J (2016). Danio rerio ABC transporter genes abcb3 and abcb7 play a protecting role against metal contamination. J. Appl. Toxicol. 36, 1551–1557. Doi: 10.1002/jat.3313

Lin HJ, Sunge T, Cheng CY, Guo HR (2013). Arsenic levels in drinking water and mortality of liver cancer in Taiwan. *J. Hazard. Mater.*, 262, 1132–1138.

Liu M, Dong F, Kang W, Sun S, Wei H, Zhang W, Liu Y (2014). Biosorption of strontium from simulated nuclear wastewater by scenedesmus spinosus under culture conditions: Adsorption and bioaccumulation processes and models. *Int. J. Environ. Res. Public Health*, 11 (6), 6099–6118. http://doi.org/10.3390/ijerph110606099

Mancuso F, Arato I, Lilli C, Bellucci C, Bodo M, Calviai M, Luca G (2018). Acute effects of lead on porcine neonatal Sertoli cells in vitro. *Toxicol. In Vitro*, 48, 45–52.

Mart´ınez-Ju´arez VM, C´ardenas-Gonz´alez JF, Torre-Bouscoulet ME, and Acosta-Rodr´ıguez (2012). Biosorption of mercury (II) from aqueous solutions onto fungal biomass. *Bioinorg. Chem. Appl.*, 18, 1–5.

Messer RL, Lockwood PE, Tseng WY, Edwards K, Shaw M, Caughman GB, Lewis JB, Wataha JC (2005). Mercury (II) alters mitochondrial activity of monocytes at sublethal doses via oxidative stress mechanisms. *J. Biomed. Mat. Res. B*, 75, 257–263.

Michalak I, Chojnacka K, Witek-Krowiak A (2013). State of the art for the biosorption process—A review. *Appl. Biochem. Biotechnol.,* 170, 1389–1416. https://doi.org/10.1007/s12010-013-0269-0

Milani PA, Debs KB, Labuto G, Vasconcelos EN, Carrilho M (2018). Agricultural solid waste for sorption of metal ions: Part I—characterization and use of lettuce roots and sugarcane bagasse for Cu (II), Fe (II), Zn (II), and Mn (II) sorption from aqueous medium. *Environ. Sci. Pollut. Res.* https://doi.org/10.1007/s11356-018-1615-0

Mohanpuria P, Rana N K, Yadav S K (2007). Cadmium induced oxidative stress influence on glutathione metabolic genes of Camella sinensis (L.). O Kuntze. *Environ. Toxicol.*, 22, 368–374.

Mullassery MD, Fernandez NB, Anirudhan TS (2014). Removal of mercury (II) ions from aqueous solutions using chemically modified banana stem: Kinetic and equilibrium modeling. *Sep. Sci. Technol.*, 49, 1259–1269.

Murtaza G, Murtaza B, Niazi KN, Sabir M, Ahmad P, Wani RM, Azooz MM, Phan Tran L-S (2014). Soil contaminants: Sources, effects, and approaches for remediation. *Improve. Crops Era Clim. Changes.*, 2, 171–196.

Mussin JE, Roldán MV, Rojas F, de los Ángeles Sosa M, Pellegri N, Giusiano G (2019). Antifungal activity of silver nanoparticles in combination with ketoconazole against Malassezia Furfur. *AMB Express*, 9, 131. https//doi.org/10.1186/s13568-019-0857-7. [PMC free article] [PubMed] [CrossRef] [Google Scholar].

Nagajyoti P, Lee K, Sreekanth T (2010). Heavy metals, occurrence and toxicity for plants: A review. *Environ. Chem. Lett.*, 8, 199–216.

O'Neal SL, & Zheng W (2015). Manganese toxicity upon overexposure: A decade in review. *Curr. Environ. Health Rep.*, 2 (3), 315–328. http://doi.org/10.1007/s40572-015-0056-x

Perpetuo EA, Souza CB, Nascimento CAO (2011). Engineering bacteria for bioremediation. In: CARPI, A. (Ed.) *Progress in Molecular and Environmental Bioengineering – From Analysis and Modeling to Technology Applications, InTech, Croatia*, 606-632.

Pichhode M, Nikhil K (2015). Effect of copper dust on photosynthesis pigments concentrations in plants species. *Int. J. Eng. Res. Manage. (IJERM)*, 2, 63–66.

Plum LM, Rink L, Haase H (2010). The essential toxin: Impact of zinc on human health. *Int. J. Environ. Res. Public Health*, 7 (4), 1342–1365. http://doi.org/10.3390/ijerph7041342

Puranik P, Modak J, Paknikar K (2005). A comparative study of the mass transfer kinetics of metal biosorption by microbial biomass. *Hydrometallurgy*, 52, 189–197.

Qin H, Hu T, Zhai Y, Lu N, Aliyeva J (2020). The improved methods of heavy metals removal by biosorbents: A review. *Environ. Pollut.*, 258, 113777.

Rani MJ, Hemambika B, Hemapriya J, Kannan VR (2010). Comparative assessment of heavy metal removal by immobilized and dead bacterial cells: A biosorption approach. *African Journal of Environmental Science and Technology*, 4 (2), 77–83.

Razzak SA, Faruque MO, Alsheikh Z, Alsheikhmohamad L, Alkuroud D, Alfayez A, Zakir Hossain SM, Hossain MM (2022). A comprehensive review on conventional and biological-driven heavy metals removal from industrial wastewater. *Environ. Adv.*, 7. 100168.

Redha AA (2020). Removal of heavy metals from aqueous media by biosorption. *Arab. J. Basic Appl. Sci.*, 27 (1), 183–193. https://doi.org/10.1080/25765299.2020.1756177

Remenar M, Kamlarova A, Harichova J, Zámocký M, Ferianc P (2018). The heavy-metal resistance determinant of newly isolated bacterium from a nickel-contaminated soil in Southwest Slovakia. Pol. J. Microbiol. 67, 191. Doi: 10.21307/pjm-2018-022

Salman M, Athar M, Farooq U (2015). Biosorption of heavy metals from aqueous solutions using indigenous and modified lignocellulosic materials. *Rev. Environ. Sci. Biotehnol.*, 14, 211–228. https//doi.org/10.1007/s11157-015-9362-x

Salman M, Athar M, Farooq U, Rauf S, Habiba U (2014). A new approach to modification of an agro-based material for PB (II) adsorption. *Korean J. Chem. Eng.*, 31 (3), 467–474. https//doi.org/10.1007/s11814-013-0264-8.

Sedlakova-Kadukova J, Kopcakova A, Gresakova L, Godany A, Pristas P. (2019). Bioaccumulation and biosorption of zinc by a novel Streptomyces K11 strain isolated from highly alkaline aluminium brown mud disposal site. *Ecotoxicol. Environ. Saf.*, 167, 204-11.

Shahin BI, Gayatri PD, Aboli VR, Sonali MA, Madhuri SC, et al. (2017). Chromium detoxification by using mango (Mangifera indica) and neem (Azadirachta indica) leaves. *J. Bioremediat. Biodegrad.,* 8, 409. https//doi.org/10.4172/2155-6199.1000409

Shen Z, Zhang Y, McMillan O, Jin F, Al-Tabbaa A (2017). Characterization and mechanism of nickel adsorption on biochar produced from wheat straw pellets and rice husk. *Environ Sci. Pollut. Res.* https//doi.org/10.1007/s11356-017-8847-8.

Shoaib NA, Tanveer F, Aslam N (2012). Removal of Ni (II) ions from substrate using filamentous fungi. *Int. J. Agric. Biol.*, 14, 831–833.

Sinha A, Pant KK, Khare SK (2012). Studies on mercury bioremediation by alginate immobilized mercury tolerant Bacillus cereus cells. *Int. Biodeterior. Biodegrad.*, 71, 1–8.

Song N, Teng Y, Wang JW, Liu Z, Orndorff W, Pan WP (2014). Effect of modified fly ash with hydrogen bromide on the adsorption efficiency of elemental mercury. *J. Therm. Anal. Calorim.*, 116, 1189–1195.

Soto DF, Recalde A, Orell A, Albers S-V, Paradela A, Navarro CA, et al (2019). Global effect of the lack of inorganic polyphosphate in the extremophilic archaeon Sulfolobus solfataricus: a proteomic approach. J. Proteomics 191, 143–152. Doi: 10.1016/j.jprot.2018.02.024

Stumm W., Morgan J.J., 1996. *Aquatic Chemistry. Chemical Equilibria and Rates in Natural Waters*, Wiley, New York.

Subhashini S, Kaliappan S, Velan M (2011). Removal of heavy metal from aqueous solution using Schizosaccharomyces pombe in free and alginate immobilized cells. 2nd International Conference on Environmental Science and Technology IPCBEE, IACSIT Press, Singapore, vol. 6.

Sulaymon AH, Mohammed AA, Al-Musawi TJ (2012). Competitive biosorption of lead, cadmium, copper, and arsenic ions using algae. *Environ. Sci. Pollut. Res. (Int.)*, 20, 3011–3023.

Sutton DJ, Tchounwou PB (2007). Mercury induces the externalization of phosphatidylserine in human proximal tubule (HK-2) cells. *Intl J. Environ. Res. Public Health*, 4(2), 138–144.

Swathy PS, Swathy MR, Anitha K (2017). Defluoridation of water using neem (Azadirachta indica) leaf as adsorbent. *IJARIIE*, 3 (4), 483–492. ISSN(O)-2395-4396.

Taghi G, Khosravi M, Rakhshaee R (2005). Biosorption of Pb, Cd, Cu and Zn from the wastewater by treated Azolla filiculoides with H2O2/MgCl2. *Int. J. Environ. Sci. Technol.*, 1 (4), 265–271.

Thelwell C, Robinson NJ, Turner-Cavet JS (1998). An SmtB-like repressor from Synechocystis PCC 6803 regulates a zinc exporter. Proc. Natl. Acad. Sci. 95, 10728–10733. Doi: 10.1073/pnas.95.18.10728

Torres E (2020). Biosorption: A review of the latest advances. *Processes*, 8, 1584. https//doi.org/10.3390/pr8121584

Tovar CT, Ortiz AV, Correa DA, Gómez NP, Amor MO (2018). Lead (II) remotion in solution using lemon peel (Citrus limonum) modified with citric acid. *Int. J. Eng. Technol. (IJET)*, 10 (1). https//doi.org/10.21817/ijet/2018/v10i1/181001046

Tseveendorj E, Enkhdul T, Lin S, Dorj D, Oyungerel S, Soyol-Erdene TO (2017). Biosorption of lead (II) from an aqueous solution using biosorbents prepared from water plants. *Mongol. J. Chem.*, 18 (44), 52–61.

Tsezos M, Remoundaki E, Hatzikioseyian A (2014). Biosorption – Principles and applications for metal immobilization from waste-water streams. *Clean Product. Nano Technol.*, 23–33. Retrieved from http://citeseerx.ist.psu.edu/viewdoc/download?doi=10.1.1.490.1498&rep=rep1&type=pdf

Usman A, Sallan A, Vithanage M, Ahmad M, Al-Farraj AOKYS, Abduljabbar A, Al-Wabel M (2016). Sorption process of date palm biochar for aqueous Cd(II) removal: Efficiency and mechanisms. *Water Air Soil Pollut.*, 227, 449.

Vijayaraghavan K, Yun YS (2008). Bacterial biosorbents and biosorption. *Biotechnol. Adv.*, 26 (3), 266–291. https//doi.org/10.1016/j.biotechadv 2008.02.002

Volesky B (2003). Biosorbent materials. *Biotechnol. Bioeng. Symp.*, 16, 121–126.

Wang C, Wang H (2018). Carboxyl functionalized Cinnamomum camphora for removal of heavy metals from synthetic wastewater-contribution to sustainability in agroforestry. *J. Clean. Product.,* 184, 921–928.

Wang J, Chen C (2009). Biosorbents for heavy metals removal and their future. *Biotechnol. Adv.*, 27, 195–226.

Wu S (1994). Effect of manganese excess on the soybean plant cultivated under various growth conditions. *J. Plant. Nutri.,* 17, 993–1003.

Xiang Y, Xu Z, Wei Y, Zhou Y, Yang X, Yang Y, Yang J, Zhang J, Luo L, Zhou Z (2019). Carbon based materials as adsorbent for antibiotics removal: Mechanisms and influencing factors. *J. Environ. Manag.*, 237, 128–138. https://doi.org/10.1016/j.jenviman2019.02.068

Yang H-C, Fu H-L, Lin Y-F, Rosen BP (2012). Pathways of arsenic uptake and efflux. Curr. Top. Membr. 69, 325–358. Doi:10.1016/b978-0-12-394390-3.00012-4

Yedjou GC, Tchounwou PB (2007). In vitro cytotoxic and genotoxic effects of arsenic trioxide on human leukemia cells using the MTT and alkaline single cell gel electrophoresis (comet) assays. *Mol. Cell Biochem.*, 301, 123–130.

Yedjou GC, Tchounwou PB (2008). N-acetyl-cysteine affords protection against lead-induced cytotoxicity and oxidative stress in human liver carcinoma (HepG2) cells. *Intl J. Environ. Res. Public Health*, 4 (2), 132–137.

Zammit CM, Weiland F, Brugger J, Wade B, Winderbaum LJ, Nies DH, et al. (2016). Proteomic responses to gold (III)-toxicity in the bacterium Cupriavidus metallidurans CH34. Metallomics 8, 1204–1216. Doi: 10.1039/c6mt00142d

Zhang J, Chen X, Zhou J, Luo X (2020). Uranium biosorption mechanism model of protonated Saccharomyces cerevisiae. *J. Hazard. Mater.*, 385, 121588.

Zhong LC, Zhang YS, Liu Z, Sui ZF, Cao Y, Pan WP (2014). Study of mercury adsorption by selected Chinese coal fly ashes. *J. Therm. Anal. Calorim.*, 116, 1197–1203.

7 Adsorption Kinetics and Isotherms

Shoomaila Latif, Muhammad Imran, M. Hassan Siddique, Muhammad Ehsan, and Shahid Ahmad

Adsorption is a significant key phenomenon that occurs in porous bodies which involves the transportation of adsorbed material (i.e., adsorbate) in the porous body that acts as an adsorbent. This adsorption process is the basis of many modern as well as classical technologies, especially in industrial processes like sorption, catalysis, chromatography, purification, and separation of mixtures, and plays a vital role in the industry as well as environmental protection. Although the pioneer work on adsorption was done by Mantell in 1951 [1], since then adsorption process and its various fundamental aspects have been under continuous investigation. In this chapter, we'll try to emphasize the understanding of the mathematical modeling of the adsorption system, especially the adsorption kinetics and isotherms. This modeling is quite helpful in the elucidation of the experimental data related to the adsorption process which provides a theoretical base for the evaluation of the feasibility of the pilot-scale adsorption process at the full-scale commercial level [2].

7.1 INTRODUCTION TO ADSORPTION KINETICS AND ISOTHERMS

Adsorption-based techniques are a vital part of many industrial processes so the basic principles of the adsorption process are well understood. However, the knowledge of adsorption kinetics is a dynamic step for the designing of operational adsorption equipment. In fact, kinetic studies provide informative data about the rate of the adsorption process i.e., they quantify the adsorption uptake with respect to time [3]. The rate of adsorption is principally allied to:

Rate of diffusion of adsorbate particles through pores of the adsorbent surface
The expanse of the naked surface of the adsorbent.

There is a wide variety of kinetic models and adsorption isotherms (AIs) that were developed for the interpretation of adsorption kinetics. However, the present chapter will provide a brief description of these models. In the first part of this chapter, AIs will be described followed by kinetic models in the later part [4].

DOI: 10.1201/9781003366058-7

7.1.1 Adsorption Isotherms

AIs are crucial for maximizing the utilization of any adsorbent because they define the interaction between the adsorbent and the adsorbate. An isotherm's shape can reveal details about the adsorption affinity of molecules as well as the quality of the relations between the adsorbent and the adsorbate. Numerous mathematical structures are used to represent AIs; some of these structures are created based on adsorption's basic principles, whereas some have an empirical approach and need the correlation of experimental data [5].

This chapter's goal is to provide a clear summary of a few commonly employed isotherms of adsorption for single- as well as multicomponent adsorption systems. The adsorption–isotherm shows a connection amid adsorbed–sorbate and the sorbate left behind in liquid at the state of equilibrium that is represented as

For a single component system:

$$m = m\left(p, T, m^{s}\right) \tag{7.1}$$

For a multi-component system:

$$m_i = m_i\left(p_1, \dots, p_N, T, m^{s}\right) \tag{7.2}$$

Here

m = mass
i = adsorbed components in the gaseous phase with a certain concentration
T = System temperature
p = Partial pressure
m^s = Mass of sorbent material.

Because there have been various isotherms, mainly for single-component systems, over the past ten years, it is impossible to describe them all in this monography. Rather, we will confine the discussion to those that have shown to be beneficial for engineering disciplines or can be so, notably at extreme pressure ranges (P = 10 mPa) and across a wide temperature range (200K < T < 400K).

AIs can be utilized in two ways:

To characterize porous substances
To design industrial adsorption systems

They are required not only for single-component gas mixture systems but also for multi-component gas mixture system applications [6]. Because of the sophistication of Ad molecule contacts with atoms and molecules of sorbent, it isn't possible to compute AIs using quantitative molecular methods or experiential methods relying on several macroscopic or atomic information of the sorptive: sorbent systems.

Even perfect molecular models of surfaces made of sorbent materials and pores must not only model functions for adsorbate-sorbent (a–s) interactions but will also in standard contribute to a so-called calibration problem which can only be fixed by incorporating observational evidence on total liabilities or masses of gas adsorbed at very fine economic circumstances. Because they are difficult to measure and frequently unavailable, it is required to evaluate gas adsorption equilibrium data, particularly for multicomponent systems, that is co-adsorption equilibrium data. These frequently demonstrate or at least describe the physical process of the patent adsorption mechanism, allowing one to make a suitable decision for a quantitative AI to associate the data and then attempt to extrapolate it to other temperatures and pressures required for methodology. Because micro- and mesopores happen to varying degrees in just about all industrial adsorbent material, a mixture of fundamentally various adsorption processes such as top layer coverage and porous humidification could perhaps happen, likely to result in fairly complex AIs whose edifice can't be predicted without observed measurements.

Adsorption equilibrium data are graphically represented as

Isotherms: m and p-diagrams when T = constant.
Isobars: m and T-diagrams when p = constant.
Isosteres: In T-diagrams when m = constant.

Adsorption enthalpies are frequently required for the design process, and they haven't been here due to space constraints, but redirect to the study [7].

7.2 TYPES OF ADSORPTION ISOTHERMS

Various types of AIs are available in Figure 7.1.

7.2.1 Type-I Adsorption Isotherm

The Langmuir equation may be used to characterize type-I isotherms. They are distinguished by a horizontal plain, which reaches and retains the limited value of the mass adsorbed even at extremely elevated gas pressures.

Such isotherms are representative of microporous materials that exhibit micropore filling but no multilayered adsorption.

For instance:

Inorganic polar molecular or zeolite sieves on water.
Organic vapors on molecular sieves and zeolite.

Comments:

Chemisorption commonly exhibits type-I AIs.
The surface of porous materials cannot be evaluated by BET i.e., Brunauer–Emmett–Teller (BET) method [8].

7.2.2 Type-II Adsorption Isotherm

Type-II isotherms often explain adsorption in mesoporous materials, exhibiting single-layer adsorption at low pressures, multilayered adsorption at high-pressure approaching diffusion, and pore concentration but no lag. These isotherms may even be seen in dispersed, non-porous, and solely macroporous substances (pore diameter greater than 50 nm). They are frequently characterized by the BET formula or modifications of it.

Vapors of polar media on molecular sieves are one example, and so are vapors of non-polar organic compounds that are certainly microporous but mostly consist of mesoporous activated carbons [9].

7.2.3 Type-III Adsorption Isotherm

Type-III isotherms arise due to weak a–s interactions in contrast to strong adsorbate–adsorbate (a–a) interactions, that is ad-molecules that are firmly associated.

Example:

Water on hydrophobic zeolites and activated carbon.

7.2.4 Type-IV Adsorption Isotherm

Specialized mesoporous materials that exhibit pore condensation behavior of hysteresis between the adsorption and desorption branches are described by type-IV isotherms for their adsorption behavior [10].

Example:

Humid air vapors on particular kinds of hydrophilic zeolites and activated carbons.

7.2.5 Type-V Adsorption Isotherm

Type-V curves deviate from type-IV isotherms by having almost upright mid regions of desorption and branches of adsorption, which are frequently around relative pressures of gas, suggesting the presence of mesopores in which the phase may shift such as condensation of pores may occur.

For example:

Water on specific carbon molecular sieves and activated carbons

7.2.6 Type-VI Adsorption Isotherm

Stepwise multilayer adsorbates are found in type-VI isotherms, with the layers seeming to be increasingly prominent with the decrease in temperature.

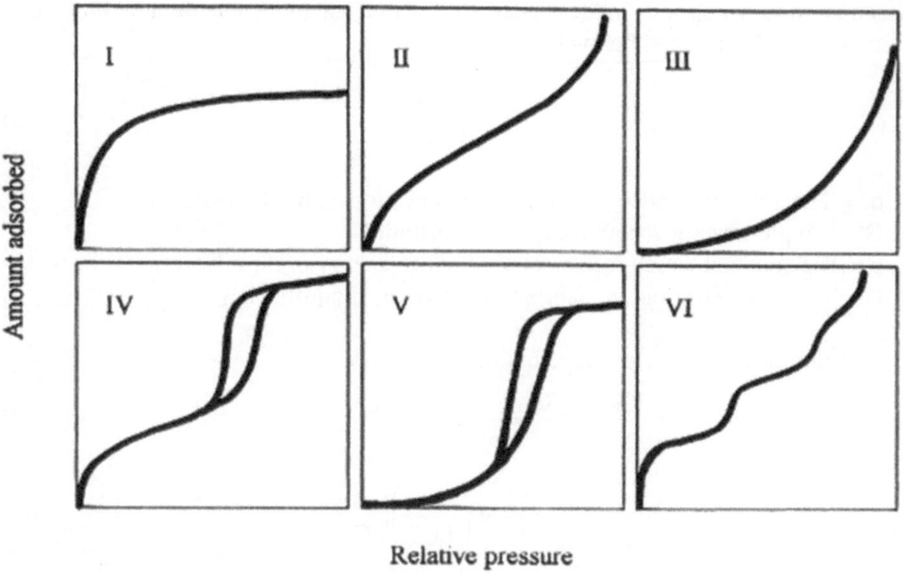

FIGURE 7.1 Types of Adsorption Isotherms (reproduced with permission under license no. 5431320424996 from the source [11]

The Langmuir AI (LAI) and its various extensions for mono- and multicomponent parameters will be covered in the chapter. LAI is quite effective in describing the phenomenon of adsorption in zeolites and metal-organic frameworks [11].

7.3 LANGMUIR ADSORPTION ISOTHERM (LAI)

The Langmuir hypothesis supposes that the phenomenon happens to the specified uniform positions inside the adsorbent, and after an adsorbate occupied the position, further no adsorption may happen to that position. It may be used to estimate the effectiveness of various adsorbents. It is known that a K_L value (related to sorption energy) showing strong affinity to adsorption and maximum adsorption capacity reflected by q_m is declined with escalating temperature [12]. Langmuir developed a link amid the quantity of adsorbed gas to the surface and the pressure of gas based on his surface chemistry findings, which is called LAI. He alleged that

Each adsorbent has a set of adsorption sites on its surface and the percentage of sites that are able to be occupied by adsorbates at a certain temperature and pressure.

Every adsorption region to the surface of the adsorbent may allow just a single entry.

Every adsorption produces heat.

Regardless of the fraction of sites covered by adsorbate, the site remains the same.

There is neither connection between adsorbates at different adsorption sites [13,14].

Based on the assumptions made earlier, the LAI is just applicable to monomolecular adsorption and explained below [15,16]:

$$q_e = \frac{bCQ^o}{1+bC} \tag{7.3}$$

q_e = The amount of adsorbate removed per unit weight of adsorbent.
Q^o = Single-layer adsorption capacity constant.
b = The constant which measures the surface energy of the adsorption process.
C = The concentration of adsorbate in solution at equilibrium.

The Langmuir equations are depicted in the equation.

$$\frac{C}{q_e} = \frac{1}{bQ^o} + \frac{C}{Q^o}\frac{1}{q_e} \tag{7.4}$$

$$\frac{C}{q_e} = \frac{1}{Q^o} + \left(\frac{1}{bQ^o}\right)\left(\frac{1}{C}\right) \tag{7.5}$$

A straight line with slope $1/bQ_o$ and intercept $1/Q^o$ would be obtained by plotting $1/q_e$ versus 1/C (Figure 7.2). The Langmuir isotherm model is used to elucidate equilibrium conditions of adsorptions and offer parameters that can be used to quantity wise match adsorption behaviors in various adsorbate–adsorbent systems [17].

The equilibrium parameter R_L will be expressed as LAI, the factor can be expressed as

$$R_L = \frac{1}{1+bC_o} \tag{7.6}$$

Where

C_o ($mg\ dm^{-3}$) = The highest initial concentration
b ($dm^3\ mg-^1$) = The Langmuir constant
RL = The nature of the isotherm shape (interpreted in Table 7.1) [5].

TABLE 7.1
Langmuir isotherm parameters

RL value	Type of isotherm
RL > 1	Unfavorable
RL = 1	Linear
RL = 0	Irreversible
0 < RL < 1	Favorable

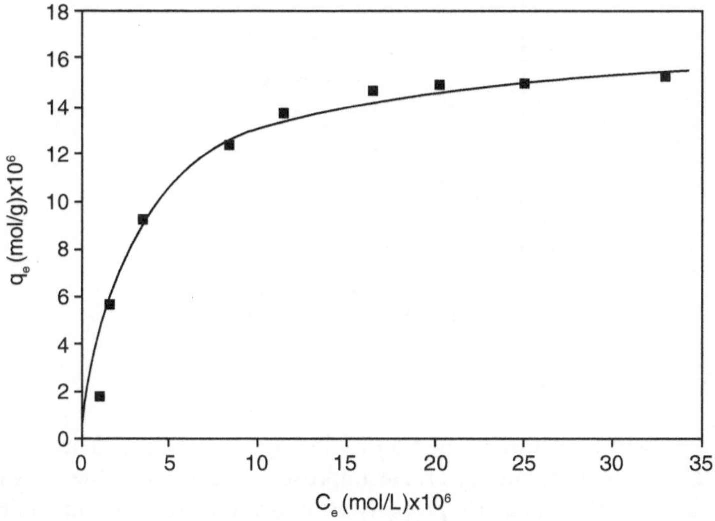

FIGURE 7.2 Langmuir isotherm (reproduced with permission under license no. 5431330031816 from the source [18])

7.4 FREUNDLICH ADSORPTION ISOTHERM

The Freundlich adsorption isotherm (FAI) is widely used to calculate the equilibrium relationship involving the quantity of adsorbate removed per weight of carbon and the concentration of adsorbate that remains in the solution [19]. FAI is expressed as

$$qe = \frac{KfC1}{n} \tag{7.7}$$

n = Sorption and sorption energy efficiency constant.
Kf = Adsorption capacity constant.
qe = Adsorbate amount removed per unit weight of carbon.
C = The equilibrium concentration of the adsorbate in the solution.

The Freundlich equation is primarily experimental; however, it is occasionally used to describe data. This is owing to the equation's ability to elucidate the non-uniform adsorption mechanism over a wide variety of adsorbate concentrations. Its numerical simplicity allows it for easy application, as does its capability to represent adsorption processes on potentially heterogeneous surface adsorption areas. By determining the logarithmic value of both the left- and right-hand sides of the equation, the Freundlich equation may be employed in linear form.

$$\log q_e = \log K_f + \frac{1}{n} \log C \tag{7.8}$$

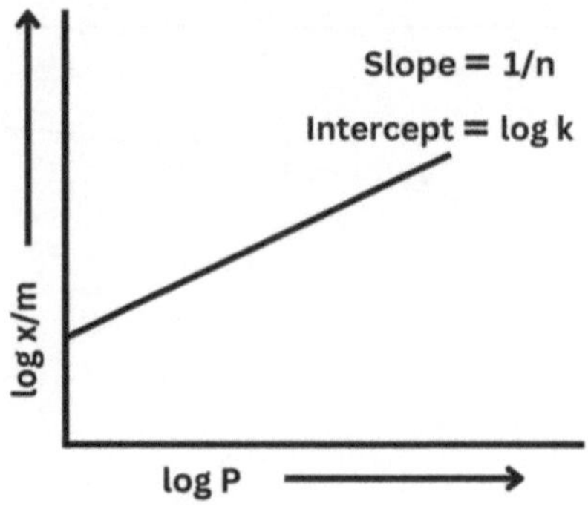

FIGURE 7.3 Freundlich isotherm

A straight line will be obtained by plotting log q_e versus log C. The ability of the activated carbon to eliminate adsorbate improves, and so does the K_f value [20]. Although the Freundlich isotherm properly establishes the relationship amid adsorption and pressures at small values, it isn't anticipated the adsorption value at extraordinary pressures (Figure 7.3).

7.5 RADLICH–PETERSON ADSORPTION ISOTHERM (RPAI)

The intermediate AI used Langmuir and Freundlich isotherm characteristics and merged them into an equation. It depicts a robust adsorption system that functions over a wide concentration range [21].

Its equation is given as

$$q_e = \frac{KRC_e}{1 + aRC_e^g} \tag{7.9}$$

The linear equation is mentioned as

$$\mathrm{Ln}\frac{C_e}{q_e} = g \ln C_e - \ln KR \tag{7.10}$$

aR ($dm^3\ mg^{-1}$) = RPAI constant
K_R ($dm^3\ g^{-1}$) = RPAI constant
g = exponent (where $0 < g < 1$).

The values of lnKR and g can be obtained via plotting $\ln(C_e / q_e)$ against ln C_e [22].

7.6 TEMKIN ADSORPTION ISOTHERM

Due to adsorbent–adsorbate interaction, Temkin isotherm forecast that the heat of sorption would be decreased with gradual coverage of sorption around adsorbent [23]. Temkin isotherm relation is mentioned as

$$q_e = \frac{R_T}{b_T} \ln(k_t C_e) C \tag{7.11}$$

The linear form equation is mentioned as

$$q_e = \frac{R_T}{b} \ln k_t + \frac{R_T}{b} \ln C_e \tag{7.12}$$

Where

K_t ($dm^3\ g^{-1}$) = Optimum binding energy at equilibrium state.
bT ($J\ mol^{-1}$) = Heat of adsorption constant.
R = Ideal gas constant
T = Absolute temperature [24,25].

7.7 DUBININ–RADUSHKEVIC (D–R) ADSORPTION ISOTHERM

The model presupposes that the adsorption feature curve is linked to the pore's characteristics of the adsorbent [26]. It helps to identify the type of adsorption that took place and to calculate the ostensible adsorption energy. The model does not presuppose that the adsorbent's surface is uniform.

The linear equation is expressed as

$$\ln q_e = \ln QD - BD + \left\{ RT \ln\left(1 + \frac{1}{C_e}\right) \right\}^2 \tag{7.13}$$

QD ($mol\ g^{-1}$) = Maximum adsorption capacity
BD ($mol^2\ KJ^{-2}$) = D–R constants
D-R constants = Determined from the slope and intercept via plotting lnqe against $RT\ln(1+(1/C_e))$.
E ($KJ\ mol^{-1}$) = Adsorption mean free energy

The following equation is used to calculate the amount of energy E ($KJ\ mol^{-1}$) needed to transport a mole of adsorbate from the solution to the surface of an adsorbent.

$$E = \frac{1}{\sqrt{2BD}} \tag{7.14}$$

BD ($mol^2\ KJ^{-2}$) = Adsorption means free energy per unit mole of adsorbate [27].

7.8 KINETICS MODELS

Various kinetic models are commonly applied to the experimental data to find the adsorption dynamics to understand the adsorption mechanism, and finally the kinetic constants for the applied models are calculated [28]. The calculated "$\mathbf{R^2}$" values (linear regression correlation coefficient) were compared for the evaluation of the best-fitted model for the adsorption process under study. Several models have been developed for kinetic studies which are difficult to summarize here so the most common models that are present in literature for the kinetics studies of adsorption systems are briefly charted in the following table (Table 7.2) along with a short description.

TABLE 7.2
List of kinetic models

Kinetic models	Equation	Explanation
Lagergren's pseudo-first-order model	$dq_t/d_t = k_t (q_e - q_t)$	Used to monitor the reaction kinetics of the adsorption process. The model assumes that the rate of change in adsorbate uptake with reaction time is directly proportional to the difference in adsorbate concentration and removal rate with time [29].
Ho's pseudo-second-order model	$dq_t/dt = k_2(q_e - q_t)^2$	This model assumes that the rate of change in adsorbate uptake at a specific reaction time is directly proportional to the square of the difference in adsorbate concentration and removal rate with time [30].
Elovich model	$dq_t/dt = \alpha\, e^{-\beta q_t}$	On the basis of the premise that the sorbent surface is energetically heterogeneous, this model is utilized to define the pseudo-second-order kinetics [31] further.
Bhattacharya and Venkobachar	$\text{Log}\,[U - U_O(T)] = -K_B t/2.303$	The first-order reversible kinetic model developed by Bhattacharya and Venkobachar is simplified. It is based on solution concentration and is helpful in understanding adsorption mechanisms and identifying sorption characteristic constants [32].
Natarajan and Khalaf	$\text{Log}\, C_o/C_t = Kt/2.303$	The link between the initial concentration of adsorbate and concentration at any time is shown by Natarajan and Khalaf's equation [33].

7.9 CONCLUSION

In this chapter, various types of adsorption kinetics and isotherms with mathematical expressions have been described. Several models on adsorption kinetics are available and few of them have been elaborated and compared in this chapter.

REFERENCES

1. Mantell, C.L. (1951), *Adsorption* (Chemical Engineering Series), PhD. 2 ed.
2. Li, S., Deng, S., Zhao, L., Zhao, R., Lin, M., Du, Y., & Lian, Y. (2018). Mathematical modeling and numerical investigation of carbon capture by adsorption: Literature review and case study. *Applied Energy*, *221*, 437–449.
3. Bonelli, B., Freyria, F. S., Rossetti, I., & Sethi, R. (Eds.). (2020). *Nanomaterials for the Detection and Removal of Wastewater Pollutants*. Elsevier.
4. Agbovi, H. K., & Wilson, L. D. (2021). Adsorption processes in biopolymer systems: Fundamentals to practical applications. In *Natural Polymers-Based Green Adsorbents for Water Treatment* (pp. 1–51). Elsevier.
5. Obaid, S. A. (2020). Langmuir, Freundlich, and Tamkin adsorption isotherms and kinetics for the removal of aartichoke tournefortii straw from agricultural waste. *Journal of Physics: Conference Series*, *1664*, 012011.
6. Guo, X., & Wang, J. (2019). A general kinetic model for adsorption: Theoretical analysis and modeling. *Journal of Molecular Liquids*, *288*, 111100.

7. McKay, G., & Al Duri, B. (1989). Prediction of multicomponent adsorption equilibrium data using empirical correlations. *The Chemical Engineering Journal*, *41*(1), 9–23.
8. Lowell, S., & Shields, J. E. (1991). Adsorption isotherms. In *Powder Surface Area and Porosity* (pp. 11–13). Springer, Dordrecht.
9. Yahia, M. B., Torkia, Y. B., Knani, S., Hachicha, M. A., Khalfaoui, M., & Lamine, A. B. (2013). Models for type VI adsorption isotherms from a statistical mechanical formulation. *Adsorption Science & Technology*, *31*(4), 341–357.
10. Giles, C. H., Smith, D., & Huitson, A. (1974). A general treatment and classification of the solute adsorption isotherm. I. Theoretical. *Journal of colloid and interface science*, *47*(3), 755–765.
11. Donohue, M. D., & Aranovich, G. L. (1998). Classification of Gibbs adsorption isotherms. *Advances in Colloid and Interface Science*, *76*, 137–152.
12. Duff, D. G., Ross, S. M., & Vaughan, D. H. (1988). Adsorption from solution: An experiment to illustrate the Langmuir adsorption isotherm. *Journal of Chemical Education*, *65*(9), 815.
13. Musah, M., Yisa, J., Suleiman, M. A. T., Mann, A., & Shaba, E.Y. (2018) A study of isotherm models for the adsorption of cr (vi) ion from aqueous solution onto Bombax buonopozense calyx activated carbon. *Lapai Journal of Science and Technology*, *4*(1), 22–32.
14. Meera, M. S. & Ganesan, T. K. (2015). Adsorption isotherm and kinetics studies of cadmium (II) ions removal using various activated carbons derived from agricultural bark wastes: A comparative study. *Journal of Chemical and Pharmaceutical Research*, *7*(4), 1194–1200.
15. Islam, A., Chowdhury, M. A., Mozumder, S. I., & Uddin, T. (2021). Langmuir adsorption kinetics in liquid media: Interface reaction model. *ACS Omega*, *6*, 14481–14492.
16. Ademiluyi, F. T., & Nze, J. C. (2016). Sorption characteristics for multiple adsorptions of heavy metal ions using activated carbon from Nigerian bamboo. *Journal of Materials Science and Chemical Engineering*, *4*, 39–48.
17. Ayawei, N., Ebelegi, A. N., & Wankasi, D. (2017). Modelling and interpretation of adsorption isotherms. *Journal of Chemistry*, *2017*, 1–11.
18. Doğan, M., Alkan, M., Demirbaş, Ö., Özdemir, Y., & Özmetin, C. (2006). Adsorption kinetics of maxilon blue GRL onto sepiolite from aqueous solutions. *Chemical Engineering Journal*, *124*(1–3), 89–101.
19. Appel, J. (1973). Freundlich's adsorption isotherm. *Surface Science*, *39*(1), 237–244.
20. Paudel, S., & Shrestha, B. (2020). Adsorptive removal of Iron (II) from aqueous solution using raw and charred Bamboo Dust. *International Journal of Advanced Social Sciences*, *3*(2), 27–35.
21. Sampranpiboon, P., Charnkeitkong, P., & Feng, X. (2014). Equilibrium isotherm models for the adsorption of Zinc (II) ion from aqueous solution on pulp waste. *WSEAS Transaction on Environment and Development*, *10*, 35–47.
22. Sheba, C. M., & Nandini, M. G. K. (2016). Heavy metal lead removal by biosorption – A review. *International Journal of Engineering Research & Technology*, *5*(11), 518–524.
23. Johnson, R. D., & Arnold, F. H. (1995). The Temkin isotherm describes heterogeneous protein adsorption. *Biochimica et Biophysica Acta (BBA)-Protein Structure and Molecular Enzymology*, *1247*(2), 293–297.
24. Edet, U. A., & Ifelebuegu, A. O. (2020). Kinetics, isotherms, and thermodynamic modeling of the adsorption of phosphates from model wastewater using recycled brick waste. *Processes*, *8*(6), 665.
25. Javadian, H., Ghorbani, F., Tayebi, H., & Hosseini, S. M. (2015). Study of the adsorption of Cd (II) from aqueous solution using zeolite-based geopolymer synthesized from coal fly ash; kinetic, isotherm, and thermodynamic studies. *Arabian Journal of Chemistry*, *8*, 837–849.

26. Lee, T. Z. E., Zhang, J., Feng, Y., Lin, X., & Zhou, J. (2021, February). Adsorption of Cd (II) ions by Coconut Copra: Isotherm and regeneration studies. In *IOP Conference Series: Earth and Environmental Science* (Vol. 657, No. 1, p. 012026). IOP Publishing.
27. Dada, A. O., Olalekan, A. P., Olatunya, A. M., & Dada, O. J. I. J. C. (2012). Langmuir, Freundlich, Temkin and Dubinin–Radushkevich isotherms studies of equilibrium sorption of Zn2+ unto phosphoric acid modified rice husk. *IOSR Journal of applied chemistry, 3*(1), 38–45.
28. Buzzi-Ferraris, G., & Manenti, F. (2009). Kinetic models analysis. *Chemical Engineering Science, 64*(5), 1061–1074.
29. Revellame, E. D., Fortela, D. L., Sharp, W., Hernandez, R., & Zappi, M. E. (2020). Adsorption kinetic modeling using pseudo-first order and pseudo-second order rate laws: A review. *Cleaner Engineering and Technology, 1*, 100032.
30. Wu, F. C., Tseng, R. L., Huang, S. C., & Juang, R. S. (2009). Characteristics of pseudo-second-order kinetic model for liquid-phase adsorption: A mini-review. *Chemical Engineering Journal, 151*(1–3), 1–9.
31. Cheung, C. W., Porter, J. F., & McKay, G. (2000). Elovich equation and modified second-order equation for sorption of cadmium ions onto bone char. *Journal of Chemical Technology & Biotechnology, 75*(11), 963–970.
32. Bhattacharya, A., & Dunson, D. B. (2011). Sparse Bayesian infinite factor models. *Biometrika, 98*(2), 291–306.
33. Natarajan, R., & Kass, R. E. (2000). Reference Bayesian methods for generalized linear mixed models. *Journal of the American Statistical Association, 95*(449), 227–237.

8 Pretreatment Methods of Biosorbents

Madhu Manikpuri, Suchitra Kumari Panigrahy, and Akanksha Jain

8.1 INTRODUCTION

For many years, environmental pollutants have posed a major hazard by contaminating the soil and water systems. The primary cause of pollution is the discharge of wastes from numerous industries. Utilizing microbes including bacteria, fungi, algae, and yeasts, biosorption has gained acceptance as an alternative eco-friendly green technology for the elimination of various contaminants (Yaashikaa et al. 2021).

Biosorbents are biomass-based substances that use preferential adsorption to move adsorbate molecules from the bulk liquid phase to their surfaces. These biosorbents are plentiful, biodegradable, and reasonably priced. They produce little to no sludge. They require only basic pretreatment procedures They are straightforward to use; and they offer highly adaptable or manipulable surface functional groups and beneficial surface-related properties. They have been recognized as viable options for eliminating pollutants from wastewater. The phytochemistry of the biomass and their pretreatment synthesis processes typically affect the physicochemical characteristics of these sorbents. In addition to these traditional classes, new-generation biosorbents that are enhanced with natural polymers employing cross-linkers and contain custom-made functional groups thanks to methods of surface modification are also on the rise. Therefore, there is an increase in research focusing on using different pretreatment techniques to create novel biosorbents (Selvasembian and Singh 2021).

8.2 PRETREATMENT OF BIOSORBENT MATERIALS

Different types of biomass cannot be used directly as biosorbent materials. There is not much sorption capacity for metal removal in the raw materials. The materials must be pretreated to increase their capacity and usefulness. One biological adsorbent substance is insufficient to meet the demands of practical applications. There are now several different biosorbent pretreatment techniques available. The sorption capacity of these materials can be improved by optimizing the functional groups on the surface of adsorbent material and enhancing the adsorption sites on the surface of biosorbents (Benis et al. 2020).

Physical pretreatment and chemical pretreatment are the two basic categories under which pretreatment techniques can be divided. Physical preparation is frequently

DOI: 10.1201/9781003366058-8

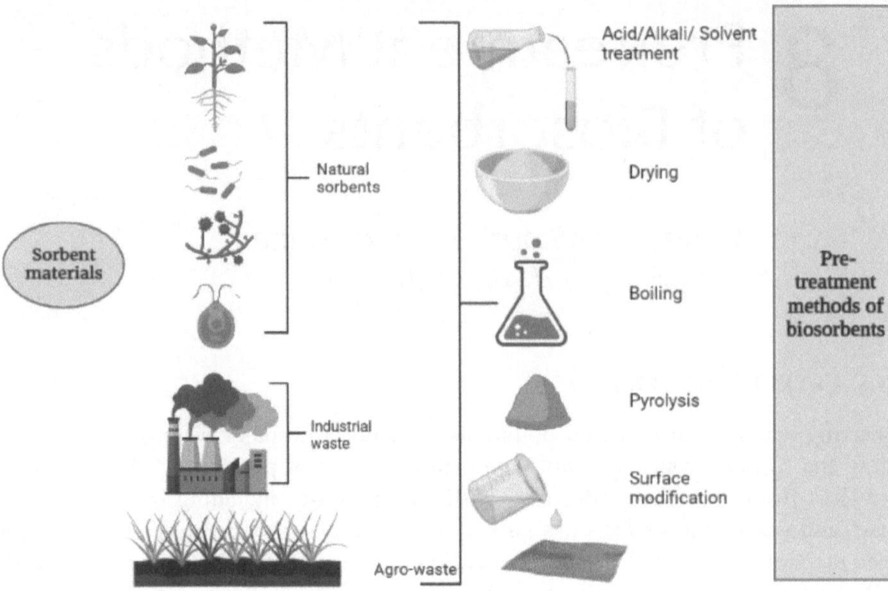

FIGURE 8.1 Pretreatment methods of biosorbents

treated with fragmentation and heat, whereas chemical pretreatment is frequently treated with acid–base treatment, oxidant, inorganic salts, and organic solvents. The physical and chemical characteristics of adsorbents, the pretreatment technique, and the type of metal ion that is adsorbed all contribute to the influence of pretreatment procedures on biosorbents. The preparation procedures for biosorbent materials are shown in Figure 8.1.

8.3 PHYSICAL PRETREATMENT

8.3.1 Surface Modification

Surface modification is the process of modifying a material's surface so that it is more suited for the intended use by changing the surface's physical, chemical, or biological properties from those that were originally there. Solid materials are frequently altered to change a variety of surface characteristics, including surface energy, surface charge, roughness, surface area, hydrophobicity, reactivity, and functional groups (Abegunde et al. 2020).

8.3.2 Surface Modification by Physical Pretreatment

The treatment of heavy metal contamination in soil and water has included the use of biochar. A carbon-rich substance called biochar is created by burning feedstocks, primarily biological waste products, without the presence of air. The efficacy of biochar as a heavy metal adsorbent is significantly influenced by the size of its pores and its surface functional groups (Fahmi et al. 2018). The adsorption of metals by biosorbents can be via complexation between the metals and various functional groups on the surface of the biosorbents, or electrostatic attractions between metal cations with negative charges and the functional groups. Functional groups can be found throughout the biochar matrix. Therefore, crushing the biochar will expose the functional groups which can adsorb the metals (Mohan et al. 2015). Crushing biochar into smaller particle sizes has been used to improve the adsorptive capacity of biochar for heavy metals.

8.3.3 Pyrolysis

This process is thought to be a crucial one for improving adsorption rate. The biomass is subjected to heat treatment under controlled oxygen content for conversion of biomass to biochar (Ahmad et al. 2014; Yaashikaa et al. 2020). In contrast, gasification is another method where an excess quantity of air is supplied and as a result syngas will be produced. Torrefaction is the method of biomass decomposition at temperature 200–300°C and heating rate <10°C per minute. The biomass from rice husk, peanut husk, sawdust, water hyacinth, etc. can be converted to biochar using the torrefaction method (Chen et al. 2021). Slow pyrolysis is a method of decomposing biomass at the temperature of 300–700°C and heating rate <10°C per minute. The various types of biomass materials that are converted to biochar using slow pyrolysis include wheat straw, orange waste, pine wood, green waste, olive pomace, etc. Fast pyrolysis is another method utilizing a temperature of 400–800°C and heating rate of 1000°C per second. The biomass from oak wood, soybean straw, oak bark, etc. can be transformed into biochar (Mohan et al. 2014). Another process, gasification disintegrates biomass from wheat straw, wood pellets, wood chunks, etc. at temperature 700–1500°C with a fast heating rate.

8.3.3.1 Over Drying

Many biosorbents have been treated by this method, mostly in combination with chemical method. For example, the biosorption of cadmium and lead ions from artificial aqueous solutions using waste baker's yeast biomass was investigated. For the enhanced biosorption along with ethanol treatment, all the treated cells were dried at 70°C for 12 hours and then ground to a gritty consistency to yield granular biosorbent samples (Göksungur et al. 2005).

8.3.3.2 Autoclaving

Another pretreatment for the enhanced absorption of pollutant is autoclaving. Many fungal biomass such as *Aspergillus flavus* is reported to be used for the removal of cromium pollutants from water and the autoclaved biomass showed the maximum adsorption capacity (Deepa et al. 2006).

8.4 CHEMICAL PRETREATMENT

8.4.1 Surface Modification by Chemical Pretreatment

Chemical pretreatment improves the specificity and surface morphology of biosorbents, improving their adsorption capabilities. In order to improve metal adsorption, chemical treatment is also used to increase porosity and surface area. When a biosorbent material is exposed to acid or alkali, its surface area is oxidized, which affects its physicochemical properties, such as hydrophilicity and hydrophobicity, and creates oxygen-containing functional groups at the surface, among other things.

8.4.1.1 Acidic Treatment

To improve the efficiency of biosorbents, acidic pretreatment uses sulfuric acid and hydrochloric acid. Other mineral acids and oxidants such as HNO_3, H_2O_2, HCl, H_3PO_4 (Shim et al. 2001; Vlasova et al. 2003; Bel'chinskaya et al. 2015). Organic acids, such as acetic, carboxylic acid, formic acid, and oxalic, are seldom used because their effect is weaker owing to their low strength (Kong et al. 2014). The surface area of sorbent material changes when biosorbent material is treated with strong acids like hydrochloric acid. When compared to untreated material, the biosorbent that has been acid pretreatment shows greater stability and regeneration potential. Increased pore structure and better sorption characteristics might be the result of acid treatment. This method improves the surface of the biosorbent material, increasing the number of acidic functional groups, especially carboxylic ones, to make it more receptive to the sorption of cation metals.Any absorbent having an acidic surface and composed of oxygen containing functional groups such as carboxyl, carbonyl, quinone, hydroxyl, lactone, and carboxylic anhydride. These functional groups play a vital role in regulating the chemical composition of the material and are typically found on the outside surfaces or edges of the basal plane of activated carbon (Ramrakhiani et al. 2016).

Additionally, they might prevent contaminants from bonding with anionic groups. The biosorbent surface's polysaccharide group is broken down by the acidic treatment, which also removes contaminants and increases the number of binding sites that are available there. The biosorbent cell wall's overall surface charge is converted to a positive or less negative charge by the acidic agents, making it possible for the sorption of negatively charged molecules. Biosorbent qualities may be lost after acid treatments. The mass of the biosorbent may also decrease, slowing the rate of biosorption.

8.4.1.2 Alkaline Modification

The surface functional groups of adsorbents can be significantly influenced by modifying with reducing agents. This process can improve the relative content of alkali groups and the non-polar surface, hence enhances the adsorption capacity of the material for non-polar substances (Abegunde et al. 2020). An adsorbent with alkali treatment adsorbs a positive charge on the surface that enhances the adsorption of negatively charged species (Rahman et al. 2019a,b). The commonly used alkali for biosorbent pretreatment is sodium hydroxide, calcium hydroxide, and sodium bicarbonate. Treatment of biosorbent material with sodium hydroxide influences the morphology of biosorbent by saponification reaction and results in conversion of esters to alcohols and carboxylates. NaOH can also be employed for the modification of activated carbon (Guo et al. 2019). However, the specific surface area and the pore volume concentration increased with the increasing alkali concentration, which resulted in a higher adsorption capacity of methane.

There was a greater number of binding sites available on the biosorbent surface produced by attaching the soluble proteins on the surface making way for passive diffusion of pollutants inside the sorbent material (Pertile et al. 2021). Sodium hydroxide eliminates the liposomes present on the cell membrane. This causes deacetylation and proteolysis of chitin aiding in binding sites for removing low molecular organic compounds and enhancing the sorption of pollutants. Calcium hydroxide also acts as saponifying agent. At high pH, the carboxylic and alcohol functional groups participate and produce the metal–calcium complex increasing the binding sites with a negative charge. The treatment of biosorbent with sodium bicarbonate increases the porosity and sorption rates but at high concentration exhibits reverse reaction by damaging the surface of biosorbents and minimizing the binding sites. The processing efficacy was noticed when the biosorbent was treated with alkali.

Przepiórski (2006) studied the influence of gaseous ammonia treatment on the CWZ (molecular name- N-[[3-[3-(6,7-dihydro-4H-[1,3]thiazolo[5,4-c]pyridin-5-ylmethyl)phenyl]phenyl]methyl]-2-pyridin-3-yl-ethanamine) -series-activated carbon toward phenol adsorption from water. A temperature range of 400–800°C was used for the treatment. The treated activated carbon's adsorption capacity was found to be 24% higher than that of the untreated CWZ-activated carbon. The author attributed the enhancement in phenol adsorption capacity of the activated carbon to the introduction of N-containing species and larger volumes of micropores in the treated material.

8.4.1.3 Salts and Organic Solvents

Salts like sodium chloride and sodium nitrate and organic solvents such as acetone and ethanol are used as biosorbent pretreatment washing materials. Compared to acid and alkali agents, inorganic salts have a less significant impact on the degradation of biomass quality. Washing with salts and organic solvents is preferable to maintain the stability and protection of biosorbent (Yaashikaa et al. 2021).

When pretreatment is done with salts, they act as ion-exchangers, and the binding site is drenched with a pollutant, the sorption efficiency depends on the affinity of the pollutant with the binding site. This method is better suited for a continuous system compared to batch as during batch process the ion-exchangers in solution may compete with the target molecules for binding, resulting in decreased sorption rate (Barquilha et al. 2017). Similar to this, removing the proteins and lipids from the biomass surface of biosorbents using solvents increases the surface area and binding sites, resulting in improved adsorption effectiveness (Bertagnolli et al. 2014). Other solvents like formaldehyde, methanol, and formic acid result in methylation and esterification of functional groups affecting the sorption efficacy.

8.4.2 Other Chemical Methods (Oxidants, Organic, and Metal Ions) for the Modification

Other than acidic and basic modes of adsorbent surface modifications, chemical modification can be done with chemical agents such oxidizing agents (potassium permanganate and hydrogen peroxide), neutral agents ($ZnCl_2$ and NaCl), metal impregnation (ferric chloride, cerium, zirconium, hydroxides, carbonates, chromates, or nitrates), and organic agents (ethanol).

8.4.2.1 Oxidizing Agents

Oxidants increase the contents of oxygen-containing functional groups on the adsorbents (Enaime et al. 2020). The instrumental analysis showed that the biochar surface was covered with MnOx ultrafine particles, with more surface oxygen-containing functional groups and much larger surface area.

Tendu leaf has been used as a biosorbent for the removal of Congo red dye from aqueous solution. The biosorbent was obtained by treating with hydrogen peroxide. The authors reported the maximum adsorption capacity of untreated tendu waste adsorbent as 46.95 mg g^{-1}, which was enhanced by 2.8 times after the adsorbent was modified with hydrogen peroxide treatment to attain 134.4 mg g^{-1}. The author concluded by recommending the use of milder chemical treatment of tendu waste to obtain a biosorbent with enhanced dye removal capacity (Nagda and Ghole 2009).

8.4.3 Internal Modification of Biosorbents

Despite the fact that chemical pretreatment procedures increase sorption rate and exhibit enhanced adsorption efficiency, the anticipated outcome was not attained. Chemical treatments caused secondary contamination because of chemical residues, and the issue was caused by final disposal. Internal alteration of biosorbent is gaining popularity among academics recently as a means of overcoming these limitations. The cells of biosorbents go through numerous stress reaction situations that change the cell's protein and enzyme content as well as its growth and structure.

By artificially regulating the expression of these proteins and enzymes found to increase the tolerance ability and accumulation of pollutants in biomass cells. The expression of various proteins and enzymes accelerates the metal removal through processes such as polymerization, reduction, oxidation, and extraction. Few compounds involved in the resistance of heavy metals include glutathione, metallothioneins, glutathione S-transferase, proline, adenosine-5'-triphosphate **sulfurylase** (ATP sulfurylase) , etc. Genetic engineering approaches can be used for cellular enzyme expression in cells. But the complexity in producing genetically engineered microbes restricts this method in a wider application. The ecosystem conditions also influence the biosorbent tolerance to pollutants. Exposure of sorbent materials to unfavorable environmental conditions induces to the production of mutants and genetically transmits the resistance to the future.

8.4.4 Hydrothermal Carbonization

In contrast to the pyrolysis procedure, this approach uses a low temperature of 175–250°C and a gradual heating rate to convert biomass. Hydrothermal carbonization can be defined as combined dehydration and decarboxy lation of a fuel to raise its carbon content with the aim of achieving a higher calorific value. It is accomplished by exposing biomass suspended in water at saturation pressure to high temperatures

(180–220°C) for several hours (Funke and Ziegler 2010). Bamboos, corn stock, pine wood, coconut fibers, etc. can be converted using this method. In this process, the raw biosorbent material is first subjected to acid/alkali treatment that aids in functionalizing the biosorbent and increasing the efficiency of interaction between the adsorbent and adsorbate. After the addition of acid, the material is treated at a low temperature of around 50°C. Then the acid-treated material is filtered and again subjected to heat 90–95°C. During drying, dehydration, carbonization, and polymerization reactions take place to obtain the final biochar product (Sizmur et al. 2017). Then the formed solid product is washed repeatedly and exposed to 120°C for complete drying. The biochar so formed is termed hydrochar used as an adsorbent or activated carbon.

8.4.4.1 Impregnation

Impregnation is a process of loading a given porous material with a metal component either in the solid state (with both components in solid state) or via wet impregnation with the support in solid state and the metal component dissolved in a solution. Impregnation is the uniform distribution of chemicals in the internal surface of porous material (Rahman et al. 2019a,b). In dry impregnation method, the just enough solvent is added to fill the pores of the adsorbent, whereas, in wet impregnation, excess solvent is added after the pores are filled. The impregnating materials can be metals or polymeric substances that generally have no significant effect on pH. The excess chemical is allowed to dry, and the loading of the chemical species can be controlled. The impregnated adsorbent has the advantage of stability and a promising regeneration capacity.

8.4.5 Nano-Biosorbents

In recent years, an increase in interest has been seen in the manufacture of innovative biosorbents due to rising air and water pollution. Researchers are working to create more novel materials that are more effective than the currently available biosorbents. These issues are resolved by the use of hybrid technology. Through this method, multicomponent bioadsorbent materials can be synthesized (Thekkudan et al. 2017). Nanoparticles are made to combine with biosorbents and materials so-termed "Nano bioadsorbents" are being produced. These innovative nano bioadsorbents have increased performance rate (Kyzas and Matis 2015). Large bioadsorbents can be broken down into smaller, more surface-area-rich ultrafine molecules. The ultrafine particles may agglomerate to form large particles producing an adverse effect. The nanoparticle present along with the bioadsorbent prevents this agglomeration and, in succession, stabilizes the porous structure of the adsorbent material (Saratale et al. 2018). This increases the availability of adsorptive sites with more surface area and high adsorption efficiency. The application of nano bioadsorbent minimizes the quantity of adsorbent material required for adsorption process. The occurrence of nanoparticles along with bioadsorbents increases the chance of transport and minimizes toxicity. This method of nano bioadsorbent utilization is a novel and effective technique for the removal of pollutants, particularly heavy metals in an aqueous solution.

8.4.6 Polymer Grafting

Grafting has been used to create a variety of novel adsorbents. For instance, grafting cotton fibers to an amino-terminated hyperbranched polymer has been used to adsorb heavy metal ions like Cu2+ and Pb2+ from aqueous solution. The prepared novel adsorbent was characterized by Fourier transform infrared spectroscopy and scanning electron microscopy. The experimental results show that the amino-terminated hyperbranched polymer was grafted to the oxidized cotton fibers, and the adsorbent with amino-terminated hyperbranched polymer was successfully obtained. The grooves on the surface of the grafted cotton fiber were filled with amino-terminated hyperbranched polymer (Zang et al. 2014).

8.5 CONCLUSION

Environmental contaminants are severe problem that provides many removal obstacles. There are numerous plant and microbial species that may break down contaminants in the ecosystem and have improved qualities. Contaminants released into the environment produce bioaccumulation and bioaugmentation, which in turn leads to major health issues in humans such as cancer, muscle weakness, nervous system issues, skin irritations, headaches, stomach infections, and dysentery. The creation of a novel technology for the extensive degradation of pollutants from the source of contamination is urgently needed. For the elimination of these pollutants, a variety of techniques have been used, including internal, physical, and chemical techniques. These sorbents must be pretreated in order to remove the pollutant more effectively. Thus, the biosorption process is feasible, simple, and flexible, and more researches are required for its wide utilization on a large scale.

REFERENCES

Abegunde SM, Idowu KS, Adejuwon OM, Adeyemi-Adejolu T. A review on the influence of chemical modification on the performance of adsorbents. Resources, *Environment and Sustainability*. 2020 Sep 1; 1: 100001.

Barquilha CE, Cossich ES, Tavares CR, Silva EA. Biosorption of nickel (II) and copper (II) ions in batch and fixed-bed columns by free and immobilized marine algae Sargassum sp. *Journal of Cleaner Production*. 2017 May 1; 150: 58–64.

Bel'chinskaya LI, Khodosova NA, Novikova LA, Strel'nikova OY, Roessner F, Petukhova GA, Zhabin AV. Regulation of sorption processes in natural nanoporous aluminosilicates. 2. Determination of the ratio between active sites. *Protection of Metals and Physical Chemistry of Surfaces*. 2016 Jul;52: 599–606.

Bertagnolli C, Uhart A, Dupin JC, da Silva MG, Guibal E, Desbrieres J. Biosorption of chromium by alginate extraction products from Sargassum filipendula: investigation of adsorption mechanisms using X-ray photoelectron spectroscopy analysis. *Bioresource Technology*. 2014 Jul 1; 164: 264–269.

Chen Y, Li M, Li Y, Liu Y, Chen Y, Li H, Li L, Xu F, Jiang H, Chen L. Hydroxyapatite modified sludge-based biochar for the adsorption of Cu2+ and Cd2+: adsorption behavior and mechanisms. *Bioresource Technology*. 2021 Feb 1; 321: 124413.

Deepa KK, Sathishkumar M, Binupriya AR, Murugesan GS, Swaminathan K, Yun SE. Sorption of Cr (VI) from dilute solutions and wastewater by live and pretreated biomass of Aspergillus flavus. *Chemosphere*. 2006 Feb 1; 62(5): 833–840.

Enaime G, Baçaoui A, Yaacoubi A, Lübken M. Biochar for wastewater treatment—conversion technologies and applications. *Applied Sciences*. 2020 Jan; 10(10): 3492.

Fahmi AH, Jol H, Singh D. Physical modification of biochar to expose the inner pores and their functional groups to enhance lead adsorption. *RSC advances*. 2018; 8(67): 38270–38280.

Funke A, Ziegler F. Hydrothermal carbonization of biomass: a summary and discussion of chemical mechanisms for process engineering. *Biofuels, Bioproducts and Biorefining*. 2010 Mar; 4(2): 160–177.

Göksungur Y, Üren S, Güvenç U. Biosorption of cadmium and lead ions by ethanol treated waste baker's yeast biomass. *Bioresource Technology*. 2005 Jan 1; 96(1): 103–109.

Guo Q, Liu Y, Qi G. Application of high-gravity technology NaOH-modified activated carbon in rotating packed bed (RPB) to adsorb toluene. *Journal of Nanoparticle Research*. 2019 Aug; 21(8): 1–4

Kong L, Enders A, Rahman TS, Dowben PA. Molecular adsorption on graphene. *Journal of Physics: Condensed Matter*. 2014 Oct 7; 26(44): 443001.

Kyzas GZ, Matis KA. Nanoadsorbents for pollutants removal: a review. *Journal of Molecular Liquids*. 2015 Mar 1; 203: 159–168.

Mohan D, Abhishek K, Sarswat A, Patel M, Singh P, Pittman CU. Biochar production and applications in soil fertility and carbon sequestration–a sustainable solution to crop-residue burning in India. *RSC Advances*. 2018 Nov; 8(1): 508–520.

Mohan D, Singh P, Sarswat A, Steele PH, Pittman Jr CU. Lead sorptive removal using magnetic and nonmagnetic fast pyrolysis energy cane biochars. *Journal of Colloid and Interface Science*. 2015 Jun 15; 448: 238–250.

Nagda G, Ghole V. Biosorption of Congo red by hydrogen peroxide treated tendu waste. *Journal of Environmental Health Science & Engineering*. 2009; 6(3): 195–200.

Pertile E, Dvorský T, Václavík V, Heviánková S. Use of different types of biosorbents to remove Cr (VI) from aqueous solution. *Life*. 2021 Mar 14; 11(3): 240.

Przepiórski J. Enhanced adsorption of phenol from water by ammonia-treated activated carbon. *Journal of Hazardous Materials*. 2006 Jul 31; 135(1–3): 453–456.

Rahman A, Hango HJ, Daniel LS, Uahengo V, Jaime SJ, Bhaskaruni SV, Jonnalagadda SB. Chemical preparation of activated carbon from Acacia erioloba seed pods using H_2SO_4 as impregnating agent for water treatment: an environmentally benevolent approach. *Journal of Cleaner Production*. 2019a Nov 10; 237: 117689.

Rahman Z, Thomas L, Singh VP. Biosorption of heavy metals by a lead (Pb) resistant bacterium, Staphylococcus hominis strain AMB-2. *Journal of Basic Microbiology*. 2019b May; 59(5): 477–486.

Ramrakhiani L, Ghosh S, Majumdar S. Surface modification of naturally available biomass for enhancement of heavy metal removal efficiency, upscaling prospects, and management aspects of spent biosorbents: a review. *Applied Biochemistry and Biotechnology*. 2016 Sep; 180(1): 41–78.

Saratale RG, Karuppusamy I, Saratale GD, Pugazhendhi A, Kumar G, Park Y, Ghodake GS, Bharagava RN, Banu JR, Shin HS. A comprehensive review on green nanomaterials using biological systems: recent perception and their future applications. *Colloids and Surfaces B: Biointerfaces*. 2018 Oct 1; 170: 20–35.

Shim T, Yoo J, Ryu C, Park YK, Jung J. Effect of steam activation of biochar produced from a giant Miscanthus on copper sorption and toxicity. *Bioresource Technology*. 2015 Dec 1; 197: 85–90.

Sizmur T, Fresno T, Akgül G, Frost H, Moreno-Jiménez E. Biochar modification to enhance sorption of inorganics from water. *Bioresource Technology*. 2017 Dec 1; 246: 34–47.

Thekkudan VN, Vaidyanathan VK, Ponnusamy SK, Charles C, Sundar S, Vishnu D, Anbalagan S, Vaithyanathan VK, Subramanian S. Review on nanoadsorbents: a solution for heavy metal removal from wastewater. *IET Nanobiotechnology*. 2017 Apr; 11(3): 213–224.

Vlasova M, Dominguez-Patino G, Kakazey N, Dominguez-Patino M, Juarez-Romero D, Méndez EY. Structural-phase transformations in bentonite after acid treatment. *Science of Sintering*. 2003 Jan; 35(3): 155–166.

Yaashikaa PR, Kumar PS, Saravanan A, Vo DV. Advances in biosorbents for removal of environmental pollutants: a review on pretreatment, removal mechanism and future outlook. *Journal of Hazardous Materials*. 2021 Oct 15; 420: 126596.

Zang C, Zhang D, Xiong J, Lin H, Chen Y. Preparation of a novel adsorbent and heavy metal ion adsorption. *Journal of Engineered Fibers and Fabrics*. 2014 Dec; 9(4): 155892501400900420.

9 Pretreatment of *Aspergillus* Mycelium for the Enhancement of Lead Biosorption

Pramod Kumar Mahish and Shailesh Kumar Jadhav

9.1 INTRODUCTION

Nowadays, it has become a challenge to solve the problem of water pollution by toxic heavy metals resulting from anthropogenic activities. In this series, biosorption can be a part of such a solution. The biosorption uses biologically derived materials as a biosorbent for the removal of heavy metal ions from wastewater (Ramírez et al., 2020). Seaweed, molds, yeasts, mushrooms, algae, bacteria, actinomycetes, crab shells, and plants are important biosorbents used for the removal of pollutants (Yaashikaa et al., 2021). These biomasses can be obtained from natural and industrial wastes (Sheth et al., 2021). Biosorption of heavy metals and other pollutants using fungal biomass has little advantage over other biosorbents because of its cell wall characteristics, easy growing, manipulation, natural availability, and eco-friendly biosorbent (Ayele et al., 2021). The fungal cell wall contains chitin, glucans, mannans, some polysaccharides, and proteins. Chitin from fungal cell walls is a good biosorbent for heavy metals and pollutants; apart from these, fungal cell walls also contain some functional groups which help to absorb pollutants (Sarode et al., 2019).

The fungi are very useful in the sorption of metal due to their cell wall characteristics, especially the chitin and chitosan compositions. The mycelial structure provides another advantage to fungi. Various functional groups take part in the adherence of metal in the cell wall like the phosphate group, amino group, carboxyl group, polysaccharides, hydroxyl group, etc. (Gahlout et al., 2021). The metal uptake capacity of biological material becomes quite useful after the physical and chemical pretreatment in comparison to the non-treated biomass because more metal-binding site is exposed after pretreatment (Chauhan et al., 2020). The pretreatment may add some functional groups to the cell wall that enhances sorption or remove some unwanted groups from the cell wall that restricts the binding of metal ions to the surface.

Aspergillus is the most common filamentous fungi ubiquitously distributed in the environment. The genus consists of about 340 officially recognized species (Osman, 2021). Decomposing organic substances; causing plant, animal, and human disease; produce toxins are the major significance of *Aspergillus* fungi (Jing and Lu,

DOI: 10.1201/9781003366058-9

2022; Taniwaki et al., 2018). In the most recent, the bio-removal of heavy metals by *Aspergillus* was studied due to their ability toward metal resistance and sorption (Acosta-Rodríguez et al., 2018). So, the present work aims to find out the biosorption ability of *Aspergillus niger* and *Aspergillus flavus* with live and pretreated biomasses and to compare their ability with different parameters.

9.2 SOURCE OF FUNGI AND GROWTH

The sample microorganisms were gathered from the wastewater of industries belonging to the metal and petrochemical works. The wastewater samples have been collected in the winter seasons because of the effect of the shower in rainy and high temperatures in summer. Wastewater has been collected in a plastic bottle sterile before in 70% alcohol. Fungi were isolated in the agar medium, and their maintenance was done as specified in our previous work (Mahish et al., 2015). Among various isolated fungi, *A. flavus var. scherotorium, A. niger,* and *A. niger var. scherotorium* are selected in the present study (Figure 9.1).

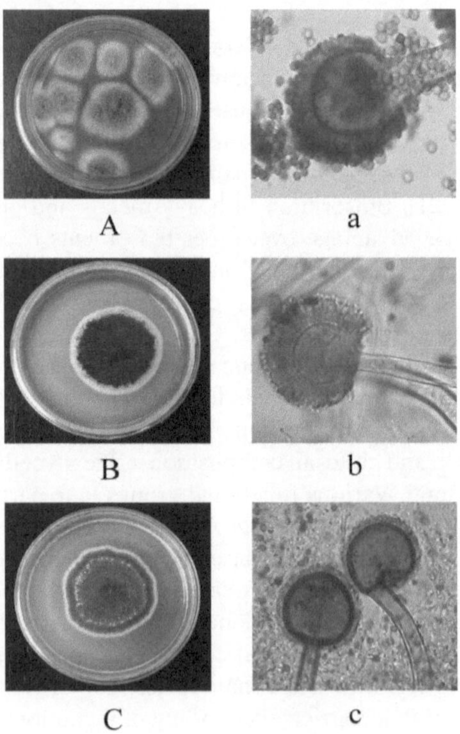

FIGURE 9.1 *Aspergillus* spp. used in the biosorption of lead – *A. flavus var. scherotorium* (A and a), *A. niger* (B and b), and *A. niger var. scherotorium* (C and c)

9.3 BIOSORPTION OF LEAD BY LIVING FUNGI

The methodology of biosorption of lead by living biomass of fungi has been adopted by Jadhav and Mahish (2020). Fungi were grown in liquid potato dextrose medium with a known concentration of lead. After the steam sterilization of the medium, 1 ml of fungal inoculums (72 hours old) was added to the medium. The growth of the microorganism was performed according to the parameter taken (temperature, pH, incubation period). After the incubation period, filtered aqueous solutions were analyzed for metal concentration. Spectroquant Nova60 is used for the assessment of lead metal ions which follows 4-(Pyridylazo)-resorcinol (PAR) methodology, followed by photometric observation. The percent of biosorption of lead by living biomass of fungi was calculated using the normal percent formula.

The incubation period, temperature, and pH were taken as parameters to find out better optimum conditions for the removal of lead from the aqueous solution. The incubation period was established from day 1 to 9. pH was adjusted from 3.0 to 8.0, similarly temperature was adjusted from 20 to 44°C.

An increase in biosorption was found during the incubation period performed with *A. niger.* A high removal efficiency is achieved from day 3 (83.89%) 85.09, 86.53, and 87.01% on days 5, 7, and 9. The optimum pH was found to be 4.0 for better absorption of lead by *A. niger.* A decrease in biosorption was recorded from pH 5 to 8. The effect of temperature on the biosorption of lead was also recorded. Maximum removal (86.20%) was found at the 26°C. Biosorption was decreased from 32 to 3826°C. Growth was completely restricted at 44°C. While studying with *A. flavus var. scherotorium*, biosorption was increased with the period of incubation. Maximum metal recovery was seen on day 9 of incubation. 81.36% lead biosorption was noted in the pH 5.0; similarly favorable temperature was recorded at 26°C (81.91%). A similar effect of various parameters was recorded with *A. niger var. scherotorium.* 11.08% of biosorption was recorded on day 1 of incubation which was found to increase to day 9. 84.53 and 87.62% of lead have been removed at days 7 and 9, respectively. While studying with the effect of pH on biosorption of lead by *A. niger var. scherotorium* the optimum pH of the medium was found 5.0. 85.71% uptake of lead was recorded as a maximum at 32°C of incubation, while 85.11 and 80.05% removals were found at 38 and 44°C (Figures 9.2–9.4).

9.4 PRETREATMENT OF MYCELIUM

Physical and chemical modifications were performed with the biomasses of all three fungi for the enhancement of metal adsorption capacity. This modification is known as the pretreatment of biomass. For this study, fungi were grown in peptone dextrose liquid medium (Mahish et al., 2018). The incubation parameter was chosen according to their performance in biosorption with living biomass. Fungi were grown in large 1,000-ml flasks (300 ml medium) in a shaker incubator at 125-rpm speed. The spherical ball-shaped microbial culture was obtained due to the fragmentation of mycelium in shaking conditions.

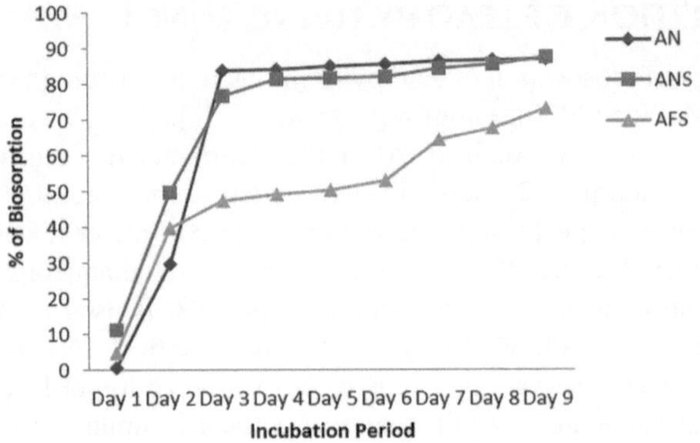

FIGURE 9.2 Effect of incubation period in the biosorption of lead by *Aspergillus* spp.

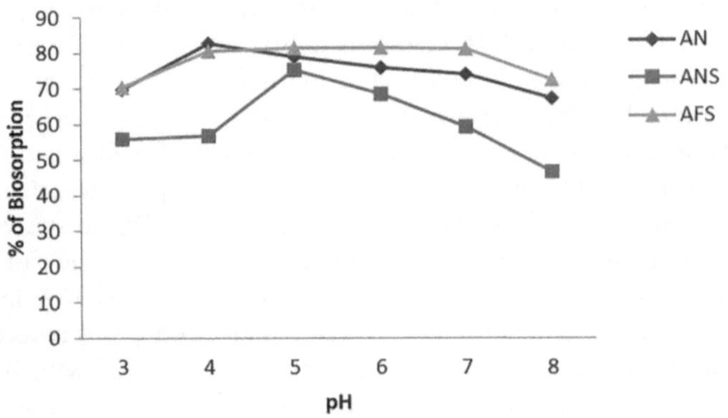

FIGURE 9.3 Effect of pH in the biosorption of lead by *Aspergillus* spp.

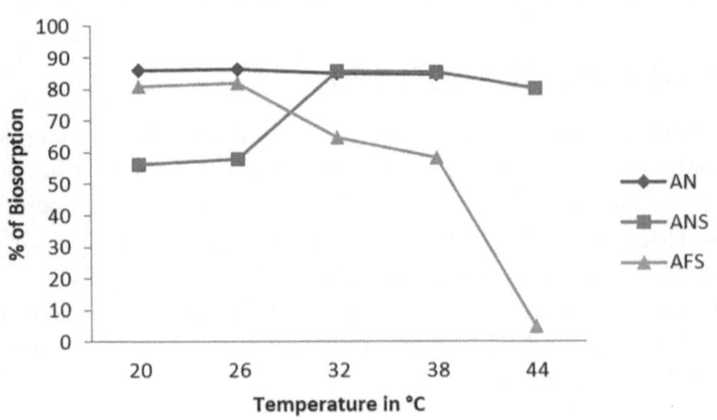

FIGURE 9.4 Effect of temperature in the biosorption of lead by *Aspergillus* spp.

TABLE 9.1
Pretreatment of fungal mycelium with chemical methods

S. no.	Chemical pretreatment	Concentration/ amount	Aqueous phase (ml)	Time of pretreatment (minutes)
1.	Sodium hydroxide	0.5N	500	15
2.	Formaldehyde	15% vol/vol	500	15
3.	Detergent	2.5 g	500	15

Biomasses have been harvested after seven days of incubation. The biomasses were washed twice with distilled water and noted as control biomass. Steam temperature (30 minutes at 15 lbs) was selected for the physical pretreatment while chemical modification of biomass was done as described in Table (9.1).

9.5 PRODUCTION OF FUNGAL BIOSORBENTS AND AQUEOUS REMOVAL OF LEAD

Washing of the pretreated biomass was done for further experiments. The biomasses were then dried in the oven for 24 hours at 60°C. After this, fine powdered particle of biomass was achieved by mortar and pestle. This fine particle is biosorbent and is ready for the biosorption of lead metal ions from an aqueous solution. The test solution is prepared in 100-ml distilled water dissolved with the known concentration of lead metal ion and 0.1 g of biosorbent. Biosorption capacity (Q) was noted after the 15-hour contact time by analyzing lead concentration in the test solution. The formula was applied to get the value of biosorption capacity.

$$Q = \frac{Ci - Cf}{M} \times V$$

where

Q = mg of lead ion absorbed by per g of biomass
Ci = initial metal ion concentration (mg l^{-1})
Cf = final metal ion concentration (mg l^{-1})
M = weight of biomass used in reaction mixture
V = volume of reaction mixture in liter

The pretreatments of biomass change the cell surface chemistry to enhance the adsorption capacity. A similar observation was noted in the present study. Untreated *A. niger* biomass has been removed 2.96 mg of lead which is lower than the physically and chemically treated biomass. 4.04, 4.00, and 4.12 mg values of Q have been obtained from NaOH, formaldehyde, and detergent pretreated mycelium, while autoclaved biomass removed 3.57 mg. While studying with different treated and untreated biomass for lead absorption, it was found that all the treated biomasses have better

FIGURE 9.5 Methodology involved in the biosorption process

absorption of lead compared to untreated biomasses. The highest lead biosorption was studied in detergent-treated biomass (Figure 9.5).

A. *flavus var. scherotorium* showed the value of Q, 2.53 mg by control (untreated) biomass. 3.39 mg of lead was absorbed per gram of physically pretreated autoclaved biomass of *A. flavus var. scherotorium*. 3.71, 3.37, and 3.46 mg of lead were absorbed by NaOH, formaldehyde, and detergent-treated biomasses of *A. flavus var. scherotorium*. The highest biosorption was found at 3.71 by NaOH pretreated biomass, while the least absorption was recorded to be 2.53 mg of lead absorption by untreated control biomass. The effect of physical and chemical modifications with *A. niger var. scherotorium* for biosorption of lead was studied. 3.41 mg of lead was absorbed per gram biomass of untreated biomass of fungus. The absorption capacity of lead 3.93, 3.86, 3.70, and 3.98 was recorded by fungal biomass treated with autoclaving, NaOH, formaldehyde, and detergent. Maximum biosorption capacity was found in detergent-treated biomass, followed by autoclaved, NaOH, formaldehyde, and untreated biomass of *A. niger var. scherotorium* (Figures 9.6 and 9.7).

Several *Aspergillus* fungi have been used for the removal of toxic metal from aqueous solution using the biosorption method. The *Aspergillus* were treated physically and chemically to improve the adsorption capacity. Among more than three hundred *Aspegillus* fungi, *A. flavus, Aspergillus fumigatus, A. niger*, and *Aspergillus terreus* have been mostly utilized for the removal of heavy metals. *A. niger* has been pretreated with physical methods like temperature treatment; chemical treatment like the use of alkaline salts, acids, and detergents for the aqueous removal of lead, cadmium, copper, nickel, chromium, iron (Albesa et al., 2019; Amini et al., 2008; Dursun, 2006; Javaid et al., 2011; Kapoor and Viraraghavan, 1998; Mukhopadhyay et al., 2007; Ren et al., 2021). It is observed that the biosorption capacity has been improved while the pretreatment of biomass as compared to the untreated. *Aspergillus nomius* has been used to remove chromium from the aqueous solution. The *A. nomius* biomass has

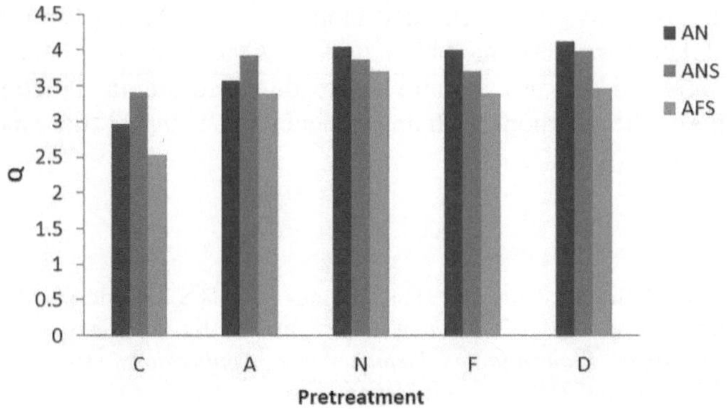

FIGURE 9.6 Effect of pretreatment in the biosorption of lead by *Aspergillus* spp.

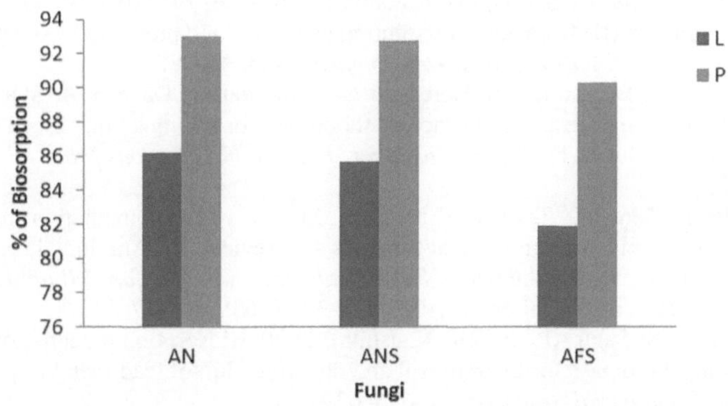

FIGURE 9.7 Comparison of biosorption of lead by live and pretreated biomasses of *Aspergillus* spp.

been pretreated with H_2SO_4, HNO_3, CH_3COOH, NaOH, and HCl in which H_2SO_4 (82%) and HCl (81%) pretreated biomass proven high metal recovery (Guha et al., 2021). *A. flavus* has been pretreated with 0.1-M sodium hydroxide for the biosorption of reactive black 5 dye. The pretreated biomass removed 91% dye from the aqueous solution (Alaguprathana et al., 2022). Copper has been recovered with improved efficiency as compared to untreated biomass of *A. flavus*, *A. fumigatus*, *A. niger*, and *A. terreus*. These fungi have been treated with NaOH (Jyoti and Nupur, 2018).

9.6 CONCLUSION

Aspergillus is an important group of fungi, generally known for the airborne allergens and production of some acids and enzymes. Their ability to control pollutants is explored for the dyes, organic contaminants, heavy metals, xenobiotics, and solid waste. In the present study, three Aspergillus fungi, *A. niger, A. niger var. scherotorium,* and *A. flavus var. scherotorium*, were investigated for their ability to remove lead metal ions. Both live and pretreated biomasses have been proven their ability to control lead. However, pretreated biomass was noted with high metal recovery as compared to untreated biomass. Sodium hydroxide-based chemical pretreatment was found better than other methods with an efficiency of 91.19% in the removal of lead from the solution.

REFERENCES

Acosta-Rodríguez I, Cárdenas-González JF, Rodríguez Pérez AS, Oviedo JT, Martínez-Juárez VM. 2018. Bioremoval of different heavy metals by the resistant fungal strain *Aspergillus niger*. *Bioinorganic Chemistry and Applications*, 3457196, https://doi.org/10.1155/2018/3457196

Alaguprathana M, Poonkothai M, Ameen F, Bhat SA, Mythili R, Sudhakar C. 2022. Sodium hydroxide pre-treated *Aspergillus flavus* biomass for the removal of reactive black 5 and its toxicity evaluation. *Environmental Research*, 214: 113859.

Albesa II JP, Alfaras AA, De Rosas AM, Catalina SB, Angeles EP. 2019. Biosorption of lead (II) and cadmium (II) from aqueous solution using raw and pretreated *Aspergillus niger*. *Journal of BIMP-EAGA Regional Development*, 5(1): 25–38.

Amini M, Younesi H, Bahramifar N, Lorestani AA, Ghorbani F, Daneshi A, Sharifzadeh M. 2008. Application of response surface methodology for optimization of lead biosorption in an aqueous solution by *Aspergillus niger*. *Journal of Hazardous Materials*, 154(1–3): 694–702.

Ayele, A, Haile, S, Alemu, D, Tesfaye, T, Kamaraj, M. 2021. Mycoremediation: fungal-based technology for biosorption of heavy metals – a review. In: Aravind, J, Kamaraj, M, Prashanthi Devi, M, Rajakumar, S (eds) *Strategies and Tools for Pollutant Mitigation*. Springer, Cham. https://doi.org/10.1007/978-3-030-63575-6_17

Chauhan J, Yadav VK, Sahu AP, Jha RK, Kaushik P. 2020. Biosorption potential of alkali pretreated fungal biomass for the removal and detoxification of lead metal ions. *Journal of Scientific & Industrial Research*, 79: 636–639.

Dursun, AY. 2006. A comparative study on determination of the equilibrium, kinetic and thermodynamic parameters of biosorption of copper (II) and lead (II) ions onto pretreated *Aspergillus niger*. *Biochemical Engineering Journal*, 28(2): 187–195.

Gahlout, M, Prajapati, H., Tandel, N., Patel, Y. (2021). Biosorption: an eco-friendly technology for pollutant removal. In: Panpatte, DG, Jhala, YK (eds) *Microbial Rejuvenation of Polluted Environment. Microorganisms for Sustainability*, vol 26. Springer, Singapore. https://doi.org/10.1007/978-981-15-7455-9_9

Guha S, Debnath S, Gayen S. 2021. Biosorption of hexavalent chromium (VI) by *Aspergillus nomius* biomass and optimization of biosorption parameters. *International Journal of Ecology and Environmental Sciences*, 3(1): 461–467.

Jadhav SK, Mahish PK. 2020. Biosorption of Lead from aqueous solution using a new fungal strain, *Cunninghamella elegans* TUFC20022. *Research Journal of Biotechnology*, 15(12): 35–42.

Javaid AM, Bajwa RU, Manzoor TA. 2011. Biosorption of heavy metals by pretreated biomass of *Aspergillus niger*. *Pakistan Journal of Botany*, 43(1): 419–425.

Jing, L., Lu, Y. (2022). Impact of Fungi on Agriculture Production, Productivity, and Sustainability. In: Rajpal, VR, Singh, I, Navi, SS (eds) *Fungal diversity, ecology and control management. Fungal Biology*. Springer, Singapore. https://doi.org/10.1007/978-981-16-8877-5_19

Jyoti C, Nupur P. 2018. Investigation on the biosorption potential of alkali pretreated *Aspergillus* sp. biomass for the removal of copper. *International Journal for Research in Applied Science and Engineering Technology*, 6(3): 2477–2483.

Kapoor, A, Viraraghavan T. 1998. Biosorption of heavy metals on *Aspergillus niger*: effect of pretreatment. *Bioresource Technology*, 63(2): 109–113. https://doi.org/10.1016/S0960-8524(97)00118-1

Mahish PK, Tiwari KL, Jadhav SK. 2015. Biodiversity of fungi from lead-contaminated industrial wastewater and tolerance of lead metal ion by dominant fungi. *Research Journal of Environmental Sciences*, 9(4): 159.

Mahish PK, Tiwari KL, Jadhav SK. 2018. Biosorption of lead by biomass of resistant *Penicillium oxalicum* isolated from industrial effluent. *Journal of Applied Sciences*, 8(1): 41–47.

Mukhopadhyay M, Noronha SB, Suraishkumar GK. 2007. Kinetic modeling for the biosorption of copper by pretreated *Aspergillus niger* biomass. *Bioresource Technology*, 98(9): 1781–1787.

Osman M, Bidon B, Abboud C, Zakaria A, Hamze B, Achcar ME, Mallat H, Dannaoui E, Dabboussi F, Papon N, Bouchara JP. 2021. Species distribution and antifungal susceptibility of *Aspergillus* clinical isolates in Lebanon. *Future Microbiology*, 16(1): 13–26.

Ramírez Calderón OA, Abdeldayem OM, Pugazhendhi A, Rene ER. 2020. Current updates and perspectives of biosorption technology: an alternative for the removal of heavy metals from wastewater. *Current Pollution Reports*, 6(1): 8–27.

Ren B, Zhao L, Wang Y, Song X, Jin Y, Ouyang F, Cui C, Zhang H. 2021. Freezing/thawing pretreatment of dormant *Aspergillus niger* spores to increase the Cr (VI) adsorption capacity: process and mechanism. *RSC Advances*, 11(13): 7704–7712.

Sarode S, Upadhyay P, Khosa MA, Mak T, Shakir A, Song S, Ullah A. 2019. Overview of wastewater treatment methods with special focus on biopolymer chitin-chitosan. *International Journal of Biological Macromolecules*. 121: 1086–1100.

Sheth Y, Dharaskar S, Khalid M, Sonawane S. 2021. An environment friendly approach for heavy metal removal from industrial wastewater using chitosan based biosorbent: a review. *Sustainable Energy Technologies and Assessments*, 43: 100951.

Taniwaki MH, Pitt JI, Magan N. 2018. *Aspergillus* species and mycotoxins: occurrence and importance in major food commodities. *Current Opinion in Food Science*, 23: 38–43.

Yaashikaa PR, Kumar PS, Saravanan A, Vo DV. 2021. Advances in biosorbents for removal of environmental pollutants: a review on pretreatment, removal mechanism and future outlook. *Journal of Hazardous Materials*, 420: 126596.

10 Pretreatment of Spent Mushroom Substrate for the Enhancement of Biosorption capacity

Anjali Kosre, Khemraj Sahu, Varsha Meshram, Deepali Koreti, Pramod Kumar Mahish, and Nagendra Kumar Chandrawanshi

10.1 INTRODUCTION

Bioadsorption is a physicochemical process that passively concentrates and binds metal ions onto specific biomass. Heavy metal contamination of industrial, mining, and agricultural land or water has increased. Heavy metals are highly toxic and, since they accumulate in tissues, irreversibly affect each link of the food chains they enter [1]. The mobility of these metals in soil occurs via the soil water, which is absorbed by fungi and plants or leached into groundwater with the subsequent heavy metals spread. Different techniques are available to minimize the effects of heavy metal pollution. However, some of them are often very costly or produce high environmental impact [2], and others are more environmental-friendly such as phytoextraction, which has been widely studied and has important limitations since it is commonly exceedingly challenging to find native hyper-accumulating plants that generate large amounts of biomass [3]. Because of these limitations, searching for new sorbents to immobilize soil contaminants or remove the contamination from water becomes necessary. Agricultural wastes are cheap materials, readily available, renewable, and show a high affinity for heavy metals [4, 5]. Agricultural wastes such as wood chips, sugar cane, and peel have been employed in lead ions biosorption. The use of agricultural wastes as biosorbents is gaining importance in the bioremediation of heavy metal-polluted water and soils due to their effectiveness and low cost. Spent mushroom substrate (SMS) is an abundant agricultural waste generated after mushroom harvest. During the growth of edible fungi, crude fibers of cellulose, hemicellulose, and lignin are converted into small molecules favorable for metallic ions biosorption [6]. It has been used to efficiently remove Cu, Zn, and Cr [7].

 DOI: 10.1201/9781003366058-10

10.2 MECHANISM OF ADSORPTION

Generally, the binding of metals, pollutants, and dyes to the mushrooms and SMS depends on the four mechanisms: adsorption, ion exchange, complexation, and precipitation. Physical adsorption is based on the electrostatic forces and van der Waals forces. Occasionally, the cation transport system transports the metal ions bearing the same charge and ionic radius along with the other required ions for metabolism [8]. It has been reported that mushroom biomass develops mechanisms to resist heavy metals through the secretion of chelating substances that can bind with metal ions. Further, metal ions accumulation is reduced due to the alterations in the metal transport system. Another mechanism to develop resistance includes binding metal ions to an intracellular molecule such as metallothionein or accumulating in intracellular organelles like vacuoles or mitochondria [9].

The principal mechanism of heavy metal binding is related to metal ions' physical binding and chemical binding to the SMS. However, ion exchange can be observed as an essential phenomenon in adsorption. Occasionally, bivalent metal ions are exchanged with counter ions of polysaccharides. As mentioned earlier, complex formation plays an essential role in adsorption. It is based on the surface charge of the SMS and mushroom. Mushroom possesses chitin, a harmful charge compound in the cell wall, which provides a negative charge to the surface of mushroom mycelium. Both possess negative charge due to the presence of carboxyl, amino, thiol, amide, imine, thioether, and phosphate, which provide the ability to make complexes during metal–ligand and adsorbate–adsorbent interactions [10].

Precipitation, a metabolically independent process, is a chemical reaction between the metal and cell surface of the mushroom and the SMS. This leads to the deposition of heavy metals into the solution and on the surface of mushroom mycelium. In the living mushroom mycelium, the adsorption process depends upon the metabolic processes, while the dead biomass of the mushroom passively binds metal ions by various physicochemical mechanisms. Nevertheless, complete knowledge of metabolism-dependent processes is required to optimize and maintain the adsorption in the living system. In the living biomass, metabolic activities are affected by respiration rate, the products formed during metabolism, and nutrient uptake, which further affects adsorption, ion exchange, complex formation, and precipitation. In the case of organic pollutants, adsorption is based on chemical structure, such as molecular size, charge, solubility, hydrophobicity, and reactivity. In addition to adsorption and complexation, the permeation of SMS biomass may contribute to the adsorption process. Hydrocarbons are hydrophobic compounds that are insoluble in water but can be associated with nonpolar environments.

They can be adsorbed on the surface of organic substances and SMS biomass. Lipophilic, hydrophobic compounds can pass through cell membranes and adsorb into the organic matrix of SMS [35]. In the case of dye adsorption, chitosan, an extracted derivative of chitin, is better than naturally occurring mushroom chitin. Dyes are adsorbed on the surface of chitosan by various mechanisms that include surface adsorption, chemisorption, film diffusion, pore diffusion, and chemical reactions like adsorption complexation and ion exchange. Amine is the leading group involved in dye adsorption; however, the hydroxyl group may also contribute to the process. Notably, the effectiveness of the substrate in adsorbing the pollutants is more important than the mechanism involved in the adsorption [8].

10.3 METHODS FOR ENHANCEMENT OF ABSORPTION USING PRETREATMENTS

To improve the biosorption capacity of SMS, chemical or physical–biological techniques (Figure 10.1) [11] have been used to pretreat the SMS biosorbents, which are discussed in the following subsections.

10.3.1 Physical Pretreatment– Biochar

Biochar is a product formed during the pyrolysis of biomass. It is a nontoxic byproduct generated from agricultural wastes [12] and can replace some more expensive adsorbents used to remove heavy metals from wastewater [13]. Its properties are similar to those of charcoal as it contains a high amount of organic carbon and includes aromatic structures that enhance the adsorption capacity of the biochar [14]. It has a lower surface area and porosity than activated carbon; however, it has a higher content of oxygen-containing acid groups that increase the metal sorption efficiency. The pore size, molecular structure, and characteristics of the biochar depend on the biomass used and the carbonization conditions during its preparation [15]. Several studies reported that spent mushroom compost is investigated as a potential biomass to prepare biochar for effective and adequate wastewater treatment by adsorption. The application of this biomass is significant due to its significantly increasing waste production, which emphasizes the need for adequate management in green wastewater treatment [16]. It can be prepared from different feedstocks and, at different carbonization conditions, possesses different structures and characteristics. The adsorption rate and capacity are a function of the properties of the biochar used, such as the aromaticity, porosity, surface area, polarity [17].

10.3.2 Chemical Pretreatment

The surface of the dried mycelium contained in the SMS was modified by chemical treatment to increase its adsorption capacity. Fourier-transform infrared spectroscopy (FTIR) analysis revealed that amine, hydroxyl, and carboxyl groups provide binding sites for adsorbate. These groups played an important role in adsorption when the modified SMS of *Pleurotuscornucopiae* was used to absorb hexavalent chromium ions. The increased dosage of this adsorbent increased the efficiency of the adsorption of chromium ions. Under the optimum controllable factors like pH, temperature, and hexavalent chromium, a removal efficiency of 75.91% was achieved [8]. Recently, an adsorbent was produced using *P. cornucopiae* and further used for the adsorption of hexavalent chromium ions from the aqueous solution. The adsorbent prepared from SMS was washed with deionized water several times, followed by drying at

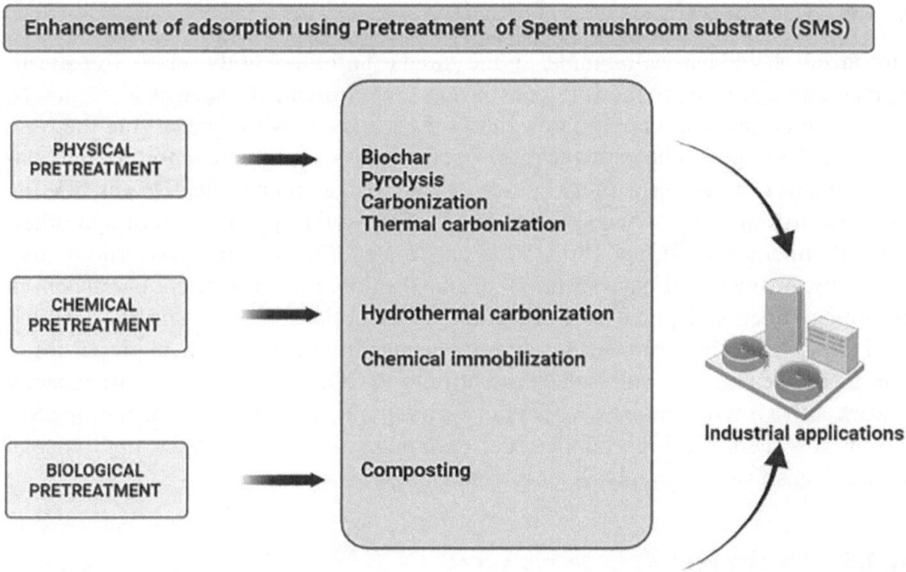

FIGURE 10.1 Methods of enhancement of adsorption using pretreatments

50°C for 3 d and grinding in a pulverized mill and sieve through 40mesh, 60mesh, and 100mesh in order to get uniform particles of specific sizes [18].

Another adsorbent treatment for SMS of *Lentinula edodes* was produced with a mixture of sodium hydroxide, ethanol, and magnesium chloride. The effect of this treatment was assessed on adsorption capacity. It was found that a considerable number of binding sites were exposed in SMS after the treatment with magnesium chloride, which helps in the adsorption of metals. In the adsorption of copper and cadmium ions, both physisorption and chemisorption were reported to involve. These processes are based on electrostatic interaction, ion exchange, and complex formation. The study of thermodynamic parameters revealed that the process was endothermic and spontaneous [19].

10.3.3 Chemical Immobilization

They recently used immobilized SMS from *Pleurotusostreatus* to remove Cd[II] from synthetic wastewater. They found that the biosorption capacity depended on pH value, initial concentration of Cd[II], and contact temperature in the batch system. In conclusion, the maximum adsorption capacity was 100 mgg^{-1} following the Langmuir isotherm model. In another study, SMS from *Auricularia auricular* with modification with cetyltrimethyl ammonium bromide and immobilized by sodium alginate was used as a novel method for removing heavy metals from industrial wastewater [20]. In addition, SMS of *Agaricus bisporus* success fully acted as a biosorbent in removing heavy metals like cadmium [Cd] and mercury [Pb]. In addition, SMS of commercialized edible mushrooms is a practical renewablebiomass resource for treating wastewater in low-cost operations [21].

10.3.4 Carbonization

Carbonization and activation make up the standard protocol of thermally transforming char into solid adsorbent. It forces the non-carbon elements, such as H, O, and N, in the form of gases and tars, to flow out of the feedstock while preserving the rigid carbon skeleton [22]. The primary role of activation is to promote a porous and continuous matrix of carbon with a high specific surface area and effective gas-binding sites. Features of chemistry and morphology of the solid byproduct from hydrothermal carbonization (HTC) at 180–250°C and 2–10 MPa support physisorption and, eventually, bifunctional physisorption–chemisorption. Injection of N2 is efficient in promoting functional pores onto the char. Other options include amines [23] and KOH [24]. These are doping agents and can change the electrochemical properties of char, thereby enhancing both selectivity/affinity to CO2/N2 and the overall capacity to work at high temperatures [25]. Transformation of the organic fraction of SMS into an adsorbent via high-temperature carbonization can offset the high cost of operating gasification, pyrolysis, or torrefaction.

10.3.5 Hydrothermal Carbonization

Hydrothermal carbonization (HTC) is a treatment technique that uses water at high temperatures (i.e., 180°C–275°C) and pressures up to around 50 bars to degrade the organic waste and mainly produce a solid product called hydrochar. It is richer in inorganic materials and less toxic than the original feedstock [26]. It can be used as a clean energy source, for soil amendment, and as a sorbent in water treatment processes. In addition, the liquid effluent of HTC is rich in phenols that have sound potential in food, cosmetics, and pharmaceutical applications [27]. Due to its low mass fraction [i.e., less than 10 wt %], the gas product is generally disposed of in nature despite its affluence in carbon dioxide (CO2) [28].

10.3.6 Autohydrolysis and Thermal Hydrolysis Pretreatment

In this pretreatment, a designated amount of crushed substrate was mixed with a proportional volume of water at room temperature. Both thermal hydrolysis and chemical hydrolysiswere conducted at 121°C for 30 min, with oxalic acid (OA) added in the latter case. The pretreatment parameters considered in this initial stage of the experiment were temperature, chemical, and various mix ratios of substrates [0–100%]. This study evaluated raw and pretreated substrates' physical, morphological, and chemical properties [29].

10.3.7 Biological Pretreatment – Composting

Composting is an aerobic self-heating oxidation process that converts the organic matter into condensed, alkalized, decomposed, humified, and stabilized materials by microbes. Compost is generally used as a polyporous and solid substrate as soil amendments. However, many researchers have found that physicochemical changes within the mixture during composting are propitious to metal removal in an aqueous solution. During composting, the organic matter is decomposed and transformed into stable humic compounds, which can interact with metal ions and buffer solution pH chelation [30]. It was found that the composted municipal solid waste had a higher biosorption capacity for Pb2+, Cu2+, and Zn2+ due to its higher cation-exchange capacity [31]. The wettability determines the affinity between materials and water molecules and subsequently influences the removal ability of materials for heavy metals [32]. During composting, hydrophilic functional groups such as carboxyl and hydroxyl were produced and increased the wettability or hydrophilicity of the materials [33].

Moreover, the increased proportion of hydroxyl, carboxyl, and aromatic hydrocarbons is also conducive to chemical precipitation, surface biosorption, and ion exchange [34]. The physicochemical properties and biosorption characteristics might change during composting, and the following features of composted spent mushroom substrate (CSMS) could be of great significance for heavy metal removal in wastewater. As far as we know,has yet to be there is no study on the effect of composting on SMS in removing lead ions from an aqueous solution, especially its characteristics and mechanisms [35, 36].

10.4 CHARACTERIZATION OF SPENT MUSHROOM SUBSTRATE ADSORBENT

The adsorbent must be sufficiently porous to adsorb pollutants from the surrounding. The SMS comprises various polymer substances such as cellulose, lignin, chitin, and hemicelluloses. These polymers are utilized as carbon and energy sources during the degradation of agro-waste by mushroom mycelia. The degradation of this agro-waste results in the formation of numerous pores in the substrate, making it a suitable substrate for adsorption [35]. These pores can be classified as micropores, mesopores, and macropores. The macroporous material has an excellent potential for adsorbing large-sized adsorbates compared to mesoporous or microporous adsorbent material. The structure of cell walls, micropores, mesopores, and macropores must be evaluated to analyze the potential of adsorbent [8]. The following methods can do the characterization of SMS for using it as an adsorbent.

10.4.1 Brunauer–Emmett–Teller Analysis

A surface area analyzer can measure the specific surface area and total pore volume of the SMS. This fully automated analyzer evaluates the material by nitrogen multilayer adsorption measured as a function of relative pressure. Nitrogen does not react chemically with the substrate and is therefore used in the analyzer. This technique determines the surface and pore area and helps acquire the necessary information for performing adsorption studies. Further, adsorbent categories can be defined based on Brunauer–Emmett–Teller analysis, such as dispersed, nonporous, macroporous (pore diameter greater than 50 nm, type-II isotherm), mesoporous (pore diameter between 2 and 50 nm, type-IV isotherm), and microporous (pore diameter lesser than 2 nm and well fitted to type-I isotherm) materials [8].

10.4.2 Scanning Electron Microscopy/Energy Dispersive X-Ray Spectroscopy Analysis

It is an analytical technique used for the elemental analysis or chemical characterization of the SMS. It is based on the interaction of X-ray and SMS. When X-rays are focused on the SMS, a unique set of peaks can be observed on its electromagnetic emission spectrum according to the atomic structure of materials in the SMS [35].

10.4.3 Fourier TransformInfrared Spectroscopy Analysis

This analysis provides insight into the adsorption process. The functional groups involved in the adsorption process can be determined by analyzing the peaks obtained by Fourier transform infrared spectroscopy (FTIR). Differences in the peaks of the SMS before and after adsorption can be investigated by FTIR. These peaks reveal the essential groups involved in the adsorption process [35].

10.4.5 SolidState CrossPolarization Magic Angle Spinning Carbon13 Nuclear Magnetic Resonance

It is used to investigate cellulose's structural changes and interactions with the pollutant during adsorption. This technique is also used in the detection of the presence of glucan and other trace compounds present in dried powder of mushroom samples. For this purpose, the high-resolution solid-state cross-polarization magic angle spinning carbon-13 nuclear magnetic resonance spectra can be recorded at the resonance frequency of approximately 100 MHz with the use of 4-mm rotors and frequency of 12 kHz and pulse duration of 1.9 microseconds (μs). A high-power proton-decoupling field of 92 kHz can be applied during data acquisition. The spectra can be obtained at room temperature averaging over 5,000–33,000 scans. Hydration of mushroom polysaccharides gives rise to conformational stabilization, which is reflected in spectra by narrowing and splitting the resonance line [37]. The evaluated strategies for potential implementation, adsorption isotherms, adsorption kinetics, and intra-particle diffusion ability can be used to explain the feasibility of the adsorption process.

10.5 APPLICATION AND FUTURE PERSPECTIVE

They were re-using SMSs as alternative low-pressure powders for making high-performance gas-adsorbing bio-interfaces. However, this carbon-to-waste concept is yet nascent, emphasizing the necessity for further investigations before venturing into an industrial scale. Thus, further directions for the improvements, expansion, and consolidation of SMS-to-adsorbent are effective to alter their selectivity to CO_2/N_2. It can be achieved by introducing high-performance doping agents, testing their ability to sequest gases other than the CO_2, such as hydrogen, methane, and secondary organic aerosols, and studying the possibility of recycling of ash from carbonizer as alkaline and porous scaffolding agent to upgrade surfacesphysicochemical and morphological properties. It is important to analyze the strengths, weaknesses, opportunities, and threats before scaling up of the SMS based bioadsorbents [25].

Biochars have been applied as sorbents in removing organic and inorganic contaminants; a little attempt has been made to employ biochars as a reactive medium associated with mining water. Based on a review study on SMS biochar, theoretically, it is an economical and practical approach to be used as filtration media to remove heavy metals from mining water. The high pH characteristic of SMS biochar will be an excellent alkaline generator to treat the low pH mining water. Its many physical and chemical composition features will provide multiple binding sites and enhance the adsorption performance. Therefore, further studies on biochar as a potential filter medium in bioretention systems could be essential, specifically for applications such as mining water treatment. It is also suggested that further study on the adsorption efficiencies, the long-term performance, and the new applications of SMS in large field-scale projects need to be carried out [38]. The SMS modified in biochar can be recycled for the adsorption of pollutants [35]. The adsorption capacity may be increased using chemical treatments and modification of SMS in the activated carbon form of soil microbes and living beings [39]. The adsorption of metal by using the SMS is a novel technique. A detailed study of the metal recovery options will make the use and disposal of the SMS safer. Furthermore, properly utilizing SMS will reduce waste generation from the mushroom industry. The SMS can be composted and stored for a longer time with maintaining its sorption capacity. It will not release any toxin which adversely affects the growth [40, 41].

10.6 CONCLUSION

SMS was used as an adsorbent to remove Pb, Cu, and Zn for wastewater treatment. It was characterized to obtain its physicochemical properties. FTIR studies showed that the functional groups on the surface of the SMS could be modified during pretreatment for adsorption. The investigated biochar has a moderate value of cation-exchange capacity. The effect of initial pH was studied, and the pH of the solutions increased at equilibrium. Composting could raise the ability of SMS to remove lead ions in the solution. The improvement in pH, cation-exchange capacity, and hydrophilicity enhanced the removal ability of materials to lead ions. Bioadsorption and precipitation were the primary mechanisms for adsorption to remove ions from an aqueous solution. The initial pH of the lead solution greatly influenced the removal efficiency, and pH was most beneficial to ions removal by SMS. Composting or adding composted cow dung could shorten the removal time, save the biosorbent dosage, and increase the removal quantity at all initial concentrations.

The possibility of using mushroom and SMS as green adsorbent has been discussed, along with their pros and cons. Mushroom and SMSs can be modified to enhance their adsorption capacities. This technology provides a good option for the adsorption of pollutants. However, the main problem of using mushroom or SMS is the generation of toxic sludge after adsorption. Researching this area is necessary to address and solve the problem of toxic sludge generation and utilization. Using SMS has a high potential to be developed into a sustainable technology. It is an environmentally friendly approach for adsorbing pollutants from industrial effluents and soil.

Moreover, the utilization of SMS will reduce the waste of mushroom farms and remove pollutants from the effluent and soil. Therefore, efforts need to be made to implement the technology in the field. Therefore, our future research will focus on the optimization of the pretreatment conditions to establish the most efficient strategy, the transformation of anSMS into value-added products via microbial fermentation, and the implementation of potential valorization techniques for residual solids to accomplish the drive toward full and sustainable utilization of biomass resources. Pretreatment temperature, hydrolysis duration, and other enzymatic saccharification conditions will be optimized accordingly.

REFERENCES

1. Esposito, A., Pagnanelli, F., Lodi, A., Solisio, C., & Veglio, F. (2001). Biosorption of heavy metals by *Sphaerotilusnatans*: an equilibrium study at different pH and biomass concentrations. *Hydrometallurgy*, *60*(2), 129–141. https://doi.org/10.1016/S0304-386X(00)00195-X
2. Khan, F. I., Husain, T., & Hejazi, R. (2004). An overview and analysis of site remediation technologies. *Journal of Environmental Management*, *71*(2), 95–122. https://doi.org/10.1016/j.jenvman.2004.02.003
3. McGrath, S. P., & Zhao, F. J. (2003). Phytoextraction of metals and metalloids from contaminated soils. *Current Opinion in Biotechnology*, *14*(3), 277–282. https://doi.org/10.1016/S0958-1669(03)00060-0
4. Abdolali, A., Guo, W. S., Ngo, H. H., Chen, S. S., Nguyen, N. C., & Tung, K. L. (2014). Typical lignocellulosic wastes and by-products for biosorption process in water and wastewater treatment: a critical review. *Bioresourcelechnology*, *160*, 57–66. https://doi.org/10.1016/j.biortech.2013.12.037
5. Medina, J., Monreal, C., Barea, J. M., Arriagada, C., Borie, F., & Cornejo, P. (2015). Crop residue stabilization and application to agricultural and degraded soils: a review. *Waste Management*, *42*, 41–54. https://doi.org/10.1016/j.wasman.2015.04.002
6. Qu, J., Zang, T., Gu, H., Li, K., Hu, Y., Ren, G.,... & Jin, Y. (2015). Biosorption of copper ions from aqueous solution by *Flammulinavelutipes* spent substrate. *BioResources*, *10*(4), 8058–8075.
7. Dong, L., Jin, Y., Song, T., Liang, J., Bai, X., Yu, S.,... & Huang, X. (2017). Removal of Cr (VI) by surfactant modified *Auricularia auricula* spent substrate: biosorption condition and mechanism. *Environmental Science and Pollution Research*, *24*(21), 17626–17641.
8. Kulshreshtha, S. (2019). Removal of pollutants using spent mushrooms substrates. *Environmental Chemistry Letters*, *17*(2), 833–847.
9. Ayangbenro, A. S., & Babalola, O. O. (2017). A new strategy for heavy metal polluted environments: a review of microbial biosorbents. *International Journal of Environmental Research and Public Health*, *14*(1), 94. https://doi.org/10.3390/ijerph14010094
10. Javanbakht, V., Alavi, S. A., & Zilouei, H. (2014). Mechanisms of heavy metal removal using microorganisms as biosorbent. *Water Science and Technology*, *69*(9), 1775–1787. https://doi.org/10.2166/wst.2013.718
11. Song, T., Yu, S., Wang, X., Teng, C., Bai, X., Liang, J.,...& Jin, Y. (2017). Biosorption of lead (II) from aqueous solution by sodium hydroxide modified *Auricularia auricular* spent substrate: isotherms, kinetics, and mechanisms. *Water, Air, & Soil Pollution*, *228*(7), 1–17.
12. Kwapinski, W., Byrne, C. M. P., Kryachko, E., Wolfram, P., Adley, C., Leahy, J. J.,... & Hayes, M. H. B. (2010). Biochar from biomass and waste. *Waste Biomass Valorization*, 1, 177–189.

13. Dunnigan, L., Morton, B. J., Ashman, P. J., Zhang, X., & Kwong, C. W. (2018). Emission characteristics of a pyrolysis-combustion system for the co-production of biochar and bioenergy from agricultural wastes. *Waste Management, 77*, 59–66. https://doi.org/10.1016/j.wasman.2018.05.004
14. Nartey, O. D., & Zhao, B. (2014). Biochar preparation, characterization, and adsorptive capacity and its effect on bioavailability of contaminants: an overview. *Advances in Materials Science and Engineering, 2014*. https://doi.org/10.1155/2014/715398
15. Wang, S., Gao, B., Zimmerman, A. R., Li, Y., Ma, L., Harris, W. G., & Migliaccio, K. W. (2015). Physicochemical and sorptive properties of biochars derived from woody and herbaceous biomass. *Chemosphere, 134*, 257–262. https://doi.org/10.1016/j.chemosphere.2015.04.062
16. Abdallah, M. M., Ahmad, M. N., Walker, G., Leahy, J. J., & Kwapinski, W. (2019). Batch and continuous systems for Zn, Cu, and Pb metal ions adsorption on spent mushroom compost biochar. *Industrial & Engineering Chemistry Research, 58*(17), 7296–7307. https://doi.org/10.1021/acs.iecr.9b00749
17. Kołodynska, D., Krukowska, J. A., & Thomas, P. (2017). Comparison of sorption and desorption studies of heavy metal ions from biochar and commercial active carbon. *Chemical Engineering Journal, 307*, 353–363. https://doi.org/10.1016/j.cej.2016.08.088
18. Xu, F., Liu, X., Chen, Y., Zhang, K., & Xu, H. (2016). Self-assembly modified-mushroom nanocomposite for rapid removal of hexavalent chromium from aqueous solution with bubbling fluidized bed. *Scientific Reports, 6*(1), 1–11.
19. Xie, H., Zhao, Q., Zhou, Z., Wu, Y., Wang, H., & Xu, H. (2015). Efficient removal of Cd (II) and Cu (II) from aqueous solution by magnesium chloride-modified *Lentinula edodes. RSC Advances, 5*(42), 33478–33488.
20. Jin, Y., Teng, C., Yu, S., Song, T., Dong, L., Liang, J.,... & Qu, J. (2018). Batch and fixed-bed biosorption of Cd (II) from aqueous solution using immobilized *Pleurotusostreatus* spent substrate. *Chemosphere, 191*, 799–808. https://doi.org/10.1016/j.chemosphere.2017.08.154
21. Zang, T., Cheng, Z., Lu, L., Jin, Y., Xu, X., Ding, W., & Qu, J. (2017). Removal of Cr (VI) by modified and immobilized *Auricularia auricula* spent substrate in a fixed-bed column. *Ecological Engineering, 99*, 358–365. https://doi.org/10.1016/j.ecoleng.2016.11.070
22. Kleszyk, P., Ratajczak, P., Skowron, P., Jagiello, J., Abbas, Q., Frąckowiak, E., & Béguin, F. (2015). Carbons with narrow pore size distribution prepared by simultaneous carbonization and self-activation of tobacco stems and their application to supercapacitors. *Carbon, 81*, 148–157. https://doi.org/10.1016/j.carbon.2014.09.043
23. Igalavithana, A. D., Choi, S. W., Dissanayake, P. D., Shang, J., Wang, C. H., Yang, X.,... & Ok, Y. S. (2020). Gasification biochar from biowaste (food waste and wood waste) for effective CO2 adsorption. *Journal of Hazardous Materials, 391*, 121–147. https://doi.org/10.1016/j.jhazmat.2019.121147
24. Nie, L., Mu, Y., Jin, J., Chen, J., & Mi, J. (2018). Recent developments and consideration issues in solid adsorbents for CO2 capture from flue gas. *Chinese Journal of Chemical Engineering, 26*(11), 2303–2317. https://doi.org/10.1016/j.cjche.2018.07.012
25. da Silva Alves, L., de Almeida Moreira, B. R., da Silva Viana, R., Dias, E. S., Rinker, D. L., Pardo-Gimenez, A., &Zied, D. C. (2022). Spent mushroom substrate is capable of physisorption-chemisorption of CO2. *Environmental Research, 204*, 111945. https://doi.org/10.1016/j.envres.2021.111945
26. Libra, J. A., Ro, K. S., Kammann, C., Funke, A., Berge, N. D., Neubauer, Y.,... & Emmerich, K. H. (2011). Hydrothermal carbonization of biomass residuals: a comparative review of the chemistry, processes and applications of wet and dry pyrolysis. *Biofuels, 2*(1), 71–106. https://doi.org/10.4155/bfs.10.81

27. Atallah, E., Kwapinski, W., Ahmad, M. N., Leahy, J. J., Ala'a, H., & Zeaiter, J. (2019). Hydrothermal carbonization of olive mill wastewater: liquid phase product analysis. *Journal of Environmental Chemical Engineering*, *7*(1), 102833.https://doi.org/10.1016/j.jece.2018.102833
28. Benavente, V., Calabuig, E., & Fullana, A. (2015). Upgrading of moist agro-industrial wastes by hydrothermal carbonization. *Journal of Analytical and Applied Pyrolysis*, *113*, 89–98. https://doi.org/10.1016/j.jaap.2014.11.004
29. Dessie, W., Luo, X., Tang, J., Tang, W., Wang, M., Qin, Z., & Tan, Y. (2021). Towards full utilization of biomass resources: acase study on industrial hemp residue and spent mushroom substrate. *Processes*, *9*(7), 1200. https://doi.org/10.3390/pr9071200
30. Liu, X., Bai, X., Dong, L., Liang, J., Jin, Y., Wei, Y.,... & Qu, J. (2018). Composting enhances the removal of lead ions in aqueous solution by spent mushroom substrate: biosorption and precipitation. *Journal of Cleaner Production*, *200*, 1–11. https://doi.org/10.1016/j.jclepro.2018.07.182
31. Paradelo, R., & Barral, M. T. (2012). Evaluation of the potential capacity as biosorbents of two MSW composts with different Cu, Pb and Zn concentrations. *Bioresource Technology*, *104*, 810–813. https://doi.org/10.1016/j.biortech.2011.11.012
32. Yalcın, E., Cavusoglu, K., & Kınalıoglu, K. (2010). Biosorption of Cu^{2+} and Zn^{2+} by raw and autoclaved *Rocellaphycopsis*. *Journal of Environmental Sciences*, *22*(3), 367–373. https://doi.org/10.1016/S1001-0742(09)60117-0
33. Simon, T. (2007). Characterisation of soil organic matter in long-term fallow experiment with respect to the soil hydrophobicity and wettability. *Soil Water Research*, *2*, 96–103.
34. Anastopoulos, I., Massas, I., &Ehaliotis, C. (2013). Composting improves biosorption of Pb2+ and Ni2+ by renewable lignocellulosic materials. Characteristics and mechanisms involved. *Chemical Engineering Journal*, *231*, 245–254. https://doi.org/10.1016/j.cej.2013.07.028
35. Javanbakht, V., Alavi, S. A., &Zilouei, H. (2014). Mechanisms of heavy metal removal using microorganisms as biosorbent. *Water Science and Technology*, *69*(9), 1775–1787. https://doi.org/10.2166/wst.2013.718
36. Ain, N. K., Tay, C. C., Amnorzahira, A., Liew, H. H., & Suhaimi, A. T. (2014). Characterization of *Pleurotus* spent mushroom compost as a potential biosorbent for Fe (III) ions removal. *Advances in Environmental Biology*, *8*(15), 1–6.
37. Fricova, O., &Kovalakova, M. (2013). Solid-state 13C CP MAS NMR spectroscopy as a tool for detection of (1→ 3, 1→ 6)-β-D-glucan in products prepared from *Pleurotusostreatus*. *International Scholarly Research Notices*, *2013*. https://doi.org/10.1155/2013/248164
38. Madzin, Z., Zahidi, I., Raghunandan, M. E., & Talei, A. (2022). Potential application of spent mushroom compost (SMC) biochar as low-cost filtration media in heavy metal removal from abandoned mining water: a review. *International Journal of Environmental Science and Technology*, *20*(6), 6989–7006.
39. Tay, C. C., Liew, H. H., Abdul-Talib, S., & Redzwan, G. (2016). Bi-metal biosorption using *Pleurotusostreatus* spent mushroom substrate (PSMS) as a biosorbent: isotherm, kinetic, thermodynamic studies and mechanism. *Desalination and Water Treatment*, *57*(20), 9325–9331. https://doi.org/10.1080/19443994.2015.1027957
40. Desa, N. M., Ab Ghani, Z., Talib, S. A., & Tay, C. C. (2016, July). Performance of spent mushroom farming waste (SMFW) activated carbon for Ni (II) removal. In *IOP Conference Series: Materials Science and Engineering* (Vol. 136, No. 1, p. 012059). IOP Publishing. https://doi.org/10.1088/1757-899X/136/1/012059
41. Yan, T., & Wang, L. (2013). Adsorptive removal of methylene blue from aqueous solution by spent mushroom substrate: equilibrium, kinetics, and thermodynamics. *BioResources*, *8*(3), 4722–4734.

11 Tools and Techniques Related to the Monitoring and Assessment of Biosorbents

Nisha Gupta, Dristi Verma, Shubhra Tiwari, Jai Shankar Paul and Shailesh Kumar Jadhav

11.1 INTRODUCTION

The anthropogenic sources contribute significantly to the deterioration of the natural quality of air, water, and soil. The worsening of these natural resources has a noxious impact on the normal functioning of living beings. The industrial sector represents a crucial source of environmental pollution. The untreated exhaust and water from various factories release toxic gases and particulate matter that deteriorate the air and water qualities. About 70% of the earth's surface is covered by water. However, despite being the most indispensable resource for every life existing on earth, the quantity and quality of potable water are unsatisfactory. Industrial activities result in the accretion of several destructive substances beyond the threshold limit recommended by the Environmental Protection Agency and World Health Organization in the environment. The main fatalistic substances of the industries include heavy metals such as mercury, cadmium, palladium, arsenic, lead, and chromium, inorganic contaminants, synthetic organic material, and nuclear waste (radionuclides) (Table 11.1) (Arshadi et al. 2015; Aggarwal and Arora 2020; Baran et al. 2020). All these pollutants demand an efficient and potential solution. Biosorption is a fascinating process which employs various biological agents as biosorbent for the removal of toxic substances (biosorbates) from air and water via different physiochemical reactions. Biosorption uses various dead or living biomass for clinging the contaminants like heavy metals on the surface of the biosorbent via several physical and chemical interactions. As it utilizes biological materials like bacteria, fungus, algae, agricultural, and industrial waste, it is cost-effective, efficient, and environment-friendly. Biosorption is a highly efficient method capable of recovery of metal ions even at a low concentration of 100 mg L^{-1} (Fuks et al. 2016).

DOI: 10.1201/9781003366058-11

TABLE 11.1
List of potential contaminants along with their source and impact on human being

S. No.	Pollutant	Source	Impact on human being	Reference
		Heavy metals		
1.	Lead[a] (Pb^{2+})	Pesticides and fertilizers industries, mining, Petroleum refining, exhaust released from smelting industries, car exhaust, batteries, paints	Acute and chronic exposure to lead can lead to severe anemia, kidney and liver malfunctioning, damage of CNS and PNS central and PNS damage	Jin et al. (2017), El-naggar et al. (2018a), Mahmoud et al. (2020)
2.	Cadmium[a] (Cd^{2+})	Textile dyeing, electroplating, pesticides and fertilizers industries	Damage to kidney, liver, nervous system disorder, cancer, endocrine dysfunction	Wang et al. (2022), El-Naggar et al. (2018b)
3.	Fluoride	Weathering, mining, pesticides	Fluorosis (dental and skeletal), bones deformity and brittleness, thyroid, skin cancer, Alzheimer's syndrome	Christina and Viswanathan (2015)
4.	Cobalt (Co)	Weathering and mining	Carcinogenic, lungs damage	Fawzy et al. (2020)
5.	Manganese (Mn^{2+})	Smelting, mining, agro-industries, batteries, ceramics	Damage to nervous and reproductive system, low immunity, pulmonary disorder	Ma et al. (2013), Deniz and Ersanli (2020)
6.	Chromium (Cr^{6+})	Ceramics manufacturing, mining, electroplating, textile and tannery industry	Causes cancer and respiratory tract infection	Aggarwal and Arora (2020), Madhuranthakam et al. (2021)
7.	Mercury[a]	Metallurgy, metal processing industries	Highly toxic, carcinogenic, damage to nervous system, cardiac failure	Soto-ríos et al. (2018), Baran et al. (2020)
8.	Lanthanum (La^{3+})[b]	Ceramics manufacturing, optical glasses, steel industry	Toxic to human blood lymphocytes	Mahmoud et al. (2020)
9.	Copper (Cu^{2+})	Battery manufacturing, electroplating, pesticides and fertilizers industries, metallurgy	Damage to kidney, liver failure, anemia, nausea	Al-Homaidan et al. (2014)

(Continued)

TABLE 11.1 (*Continued*)
List of potential contaminants along with their source and impact on human being

S. No.	Pollutant	Source	Impact on human being	Reference
10.	Arsenic (As^{3+})	Metallurgical and mining industries, arsenic rich pesticides	Various types of cancer including skin, lungs, kidney, liver, blackfoot disease	Zhang et al. (2016), Shakoor et al. (2019)
11.	Zinc (Zn^{2+})	Electroplating, dyeing, automotive and rubber industries, batteries	Reproductive malfunctioning, disturbance in normal cell division process	Jagaba et al. (2020), Maisarah et al. (2021)
		Dyes		
12.	Azo dyes	Textile industries	Cytotoxic, genotoxic, carcinogenic and causes mutagenesis	Cossolin et al. (2019)
13.	Methylene blue	Textile industries	Toxic, carcinogenic, cause mental retardation, breathing problems, irregular heart rhythm, vomiting, nausea, skin and eye irritation, mutagenesis	Amar et al. (2022), Holliday et al. (2022)
14.	Malachite green	Textile industries	Highly toxic, mutagenic, carcinogenic	Chen et al. (2014)
15.	Metanil yellow	Textile industries	Toxic, cancerous, skin and eye irritant	Tarhan et al. (2019)
16.	Reactive Black B (RBB), Reactive Red 239(RR239), and Direct Blue 85(DB85)	Textile industries	Highly toxic, mutagenic, carcinogenic	Cossolin et al. (2019)
17.	Bromophenol blue	Textile industries	Toxic to brain, liver, kidney, carcinogenic, damage CNS	de Souza et al. (2020)
18.	Reactive Orange 16(RO16)	Textile industries	Highly toxic, mutagenic, carcinogenic	Malakootian and Heidari (2018)
19.	Acid Blue 161	Textile industries	Highly toxic, mutagenic, carcinogenic	Giwa et al. (2016)
20.	Congo red	Textile industries	Carcinogenic, allergenic and causes mutagenesis	Wong et al. (2020)

(*Continued*)

TABLE 11.1 (*Continued*)
List of potential contaminants along with their source and impact on human being

S. No.	Pollutant	Source	Impact on human being	Reference
21.	Reactive Black 5(RB5)	Textile industries	Carcinogenic and causes mutagenesis	Tarhan et al. (2019), Wong et al. (2020)
		Radionuclides		
22.	Uranium (U^{6+})	Nuclear industries	Cause cancer and damages kidney, DNA, bone and reproductive diseases	Monji et al. (2016), Su et al. (2021)
23.	Thorium (Th^{4+})	Nuclear industries	Recalcitrant to human body, cause damage to liver, kidney and heart, nervous disorder, cancer	Monji et al. (2016)
24.	Cesium ($^{137}Cs^{+}$)	Nuclear industries	Nausea, CNS damage, cardiovascular disease	Yu et al. (2020)
25.	Strontium-85 (Sr^{2+})	Nuclear industries	Bone cancer and leukemia, damage to lungs, kidney and heart	Fuks et al. (2016)
26.	Americium-241 (Am^{3+})	Nuclear industries	Bone cancer and leukemia, damage to lungs, kidney and heart	Fuks et al. (2016)

[a] Big three heavy metals.
[b] REE – Rare Earth Elements, CNS – Central Nervous System, PNS – Peripheral Nervous System.

11.2 BIOSORBENT

Biosorption is the process of recovery or removal of pollutants from an aqueous solution. Different materials (inorganic and organic) are used for the biosorption process either in a soluble or insoluble form. The utilization of microorganisms, agricultural by-products, and various organic wastes released from different industries for dismissing heavy metals and other pollutants has opened a new horizon toward bioremediation tools. Biosorbents are safe, easily available, cost-effective, highly efficient, and eco-friendly. The biosorbent removes or recovers the pollutants via various methods like the ion-exchange method, complexation, adsorption, absorption, and precipitation (Aggarwal and Arora 2020).

The qualitative and quantitative analysis of the functional groups like hydroxyl, carboxyl, sulfate, and amino groups present on the biosorbent reveals the extent of the biosorption process. The adsorption % and the adsorption capacity of a biosorbent (mg g^{-1}) are calculated by using Equations (11.1) and (11.2), respectively (Bertagnolli et al. 2014; Arshadi et al. 2015; Ahsan et al. 2018; Baran et al. 2020).

$$Adsorption\ (\%) = \frac{C_0 - C_e}{C_0} \times 100 \tag{11.1}$$

$$Q_e = \frac{C_0 - C_e \times V}{m} \tag{11.2}$$

Where Q_e represents the adsorption capacity of a biosorbent, C_0 is the initial concentration, C_e is the equilibrium metal ion concentration, V is the adsorbate volume in mL, and m is the mass of the biosorbent in mg.

11.3 CONVENTIONAL METHODS OF METAL ION REMOVAL

The presence of the chemical groups facilitates the contaminants (heavy metals and dye) binding through various physiochemical reactions like chelation, adsorption, absorption, precipitation, oxidation, ion exchange, complexation, precipitation, flocculation, and coagulation (Fuks et al. 2016; Ren et al. 2016; Fawzy et al. 2020; Holliday et al. 2022).

11.3.1 Precipitation

Precipitation is the conventional method of separation of impurities either in the suspended or soluble form as sediment or precipitate based on their solubility. Filtration or centrifugation is used to remove the settled sediment. It is the widely used method of heavy metal removal due to its low operational cost and simplicity. Heavy metal ions are separated from an aqueous solution via the chemical precipitation process, which requires the addition of precipitants. The precipitate of carbonates, sulfides, and insoluble metal hydroxide is produced by adding the precipitants to the aqueous solution of heavy metals.

11.3.2 Adsorption

The physical adsorption is due to the van der Waals forces and electrostatic interactions between the adsorbate molecules and the atoms of the adsorbent. Therefore, adsorbents are categorized by their surface characteristics, such as surface area and polarity. Panda et al. (2017) studied industrial waste as a biosorbent for chromium removal from aqueous solutions. About 95% chromium adsorption capacity was recorded at pH 2 with 20 g L^{-1} of adsorbent and 5 mg L^{-1} of the adsorbate. The membrane adsorption techniques perform both membrane filtration and adsorption to effectively remove trace levels of contaminants like cationic heavy metals, anionic phosphates, and nitrates (Khulbe and Matsuura 2018).

11.3.3 Membrane Filtration

Membrane filtration is a method which uses different membranes carrying several functional groups for the binding of solutes of water via adsorption. There are several chemical moieties available on the membrane which selectively adsorb solute. In membrane filtration, a continuous liquid is allowed to pass through membranes having two distinct streams (permeate and retentate). Only a few solutes can pass through the membrane owing to their physical properties and high level of specialization. Membrane filtration is categorized into four different varieties viz. reverse osmosis, nanofiltration, ultrafiltration, and microfiltration the pore size of the membrane (Schlosser 2014).

For heavy metal removal, various kinds of membranes are used. The most commonly used adsorbent (as membrane) for wastewater purification includes charcoal, zeolites, and organic polymers. However, in recent decades, nanotechnology is widely incorporated into wastewater treatment due to its high surface-to-volume ratio and sturdiness. Various nanostructured materials like nanorods, nanospheres, and nanosheets are employed for the removal of pollutants from wastewater.

11.3.4 Liquid-Liquid Extraction

Transferring a solute from one solvent to another while ensuring that the two solvents are insoluble or only partially miscible with one another is known as liquid-liquid extraction (LLE), often referred to as partitioning. It is an essential enrichment method. LLE is based on the dispersion of an analyte between two completely immiscible solvents. The total concentration of an analyte in the organic phase divided by the total concentration in the aqueous phase at equilibrium is known as the distribution ratio. The greater the analyte-to-matrix distribution ratio, the higher the recovery and enrichment factor of the organic compound. The enrichment factor is improved after the extraction by reverse washing the matrix components selectively from the organic phase into the aqueous phase (Berk 2018). Various ionic liquids are used for the removal of heavy metals from aqueous solutions due to their less toxicity.

11.3.5 Ion Exchange

Ion exchange is one of the popular methods in wastewater treatment for the removal of various charged particles. An ion from a diluted water solution is replaced by a similarly charged ion immobilized onto solid support in a reversible chemical reaction known as ion exchange. These inorganic zeolites found naturally, or the organic resins, created synthetically, comprise these solid ion-exchange particles. Because their properties may be customized to specific uses, synthetic organic resins are the most common type utilized nowadays. Ion exchange contributes substantially to wastewater treatment by the removal of dye components and other pollutants (Khan and Mondal 2021).

11.3.6 Coagulation-Flocculation

In coagulation, the physical and chemical properties of the dissolved and suspended particles are altered through the addition of coagulants to form a solid aggregate via various electrostatic and ionic interactions. The suspended solids fraction of the water is separated from it using flocculation and coagulation. The successive coagulation and flocculation allow particle collision and flocs formation. The formed flocs are separated via sedimentation. Various coagulants such as alum, ferric chloride, and ferrous sulfate are used for the treatment of seawater and the removal of impurities (Prakash et al. 2014).

11.3.7 Electrodialysis

Ions are transferred across ion-exchange membranes in a selective manner during the membrane separation process known as electrodialysis (ED). Desalination and other treatments with considerable environmental advantages can be performed with ED techniques because of their selectivity, high separation efficiency, and chemical-free treatment. To reuse wastewater and recover water and/or other products, such as heavy metal ions, salts, acids/bases, nutrients, and organics, or electrical energy, ED technologies can be utilized in processes of concentration, dilution, desalination, regeneration (Gurreri et al. 2020).

However, these methods are not very sensitive in removing low metal ion concentrations. Moreover, they demand harsh chemicals and sophisticated equipment which increases the overall treatment cost. Therefore, a cost-effective and efficient approach like a biosorbent is required. Biosorbents are safe, environment-friendly, efficient, easily available, and cost-effective. Hence, biosorbents provide a promising alternative to the conventional techniques for the removal of heavy metals and other toxic components from the environment.

11.4 ELECTROCHEMICAL TOOLS AND TECHNIQUES FOR MONITORING BIOSORPTION PROCESS

Although electrochemical methods are very traditional but are facile, low-budget, fast, and are capable of multiplexed detection of heavy metal ions. Voltammetry and potentiometric titration are widely used electrochemical tools for metal ion concentration determination.

11.4.1 Voltammetry

It is used to study the current flow as a result of applied potential. The current flow is generally due to Faraday's reaction by the analytes. The graph between current vs. voltage (voltammogram) is plotted. Several variants of voltammetry include pulse, cyclic, stripping, and square wave voltammetry (Devaramani et al. 2021). The concentration of heavy metals is detected through the changes in current flow due to

Faraday's electrochemical reaction through oxidation or reduction at a particular voltage. Through the voltammogram, the generated current and the voltage required to bring about oxidation and reduction of a particular metal ion can be determined.

11.4.2 Potentiometric Titration

Potentiometric titration is a chemical analysis method for the detection of functional groups present on the surface of the biosorbent. It uses an electrode indicator for analyzing the end point of a titration that works as a function of the addition of known concentration of titrant. Various functional groups present on the surface of the biosorbent like carboxyl, amine, phosphoric, sulfonate, and hydroxyl offer potential binding sites for heavy metal binding through different interactions. Therefore, it is of utmost requirement to investigate the number of binding sites available on the biosorbent to determine the extent of biosorption. Potentiometric titration is used to reveal the type and number of aforementioned binding sites (functional groups). Potentiometric titration measures the amount of acidic and basic charges on the biosorbent. The pK_a of the biosorbent determined via potentiometric titration curves shows the type of functional groups present. Furthermore, the number or the quantity of functional group is determined via a mathematical model equation (proton-binding model) (Kim et al. 2018). As per several reported literature, potentiometric titration with a proton-binding model has been used for quantifying the functional groups of the biosorbent including (Hadjoudja et al. 2010; Won et al. 2013).

Potentiometric titration is done to investigate the number of functional groups on the sorbent. In a potentiometric titration, the known amount of biomass (in liquid form) is titrated against 0.1 M HNO_3 or HCl to pH 3 and then against 0.1 M NaOH to pH 10 in CO_2 free environment through constant N_2 gas purging to avoid pH fluctuations. Finally, the equilibrium pH is measured and compared with a reference electrode (Kim et al. 2018). Usually, the potentiometric data is combined with various mathematical software including a proton-binding model to determine the presence of cationic and anionic sites on the biosorbent (Hadjoudja et al. 2010).

11.5 ANALYTICAL TECHNIQUES FOR MONITORING AND ASSESSING THE BIOSORPTION PROCESS

Characterization of biosorbent is the key to unlocking the understanding of the whole biosorption process. Over the past few decades, the development of science and technology has enriched us with countless vital tools and techniques. These modern inventions are a boon for researchers as their research have become more reliable with an explicit reason behind every scientific experimentation. Biosorbents are a crucial tool in combating the noxious pollutant from the environment. The structural features and chemical composition of the biosorbent play a remarkable role in the biosorption process. Therefore, it is equally essential to unveil the morphology, surface feature, and elemental composition of biosorbents. There are several advanced instrumentation techniques (Figure 11.1) to monitor and assess the

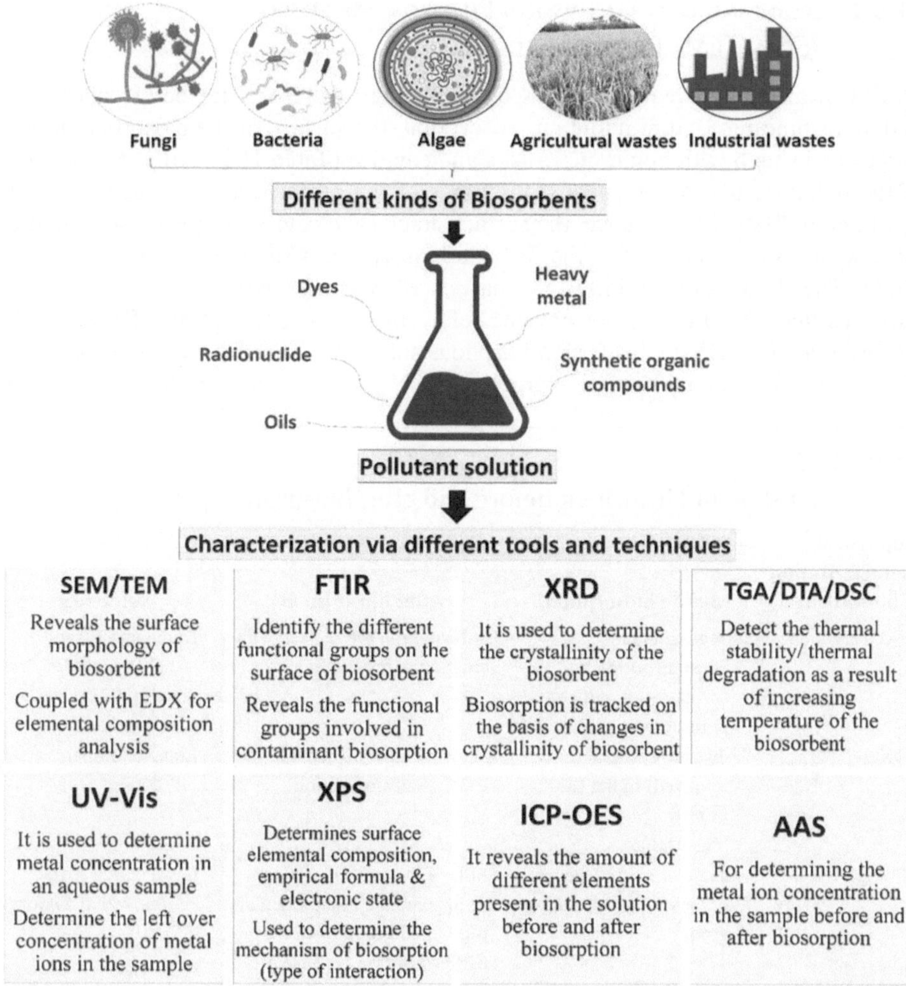

FIGURE 11.1 Various advanced instrumentation techniques for monitoring biosorbent structure and biosorption mechanism

biosorbents including scanning or transmission electron microscopy Coupled with energy dispersive X-ray spectroscopy (SEM/TEM-EDX), Fourier-transform infrared (FTIR) spectroscopy, atomic absorption spectrophotometry (AAS), UV-visible (UV-vis) spectrophotometry, X-ray diffraction (XRD), differential scanning calorimetry, thermogravimetric analysis (TGA), inductively coupled plasma optical emission spectroscopy (ICP-OES), X-ray photoelectron spectroscopy (XPS), X-ray fluorescence spectrometer, and nuclear magnetic resonance (Christina and Viswanathan 2015; Aggarwal and Arora 2020; Deniz and Ersanli 2020; Jagaba et al. 2020; Wong et al. 2020).

11.5.1 Scanning or Transmission Electron Microscopy (SEM/TEM) Coupled with EDX

The SEM analysis reveals the surface morphology of biosorbent before and after pollutants binding. Before pollutant biosorption, the surface of the biosorbent looks porous and rough with numerous cracks and cavities (Table 11.2). After the binding of the pollutant like heavy metals, the pores of the biosorbent are clogged (Deniz and Ersanli 2020). For instance, the surface cracks and cavities of the biosorbent are blocked and smoothened after the fluoride biosorption (Christina and Viswanathan 2015). The TEM along with EDS analysis of *Serratia marcescens* cell surface indicates that Yttrium was not present before biosorption (Liang and Shen 2022). However, as per EDS analysis, an amorphous substance detected on the cell surface of *S. marcescens* after biosorption process was Y^{3+}. Furthermore, SEM analysis also

TABLE 11.2
Characterization of biosorbent before and after biosorption process

Biophysical characterization of biosorbent	Before biosorption	After biosorption	Reference
SEM/TEM	Irregularity and non-uniformity on surface with numerous pores, crack, canals	Uniformity in the surface with clogged pores and cracks	Deniz and Ersanli (2020), Holliday et al. (2022)
EDX	Absence of metal (to be removed) in the EDX of biosorbent before biosorption	A new metal is detected in the EDX after biosorption	Liang and Shen (2022)
FTIR	Characteristic IR peaks of various functional group	Shifts, appearance/ disappearance of a particular functional group from the earlier position because of involvement of that functional group in metal biosorption	Giwa et al. (2016), Aggarwal and Arora (2020)
XRD	Broad peaks without sharpness indicating the amorphous nature of the biosorbent	Sharp peak after contaminant (metal ions) binding indicating the crystallinity of the biosorbent	Jaafar et al. (2016), Kofa et al. (2019)
XPS	Presence of binding energies of the normal atom present in the biosorption	Introduction of new binding energies due to pollutant binding, shortening or broadening of peaks at different binding energies	Ren et al. (2016), Liang and Shen (2022)
UV-vis	Appearance of characteristic peak due to the metal ion	Disappearance of characteristic peak due to the metal ion	Kara et al. (2012), Ahsan et al. (2018), Holliday et al. (2022)

confirms that the amount of Y^{3+} changes from 0% (before biosorption) to 0.77% (after biosorption) on *S. marcescens* cell surface. There was also a reduction in the overall metal composition especially Na^+, Mg^{2+}, and K^+ on the biosorbent surface after biosorption process which indicates that ion-exchange method was involved in Y^{3+} biosorption (Liang and Shen 2022). TEM/SEM along with EDS is a vital tool for investigating the elemental composition of the biosorbent surface before and after biosorption process.

11.5.2 Fourier Transform Infrared Spectroscopy of Biosorbents

For analyzing the changes in the surface chemistry of the biosorbent, FTIR spectroscopy is used before and after contaminants binding. FTIR study helps to identify the different functional groups present on the surface of biosorbent that might be responsible for the contaminants biosorption process. The IR peaks reveal the interaction of pollutants like heavy metals onto the surface of the biosorbents. The binding of the metal ions is due to the interaction with the functional moieties like carboxyl, hydroxyl, phosphate, and the amino group that occurs in the macromolecules of the microbial cell acting as a biosorbent. For instance, the interaction of cadmium (Cd^{2+}) on the surface of dry *Ulva fasciata* biomass through the ion-exchange method was detected via sharp IR bands observed at 3,548, 3,408, 1,624, 1,422, 1,147, 662, and 602 cm^{-1} in FTIR study (El-Naggar et al. 2018b). Maximum biosorption of Chromium (Cr^{6+}) was reported by Aggarwal and Arora (2020) through Neem leaves. The removal and biological reduction of Cr^{6+} into Cr^{3+} were traced via FTIR spectra. The functional moieties like hydroxyl, amino, nitro, and ether group of Neem phytoconstituents were involved in the biosorption of chromium as revealed via FTIR absorption spectra (Aggarwal and Arora 2020). The results of FTIR spectroscopy unveil the functional group which is actually involved in the biosorption of contaminants.

Huge adsorption sites are present as various functional groups in the biosorbent before the biosorption process. The number of the functional groups decreases after the binding of contaminants (Table 11.2). The reduction of the functional group indicates that they are involved in the biosorption process. For instance, before the biosorption process, characteristic peaks of methyl and methylene groups were observed at 2,924, 2,869, 2,852, and 2,869 cm^{-1} (Chen et al. 2014). However, these peaks were absent in the FTIR spectra of biosorbent after the absorption of malachite green. The absence of methyl and methylene groups designates that these groups were involved in the malachite green biosorption (Chen et al. 2014).

11.5.3 Inductively Coupled Plasma Optical Emission Spectroscopy (ICP-OES)

ICP-OES is an analytical technique to determine the type and quantity of elements present in a sample. Unlike SEM, the internal elemental composition of the liquid sample is detected via ICP-OES. ICP-OES reveals the amount of different elements present in the solution after the biosorption process to reveal the biosorption potential. For instance, Jamoussi et al. (2020) performed ICP-OES of the liquid sample after the biosorption process using *Acacia gummifera* gum powder to detect the left out concentration of lead (Pb^{2+}) and cadmium (Cd^{2+}). The remaining Cr^{6+} ions after adsorption via sulfonated spent coffee waste (SCW-SO_3H) were analyzed via ICP-OES (Ahsan et al. 2018).

11.5.4 Thermogravimetric Analysis (TGA)

TGA is an analytical technique to detect the thermal stability of different materials with respect to increasing temperature at a constant rate. TGA is used to detect the % weight loss of the sample as a result of increasing temperature. The change in the sample weight as a function of increasing temperature is recorded via a thermogram generated through TGA. The change in the weight is due to the physicochemical properties of the sample including absorption, adsorption, desorption, phase transition (vaporization and sublimation), and decomposition (Ammar et al. 2021). TGA is a cost-effective and most commonly employed technique for studying the thermal stability of a sample.

Thermal degradation of biosorbent (*A. gummifera*) was studied from 30 to 800°C with a constant heating rate of 10°C min^{-1} (Jamoussi et al. 2020). The TGA analysis has generated three decomposition peaks in three phases. The first phase was observed at 30–120°C with a total weight loss of about 5% due to the elimination of adsorbed water molecules. The second phase was obtained at 230–330°C with 53.74% weight loss of the biosorbent due to the absorption of heat for the decomposition of polysaccharides. The third phase was observed at 400–600°C with 6.95% weight loss due to the carbonization of the sample (Jamoussi et al. 2020). The study of weight-loss profiles of biosorbents as a consequence of increasing temperature helps to know the thermal characteristics, chemical composition, and decomposition pattern. TGA curves are useful in estimating the composition of the biosorbent. For instance, TGA curve was used to determine the biosorbent (pea haulm) composition by Holliday et al. (2022). The peak obtained around 160°C depicts moisture loss. About 22% lignin degradation was achieved in the range between 160–220 and 400–800°C. About 37% cellulose and 18% hemicellulose degradation were recorded between 315–400 and 220–315°C, respectively (Holliday et al. 2022).

TGA and differential thermal analysis (DTA) curves help to analyze the thermal behavior of the biosorbent after adsorption. As per literature, organic compound on thermal degradation splits into small compounds. The major carbohydrate thermal decomposition products are carbon monoxide, carbon dioxide, and less quantity of aldehydes, acids, volatile solids, and water molecules. The splitting of these compounds at various temperature ranges is helpful in metal ion removal. For instance, the TGA-DTA curve was used to analyze the thermal degradation of raw material and Dandelion root biosorbent (DRB) saturated with water and DRB saturated with europium (III) in different phases at 20–950°C (Fuks et al. 2016). The initial weight loss in raw material and DRB saturated with water at 220°C were due to the moisture loss, followed by 30% weight loss due to CO_2 at 450°C. On the other hand, a more intense water loss and less degradation along with a shift of about 20°C in every decomposition temperature were observed in DRB loaded with europium (III) (Fuks et al. 2016). The less efficient degradation and shift in the decomposition temperatures were due to europium (III) adsorption.

11.5.5 X-Ray Diffraction (XRD)

XRD is an analytical technique used to determine the crystallinity of the sample. XRD is widely employed for the characterization and monitoring of the mechanism of the biosorption process. In a biosorption study of Gebrezgiher and Kiflie (2020) using a cactus peel as a biosorbent was characterized via XRD to determine the crystallinity. The absence of any sharp peaks in the XRD diffractogram indicates the amorphous nature of the biosorbent. The amorphous nature of the biosorbent offers porosity for the binding of heavy metals (Gebrezgiher and Kiflie 2020). The absence of crystallinity in the structure of the *Grewia* spp. biopolymer is indicative of porous nature for binding of Cr^{6+} (Kofa et al. 2019). The amorphous nature of the biosorbent is necessary for the binding of contaminants. The broad diffraction or Bragg's peaks of *Pinus densiflora* cone powder were obtained in the 2θ range indicating the amorphous nature of the biosorbent (Shrestha 2016). After the binding of contaminants, the porosity or the amorphous nature of the biosorbent decreases as confirmed via sharp peaks in the XRD diffractogram. The decrease in the crystallinity of the biosorbent after biosorption is due to the masking of the sharp crystal peak by the metal ion (Table 11.2). The binding of the metal ions reduces the crystallinity of the biosorbent. For instance, the masking of $CaCO_3$ by the pollutants was responsible for the decreased intensity of the crystallinity (Jana et al. 2016).

11.5.6 Atomic Absorption Spectroscopy (AAS)

For determining the metal ion concentration in the sample, AAS is used. Because of the higher sensitivity and simplicity, AAS is the preferred choice for the quantification of metal ions present in a sample. For analyzing the biosorption potential of a biosorbent, AAS is done. The liquid sample after the biosorption process is analyzed via AAS to determine the non-bonded concentration of the metal ion. In AAS, the analyte in the solution phase is converted into gaseous phase atoms via an atomizer. Flame atomic absorption spectroscopy (FAAS) is the most readily and well-established variant of AAS for determining the presence of metal in a sample. It is sensitive and quick. Furthermore, the charge for characterizing a sample via FAAS is not so high. FAAS utilizes a flame atomizer for atomizing the analyte from liquid to gaseous phase. The metal ion concentration (10 mg L^{-1}) after biosorption was analyzed via FAAS (Salehi et al. 2021).

11.5.7 UV-vis Spectroscopy

UV-vis spectroscopy is a simple and readily used technique for estimating the metal ion concentration in a liquid sample. To determine the amount of metal ion adsorbed on the biosorbent, a standard calibration curve of that metal is plotted between absorbance and a known concentration of metal ion (Holliday et al. 2022). The collected sample/filtrate after the biosorption mechanism is generally analyzed via UV-vis spectroscopy to determine the leftover concentration of metal ions in the sample. As the biosorbent successfully adsorbs the metal ion from the aqueous solution, there is always a reduction in the absorption intensity after the biosorption process (Kara et al. 2012). The extent of MB dye removal after biosorption by the sulfonated spent coffee waste in the filtrate was investigated via UV-vis spectroscopy (Ahsan et al. 2018). As per data obtained via UV-vis spectroscopy, modified spent coffee waste was capable of removing 100% MB dye (20 ppm) from water (Ahsan et al. 2018).

11.5.8 X-Ray Photoelectron Spectroscopy (XPS)

XPS is a crucial technique for the determination of the surface elemental composition of a material. It also reveals the binding sites of an element. The elemental composition, empirical formula, and chemical and electronic state of the biosorbent are investigated via XPS. Being an ultra-high vacuum technique, the sample needs prior evacuation for analysis by single energy X-ray photons. XPS is a crucial technique to determine the surface composition of material through the binding energies of the photoelectrons (Ren et al. 2016). To determine the mechanism of chromium binding by alginate extraction product from brown seaweed *Sargassum filipendula*, XPS was performed (Bertagnolli et al. 2014). The main functional groups involved in the Cr^{6+} removal are carboxyl, amine, and sulfonic groups with a reduction in the oxidation state of chromium from Cr^{6+} to Cr^{3+} confirmed via a high-intensity peak at Cr $2p_{3/2}$ peak. A peak with a binding energy of 530.7 eV was observed after chromium binding (Bertagnolli et al. 2014).

As per the XPS data obtained, the amount of C, N, and O on the *S. marcescens* cell surface (biosorbent) before the biosorption process was about 67.74, 4.68, and 26.39%, respectively (Liang and Shen 2022). After the biosorption process, there was a decrease in the carbon content (53.56%) and an increase in the N (8.72%) and O (33.26%) content. Furthermore, 4.47% yttrium (Y^{3+}) was detected on the biosorbent surface after the biosorption process. The changes in the amount of C, N, and O indicate that these atoms are involved in the biosorption of Y^{3+} (Liang and Shen 2022). XPS analysis revealed two peaks at 157.23 eV and 159.28 eV corresponding to $Y3d_{5/2-3/2}$ doublet of the Y-OH group with 72.85% atomic composition (Liang and Shen 2022). The ratio of $Y3d_{5/2}$ and $Y3d_{3/2}$ was approximately 1.497, which was the same as the theoretical ratio of 1.5 indicating that the –OH group was involved in the biosorption process. Furthermore, two more peaks of $Y3d_{5/2-3/2}$ doublet of the yttrium hydroxycarboxylate were obtained at 158.98 eV and 161.08 eV with low intensities (27.15%). This indicates that hydroxyl, as well as carboxyl group, was also involved in the biosorption of Y^{3+}. Moreover, there was a shift in the binding energies of C=O and COO– after the biosorption process which confirms their involvement in Y^{3+} biosorption (Liang and Shen 2022).

XPS was performed to determine the possible biosorption mechanism of amine cross-linked reed (ACR) for the biosorption of chromium (Cr^{6+}), nitrate, and phosphate (Ren et al. 2016). Two definite peaks at binding energies of 398 eV (C–N) and 401.8 eV ($-NH_3^+$) were observed when N1s core region was scanned from 394 to 408 eV. A wide scan XPS spectrum of N1s core region of nitrate-laden ACR indicates an additional peak of $-NH_3^+$ at a binding energy of 404.9 eV which shows that the tertiary $-NH_3^+$ group was involved in nitrate removal via electrostatic interaction (Ren et al. 2016). Similarly, the analysis of N1s and P2p of phosphate-laden ACR reveals that the removal of phosphate is mediated by $-NH_3^+$ group through electrostatic interaction. The highly toxic Cr^{6+} was reduced to less toxic Cr^{3+} before adsorption on the ACR with two characteristic peaks at 585.9 (Cr $2p_{1/2}$) and 576.2 eV (Cr $2p_{3/2}$). XPS analysis reveals that after the adsorption of Cr^{3+} on the surface of ACR, there was a splitting of the Cr $2p_{3/2}$ peak at 576.2 eV into two peaks of binding energies 576.0 corresponding to Cr(III)–OH and 577.9 eV corresponding to Cr(VI)–O. The obtained XPS data confirms that adsorption phenomena were mainly facilitated via electrostatic interaction (Ren et al. 2016).

11.6 CONCLUSION

Water is an utmost necessity for every organism surviving on earth. However, human civilization and industrialization have led to the deterioration of water in terms of its quality and quantity. The consumption of deteriorated water is catastrophic. However, the advancement in science and technology has enabled us to solve this concern by using microorganisms and other biomasses. Biosorbent is an essential tool for the removal of pollutants from the water via various physicochemical processes. Biosorbents are very efficient, environment-friendly, and easy to modify via several chemical treatments. The crucial factor in understanding the biosorption process is the monitoring and biophysical characterization. The modern-day instrumentation has enabled the scientific community to monitor the biosorbent structure and its biosorption mechanism. Various spectroscopic and microscopic techniques have enabled monitoring of the biosorbent structure and therefore assessing the biosorption mechanism. Many instrumentation techniques are performed before and after the biosorption mechanism to track the changes occurring after the biosorption. The research and development in biosorption are increasing day by day. However, more studies are still required to find various new biomass as a potential biosorbent and investigation of new tools and techniques for characterization.

REFERENCES

Aggarwal, Rhythm, and Geeta Arora. 2020. "Assessment of Biosorbents for Chromium Removal from Aqueous Media." *Materials Today: Proceedings* 28: 1540–1545. doi:10.1016/j.matpr.2020.04.837.

Ahsan, Md Ariful, Vahid Jabbari, Md Tariqul Islam, Hoejin Kim, Jose Angel Hernandez-Viezcas, Yirong Lin, Carlos A. Díaz-Moreno, Jorge Lopez, Jorge Gardea-Torresdey, and Juan C. Noveron. 2018. "Green Synthesis of a Highly Efficient Biosorbent for Organic, Pharmaceutical, and Heavy Metal Pollutants Removal: Engineering Surface Chemistry of Polymeric Biomass of Spent Coffee Waste." *Journal of Water Process Engineering* 25: 309–319. doi:10.1016/j.jwpe.2018.08.005.

Al-Homaidan, Ali A., Hadeel J. Al-Houri, Amal A. Al-Hazzani, Gehan Elgaaly, and Nadine M.S. Moubayed. 2014. "Biosorption of Copper Ions from Aqueous Solutions by *Spirulina platensis* Biomass." *Arabian Journal of Chemistry* 7 (1): 57–62. doi:10.1016/j.arabjc.2013.05.022.

Amar, Ibrahim A., Esra A. Zayid, Seada A. Dhikeel, and Mohamed Y. Najem. 2022. "Biosorption Removal of Methylene Blue Dye from Aqueous Solutions Using Phosphoric Acid-Treated Balanites *Aegyptiaca* Seed Husks Powder." *Biointerface Research in Applied Chemistry* 12 (6): 7845–7862. doi:10.33263/BRIAC126.78457862.

Ammar, Chiraz, Yassine El-ghoul, and Mahjoub Jabli. 2021. "Characterization and Valuable Use of *Calotropis gigantea* Seedpods as a Biosorbent of Methylene Blue." *International Journal of Phytoremediation* 23 (10): 1085–1094. doi:10.1080/15226514.2021.1876629.

Arshadi, Mohammad, J. Etemad Gholtash, H. Zandi, and S. Foroughifard. 2015. "Phosphate Removal by a Nano-Biosorbent from the Synthetic and Real (Persian Gulf) Water Samples." *RSC Advances* 5 (54): 43290–43302. doi:10.1039/c5ra03191e.

Baran, M. Firat, Ayfer Yildirim, Hilal Acay, Cumali Keskin, and Husamettin Aygun. 2020. "Adsorption Performance of Bacillus Licheniformis Sp. Bacteria Isolated from the Soil of the Tigris River on Mercury in Aqueous Solutions." *International Journal of Environmental Analytical Chemistry* 102 (9): 2013–2028. doi:10.1080/03067319.2020.1746779.

Berk, Zeki. 2018. "Extraction." In *Food Process Engineering and Technology*, edited by Zeki Berk, 307. Academic Press (Imprint of Elsevier).

Bertagnolli, Caroline, Arnaud Uhart, Jean Charles Dupin, Meuris Gurgel Carlos da Silva, Eric Guibal, and Jacques Desbrieres. 2014. "Biosorption of Chromium by Alginate Extraction Products from *Sargassum filipendula*: Investigation of Adsorption Mechanisms Using X-Ray Photoelectron Spectroscopy Analysis." *Bioresource Technology* 164: 264–269. doi:10.1016/j.biortech.2014.04.103.

Chen, Zhengsuo, Hongbo Deng, Can Chen, Ying Yang, and Heng Xu. 2014. "Biosorption of Malachite Green from Aqueous Solutions by *Pleurotus ostreatus* Using Taguchi Method." *Journal of Environmental Health Science and Engineering* 12 (1): 1–10. doi:10.1186/2052-336X-12-63.

Christina, Evangeline, and Pragasam Viswanathan. 2015. "Development of a Novel Nano-Biosorbent for the Removal of Fluoride from Water." *Chinese Journal of Chemical Engineering* 23 (6): 924–933. doi:10.1016/j.cjche.2014.05.024.

Cossolin, Aline Silva, Hélen Cristina Oliveira dos Reis, Ketinny Camargo de Castro, Bruna Assis Paim dos Santos, Matheus Zimermann Marques, Carlos Adriano Parizotto, Leonardo Gomes de Vasconcelos, and Eduardo Beraldo de Morais. 2019. "Decolorization of Textile Azo Dye Reactive Red 239 by the Novel Strain *Shewanella xiamenensis* G5-03 Isolated from Contaminated Soil." *Revista Ambiente & Água* 14 (6): 1–11.

de Souza, Pablo Rodrigues, Thayannah Moreira do Carmo Ribeiro, Ailton Pinheiro Lôbo, Miriam Sanae Tokumoto, Raildo Mota de Jesus, and Ivon Pinheiro Lôbo. 2020. "Removal of Bromophenol Blue Anionic Dye from Water Using a Modified Exuviae of *Hermetia illucens* Larvae as Biosorbent." *Environmental Monitoring and Assessment* 192 (3): 1–16. doi:10.1007/s10661-020-8110-z.

Deniz, Fatih, and Elif Tezel Ersanli. 2020. "An Effectual Biosorbent Substance for Removal of Manganese Ions from Aquatic Environment: A Promising Environmental Remediation Study with Activated Coastal Waste of *Zostera marina* Plant." *BioMed Research International* 2020. doi:10.1155/2020/7806154.

Devaramani, Samrat, G. Banuprakash, and B. H. Doreswamy. 2021. "Electrochemical and Optical Methods for the Quantification of Lead and Other Heavy Metal Ions in Liquid Samples." In *Heavy Metals-Their Environmental Impacts and Mitigation*, edited by Mazen Khaled Nazal and Hongbo Zhao, 1–21. IntechOpen. doi: 10.5772/intechopen.95085.

El-Naggar, Noura El-ahmady, Ragaa A Hamouda, Ibrahim E Mousa, Marwa S Abdel-hamid, and Nashwa H Rabei. 2018a. "Biosorption Optimization, Characterization, Immobilization and Application of *Gelidium amansii* Biomass for Complete Pb^{2+} Removal from Aqueous Solutions." *Scientific Reports* 8(1): 1–19. doi:10.1038/s41598-018-31660-7.

El-Naggar, Noura El Ahmady, Ragaa A. Hamouda, Ibrahim E. Mousa, Marwa S. Abdel-Hamid, and Nashwa H. Rabei. 2018b. "Statistical Optimization for Cadmium Removal Using *Ulva fasciata* Biomass: Characterization, Immobilization and Application for Almost-Complete Cadmium Removal from Aqueous Solutions." *Scientific Reports* 8 (1): 1–17. doi:10.1038/s41598-018-30855-2.

Fawzy, Mustafa A., Awatief F. Hifney, Mahmoud S. Adam, and Arwa A. Al-badaani. 2020. "Environmental Technology & Innovation Biosorption of Cobalt and Its Effect on Growth and Metabolites of *Synechocystis pevalekii* and *Scenedesmus bernardii* : Isothermal Analysis." *Environmental Technology & Innovation* 19: 100953. doi:10.1016/j.eti.2020.100953.

Fuks, L., A. Oszczak, J. Dudek, M. Majdan, and M. Trytek. 2016. "Removal of the Radionuclides from Aqueous Solutions by Biosorption on the Roots of the Dandelion (*Taraxacum officinale*)." *International Journal of Environmental Science and Technology* 13 (10): 2339–2352. doi:10.1007/s13762-016-1067-3.

Gebrezgiher, Mebrahtu, and Zebene Kiflie. 2020. "Utilization of Cactus Peel as Biosorbent for the Removal of Reactive Dyes from Textile Dye Effluents." *Journal of Environmental and Public Health* 2020. doi:10.1155/2020/5383842.

Giwa, Abdur Rahim Adebisi, Khadijat Ayanpeju Abdulsalam, Francois Wewers, and Mary Adelaide Oladipo. 2016. "Biosorption of Acid Dye in Single and Multidye Systems onto Sawdust of Locust Bean (*Parkia biglobosa*) Tree." *Journal of Chemistry* 2016: 1–11. doi:10.1155/2016/6436039.

Gurreri, Luigi, Alessandro Tamburini, Andrea Cipollina, and Giorgio Micale. 2020. "Electrodialysis Applications in Wastewater Treatment for Environmental Protection and Resources Recovery: A Systematic Review on Progress and Perspectives." *Membranes* 10 (7): 1–93. doi:10.3390/membranes10070146.

Hadjoudja, S., V. Deluchat, and M. Baudu. 2010. "Cell Surface Characterisation of *Microcystis aeruginosa* and *Chlorella vulgaris*." *Journal of Colloid And Interface Science* 342 (2) : 293–299. doi:10.1016/j.jcis.2009.10.078.

Holliday, Mathew C., Daniel R. Parsons, and Sharif H. Zein. 2022. "Agricultural Pea Waste as a Low-Cost Pollutant Biosorbent for Methylene Blue Removal: Adsorption Kinetics, Isotherm and Thermodynamic Studies." *Biomass Conversion and Biorefinery*. doi:10.1007/s13399-022-02865-8.

Jaafar, Raghad, and A. Al Sulami. 2016. "Biosorption of Some Heavy Metals by *Deinococcus radiodurans* Isolated from Soil in Basra Governorate-Iraq." *Journal of Bioremediation & Biodegradation* 07 (02). doi:10.4172/2155–6199.1000332.

Jagaba, A. H., S. R. M. Kutty, S. G. Khaw, C. L. Lai, M. H. Isa, L. Baloo, I. M. Lawal, S. Abubakar, I. Umaru, and Z. U. Zango. 2020. "Derived Hybrid Biosorbent for Zinc(II) Removal from Aqueous Solution by Continuous-Flow Activated Sludge System." *Journal of Water Process Engineering* 34: 101152. doi:10.1016/j.jwpe.2020.101152.

Jamoussi, Bassem, Radhouane Chakroun, Cherif Jablaoui, and Larbi Rhazi. 2020. "Efficiency of *Acacia gummifera* Powder as Biosorbent for Simultaneous Decontamination of Water Polluted with Metals." *Arabian Journal of Chemistry* 13 (10): 7459–7481. doi:10.1016/j.arabjc.2020.08.022.

Jana, Animesh, Priyankari Bhattacharya, Sandeep Sarkar, Swachchha Majumdar, and Sourja Ghosh. 2016. "An Ecofriendly Approach towards Remediation of High Lead Containing Toxic Industrial Effluent by a Combined Biosorption and Microfiltration Process: A Total Reuse Prospect." *Desalination and Water Treatment* 57 (12): 5498–5513. doi:10.1080/19443994.2015.1004596.

Jin, Yu, Sumei Yu, Chunying Teng, Tao Song, and Liying Dong. 2017. "Biosorption Characteristic of *Alcaligenes* sp. BAPb. 1 for Removal of Lead (II) from Aqueous Solution." *3 Biotech* 7 (2): 1–12. doi:10.1007/s13205-017-0721-x.

Kara, Ilknur, Sibel Tunali Akar, Tamer Akar, and Adnan Ozcan. 2012. "Dithiocarbamated Symphoricarpus Albus as a Potential Biosorbent for a Reactive Dye." *Chemical Engineering Journal* 211–212: 442–452. doi:10.1016/j.cej.2012.09.086.

Khan, Anoar Ali, and Madhumanti Mondal. 2021. "Effective Materials in the Photocatalytic Treatment of Dyestuffs and Stained Wastewater." In *Photocatalytic Degradation of Dyes*, edited by Maulin Shah, Sushma Dave and Jayashankar Das, 91–111. Elsevier.

Khulbe, K. C., and T. Matsuura. 2018. "Removal of Heavy Metals and Pollutants by Membrane Adsorption Techniques." *Applied Water Science* 8 (1): 1–30. doi:10.1007/s13201-018-0661-6.

Kim, Sok, Chul Woong, Cho Myung, Hee Song, John Kwame, Bediako Yeoung, Sang Yun, and Yoon E. Choi. 2018. "Potentiometric Titration Data on the Enhancement of Sorption Capacity of Surface-Modified Biosorbents : Functional Groups Scanning Method." *Clean Technologies and Environmental Policy* 20 (10): 2191–2199. doi:10.1007/s10098-018-1542-2.

Kofa, G. P., G. R. Nkoue Ndongo, M. B. Kameni Ngounou, M. N. Nsoe, E. V. Amba, and S. Ndi Koungou. 2019. "*Grewia* spp. Biopolymer as Low-Cost Biosorbent for Hexavalent Chromium Removal." *Journal of Chemistry* 2019. doi:10.1155/2019/6505731.

Liang, Chang li, and Ji li Shen. 2022. "Removal of Yttrium from Rare-Earth Wastewater by *Serratia marcescens*: Biosorption Optimization and Mechanisms Studies." *Scientific Reports* 12 (1): 1–14. doi:10.1038/s41598-022-08542-0.

Ma, Lan, Yuhong Peng, Bo Wu, Daiyin Lei, and Heng Xu. 2013. "*Pleurotus ostreatus* Nanoparticles as a New Nano-Biosorbent for Removal of Mn(II) from Aqueous Solution." *Chemical Engineering Journal* 225: 59–67. doi:10.1016/j.cej.2013.03.044.

Madhuranthakam, Chandra Mouli R., Archana Thomas, Zhainab Akhter, Shannon Q. Fernandes, and Ali Elkamel. 2021. "Removal of Chromium(VI) from Contaminated Water Using Untreated *Moringa* Leaves as Biosorbent." *Pollutants* 1 (1): 51–64. doi:10.3390/pollutants1010005.

Mahmoud, Mohamed E., Nesma A. Fekry, and Amir M. Abdelfattah. 2020. "A Novel Nanobiosorbent of Functionalized Graphene Quantum Dots from Rice Husk with Barium Hydroxide for Microwave Enhanced Removal of Lead (II) and Lanthanum (III)." *Bioresource Technology* 298 (Ii): 122514. doi:10.1016/j.biortech.2019.122514.

Maisarah, S., Adityosulindro, and D. Wulandari. 2021. "Utilization of Wild Algae Biomass as Biosorbent for Removal of Heavy Metal Zinc (Zn^{2+}) from Aqueous Solution." *IOP Conference Series: Earth and Environmental Science* 824 (1): 2–9. doi:10.1088/1755-1315/824/1/012017.

Malakootian, Mohammad, and Mohammad Reza Heidari. 2018. "Reactive Orange 16 Dye Adsorption from Aqueous Solutions by Psyllium Seed Powder as a Low-Cost Biosorbent: Kinetic and Equilibrium Studies." *Applied Water Science* 8 (7): 1–9. doi:10.1007/s13201-018-0851-2.

Monji, Akbar Boveiri, Vanik Ghoulipour, and Mohammad Hassan Mallah. 2016. "Biosorption of Toxic Transition Metals and Radionuclides from Aqueous Solutions by Agro-Industrial Byproducts". *Journal of Hazardous, Toxic, and Radioactive Waste* 20 (2): 04015016-1–04015016-9. doi:10.1061/(ASCE)HZ.2153-5515.0000296.

Panda, H., N. Tiadi, M. Mohanty, and C. R. Mohanty. 2017. "Studies on Adsorption Behavior of an Industrial Waste for Removal of Chromium from Aqueous Solution." *South African Journal of Chemical Engineering* 23: 132–138. doi:10.1016/j.sajce.2017.05.002.

Prakash, N. B., Vimala Sockan, and P. Jayakaran. 2014. "Waste Water Treatment by Coagulation and Flocculation." *International Journal of Engineering Science and Innovative Technology* 3 (2): 479–484.

Ren, Zhongfei, Xing Xu, Xi Wang, Baoyu Gao, Qinyan Yue, Wen Song, Li Zhang, and Hantao Wang. 2016. "FTIR, Raman, and XPS Analysis during Phosphate, Nitrate and Cr(VI) Removal by Amine Cross-Linking Biosorbent." *Journal of Colloid and Interface Science* 468: 313–323. doi:10.1016/j.jcis.2016.01.079.

Salehi, Narges, Ali Moghimi, and Hamidreza Shahbazi. 2021. "Magnetic Nanobiosorbent (MG-Chi/Fe3O4) for Dispersive Solid-Phase Extraction of Cu(II), Pb(II), and Cd(II) Followed by Flame Atomic Absorption Spectrometry Determination." *IET Nanobiotechnology* 15 (6): 575–584. doi:10.1049/nbt2.12025.

Schlosser, Štefan. 2014. "Membrane Filtration." In *Engineering Aspects of Food Biotechnology*, edited by José A. Teixeira and António A. Vicente, 145. CRC Press.

Shakoor, Muhammad Bilal, Nabeel Khan Niazi, Irshad Bibi, Muhammad Shahid, Zulfiqar Ahmad Saqib, Muhammad Farrakh Nawaz, Sabry M. Shaheen, et al. 2019. "Exploring the Arsenic Removal Potential of Various Biosorbents from Water." *Environment International* 123: 567–579. doi:10.1016/j.envint.2018.12.049.

Shrestha, Sohan. 2016. "Chemical, Structural and Elemental Characterization of Biosorbents Using FE-SEM, SEM-EDX, XRD/XRPD and ATR-FTIR Techniques." *Journal of Chemical Engineering & Process Technology* 7 (3): 2–11. doi:10.4172/2157-7048.1000295.

Soto-ríos, Paula Cecilia, and Marco Antonio León-romero. 2018. "Biosorption of Mercury by Reed (*Phragmites australis*) as a Potential Clean Water Technology." *Water, Air, & Soil Pollution* 229 (10): 1–11. doi:10.1007/s11270-018-3978-8

Su, Yi, Marco Wenzel, Silvia Paasch, Markus Seifert, Wendelin Böhm, Thomas Doert, and Jan J. Weigand. 2021. "Recycling of Brewer's Spent Grain as a Biosorbent by Nitro-Oxidation for Uranyl Ion Removal from Wastewater." *ACS Omega* 6 (30): 19364–19377. doi:10.1021/acsomega.1c00589.

Tarhan, Tuba, Bilsen Tural, Kenan Boga, and Servet Tural. 2019. "Adsorptive Performance of Magnetic Nano-Biosorbent for Binary Dyes and Investigation of Comparative Biosorption." *SN Applied Sciences* 1 (1): 1–11. doi:10.1007/s42452-018-0011-1.

Wang, Qian, Yunlong Wang, Zi Yang, Wenqing Han, Lizhu Yuan, Li Zhang, and Xiaowu Huang. 2022. "Efficient Removal of Pb(II) and Cd(II) from Aqueous Solutions by Mango Seed Biosorbent." *Chemical Engineering Journal Advances* 11: 100295. doi:10.1016/j.ceja.2022.100295.

Won, Sung Wook, Sun Beom Choi, and Yeoung-Sang Yun. 2013. "Binding Sites and Mechanisms of Cadmium to the Dried Sewage Sludge Biomass." *Chemosphere* 93 (1): 146–151. doi:10.1016/j.chemosphere.2013.05.011.

Wong, Syieluing, Nawal Abd Ghafar, Norzita Ngadi, Fatin Amirah Razmi, Ibrahim Mohammed Inuwa, Ramli Mat, and Nor Aishah Saidina Amin. 2020. "Effective Removal of Anionic Textile Dyes Using Adsorbent Synthesized from Coffee Waste." *Scientific Reports* 10 (1): 1–13. doi:10.1038/s41598-020-60021-6.

Yu, Runlan, Hongsheng Chai, Zhaojing Yu, Xueling Wu, Yuandong Liu, and Li Shen. 2020. "Behavior and Mechanism of Cesium Biosorption from Aqueous Solution by Living *Synechococcus* PCC7002." *Microorganisms* 8 (4): 491. doi:10.3390/microorganisms8040491.

Zhang, Lingfan, Tianyi Zhu, Xin Liu, and Wenqing Zhang. 2016. "Simultaneous Oxidation and Adsorption of As(III) from Water by Cerium Modified Chitosan Ultrafine Nanobiosorbent." *Journal of Hazardous Materials* 308: 1–10. doi:10.1016/j.jhazmat.2016.01.015.

12 Aqueous Removal of Heavy Metals Using Biosorbents

Tarun Kumar Patle, Ravishankar Chauhan, Alka Patle, and Pramod Kumar Mahish

12.1 INTRODUCTION

Environmental pollution such as soil, water, air, etc. have an immense consequence on all living beings. Among, the environmental issues, water pollution from toxic heavy metals such as Hg, Cr, Pb, Zn, Cu, Ni, Cd, As, Co, Sn, etc. is a major issue [1–3]. Among these, Pb, Hg, Cd, and Cr(VI) are the most toxic heavy metals; these have a major impact on the environment and human health [3]. The major sources of heavy metal contamination are industrial influents coming from mining, paints, fertilizers, pesticides, leather, iron, steel, electroplating, photography, aerospace, atomic energy, etc. [4]. Contamination of heavy metal is non-degradable; many heavy metals such as Hg, Cr, Pb, Cd, As, etc. are frequently quantified in industrial wastewaters [5]. These heavy metals are very toxic for humans in ppb levels; for example, Pb can damage our body by attachment with specific cell components, compartmentalization, breakdown of cellular process, oxidative damage, and transport [6, 7]. Further, industrial wastewater has been drained in the rivers that further affects the other sources of drinking water too. So, the treatment of wastewater from industries is necessary for the removal of heavy metals to protect the environment as well as human health.

Several methodologies such as physical, chemical, and biological are employed in the removal of heavy metals from aqueous media [8–10]. Some conventional techniques were frequently used for the removal of heavy metals with certain drawbacks including less effective, generation of a large amount of waste, time taking, high energy demand, high cost, etc. [11, 12]. Recently, biosorbents have gained more attention for developing cost-effective and eco-friendly removal and control of heavy metal pollution in aqueous media [13, 14]. Biosorbents are natural biological materials such as plants, bacteria, fungi, algae, etc. have a tremendous property to accumulate heavy metals from aqueous media [14]. These biosorbents have advantages over conventional methods in terms of low cost, higher efficiency, nominal waste, recovery, etc. [15].

Biosorption is a physico-chemical phenomenon occurring biologically in plants, and microbes, in which absorption or adsorption of targeted heavy metals or other

 DOI: 10.1201/9781003366058-12

contamination in aqueous media takes place [16]. There are two types of phases usually present in biosorption; the first one is a solid phase which is also known as sorbent containing plants, fungi, algae, bacteria, etc. and the second one is an aqueous phase containing heavy metal contamination which is also known as sorbate [17]. The affinity of sorbents to the sorbate is higher involving different mechanisms and the process continues till the equilibrium stabilizes between biosorbent-bound heavy metal and the amount of heavy metal reaming in aqueous media [17]. The mechanism involved in biosorption of contamination of heavy metal from aqueous media via few processes such as physical absorption, precipitation, complexation, ion exchange, oxidation-reduction, etc. [18].

The cell wall is a major part of the biosorption process, which contains a variety of functional groups such as hydroxyl (–OH), carboxyl (–COOH), esters (–COOR), amino ($-NH_2$), carbonyl (–C=O), phosphate group, etc. which contributed in biosorption process [19]. These functional groups are directly involved in the removal of heavy metals through the biosorption process which can later examine the efficiency of removal of heavy metals by different analytical techniques such as infrared spectroscopy, Raman spectroscopy, X-ray photoelectron spectroscopy, energy dispersive X-ray spectroscopy, etc. [20]. This chapter deals with the efficiency of biosorbents (plant, bacteria, algae, fungi, etc.) for the removal of heavy metals from aqueous solutions. Additionally, the chapter highlights the challenges, significance as well as prospects of biosorbents for the removal of contamination from aqueous solution.

12.2 PLANTS BASED BIOSORBENT FOR THE REMOVAL OF HEAVY METALS

Different parts of a plant such as leaves, peel, stem, bark, seed, flower, cortex, etc. have been widely used as biosorbents for the removal of heavy metals from aqueous solution. Plant cellulose has different functional groups such as carboxylic, phenolic, hydroxyl, and carbonyl [21]. A study reported that the hydroxyl functional group present in aloe vera is associated with metal binding for the removal of heavy metals such as uranium and cadmium [22]. Various experimental parameters such as pH, temperature, time, concentration, particle size, etc. were optimized to achieve the highest biosorption condition [23]. The kinetic of isotherm models such as the Langmuir, Freundlich, Sips, Redlich-Peterson, Toth, Frenkel Halsey-Hill, and Dubinin-Radushkevich (D-R) were studied well by researchers [24]. The basic procedure of making plant materials as biosorbent is, first, the plant part is collected and washed properly, after that oven dried part is powdered; this powdered plant sample is treated chemically with NaOH for alkanization [25]. This sample is again oven dried and stored in an air-tight container till used as a biosorbent.

Leaves are the most common and abundant plant part which are obtained very simply in almost all environmental seasons. The leaf is studied for its potential as a biosorbent due to its low cost and natural abundance property for the removal of toxic heavy metals from aqueous solution [25]. Leaves of *Ulmus carpinifolia* and

Fraxinus excelsior were used for the removal of Pb(II), Cd(II), and Cu(II) from an aqueous solution [26]. The maximum absorption potential of plant samples with different optimization analytical parameters was studied. They found that the kinetic of the Langmuir adsorption isotherm model was second order and biosorption capacity was Pb(II) > Cd(II) > Cu(II) in order. The absorption capacity of Pb(II), Cd(II), and Cu(II) by *U. carpinifolia* was 201 mg g^{-1}, 80 mg g^{-1}, and 69 mg g^{-1} respectively. Similarly, the capacity of *F. excelsior* for removal of Pb(II), Cd(II), and Cu(II) was 172 mg g^{-1}, 67 mg g^{-1}, and 33 mg g^{-1} respectively observed. They observed that the concentration of metal ions and quantity of biosorbent play a vital role in the biosorption of heavy metal from the aqueous solution. The lower concentration of metal ion and the higher amount of biosorbent is more significant for the removal of Pb(II), Cd(II), and Cu(II) in an acidic medium. The natural and treated form of *Phragmites karka* was used for the removal of Hg(II) from an aqueous solution [27]. They used D-R model, Freundlich isotherm, and Langmuir isotherm model for their study. The optimum pH condition was 4 with a biosorption capacity of 1.79 mg g^{-1} for untreated and 2.27 mg g^{-1} for treated plant samples using the Langmuir adsorption isotherm model. The D-R model and Freundlich isotherm model show lower absorption of Hg(II) as compared to the Langmuir adsorption isotherm model. The contact time, agitation, and the temperature were optimized for better biosorption condition. They reported that 40 min (untreated) and 50 min (treated) of time, 150 rpm (untreated) and 100 rpm (treated) of agitation speed, 313K temperature for both untreated and treated biosorbent were an optimum condition for removal of Hg(II) from an aqueous sample.

The bark of the plant is the second most abundant part and can be used as a biosorbent for the removal of heavy metal from an aqueous solution. The bark is one of the most waste materials of wood industries. Usually, they are left in the forest after cutting the tree or used as fuel. The large biomass concentration of bark is available as the cheapest option as a biosorbent material. The use of bark in heavy metal removal from aqueous solution is a promising area for researchers and reports have already shown the capability of plant bark as biosorbent material [28]. The barks can be used in ion-exchange material and activate charcoal form for various applications with great absorption capacity [29]. The major challenge in biosorption using the bark is its floatability; the low density of bark can reduce the potential of metal ion removal from aqueous solution, however, the packed column method can be employed to overcome this problem [29]. A charcoal form of the wood disk of *Eucalyptus urograndis* was taken for efficient removal of Cu(II), Cd(II), and Ni(II) in aqueous solutions [30]. The maximum absorption capacity was ranging from 114.27 to 310.53 mg g^{-1} applying the Langmuir and Freundlich adsorption isotherm models. The precise result was obtained using the Langmuir adsorption isotherm model for the removal of Cu(II) with a removal efficiency of 96%. The ion exchange and electrostatic interaction mechanism have occurred for removal in Cu(II), Cd(II), and Ni(II) from an aqueous solution.

Fruit processing industries produce a large amount of fruit cortex as waste. The modified fruit cortex is a good source of biomass for binding and removal of heavy metals from water samples [31]. Modification of fruit cortex to biosorbent material is started with proper cleaning to remove excess pulp and then dried at 40°C. After drying, the sample is ground to a fine powder using mortar followed by chemical treatment to remove impurities [32]. Bananas, lemons, oranges, etc. are widely used in fruit processing industries as well as local juice vendors, and the waste cortex can be used as a biosorbent. A report has already shown that the modified cortex of bananas, lemons, and oranges as biosorbent can remove Cu(II), Pb(II), and Cd(II) from an aqueous solution [33]. The efficiency of the banana cortex for the removal of Cd(II) is greater than the lemon and orange cortex. This is due to the natural components such as starch, cellulose, pectin, etc. which are higher in the banana cortex as compared to the lemon and orange cortex. These types of natural components play a very important role to generate negative charges after alkaline treatment to bind with metallic cations [33].

The low-cost biosorbent material can be made from fruit peels like banana peel, orange peel, lemon peel, etc. Peel of fruits is excellent biosorbent material due to the presence of high content of cellulose, galacturonic acid, hemicelluloses, and lignin. A surface-modified co-polymerized orange peel was employed for the removal of Pb(II), Cd(II), and Ni(II) from an aqueous solution [34]. Langmuir and Freundlich adsorption isotherm models were used to measure the absorption efficiency of unmodified and modified orange peel. They found that the Langmuir adsorption isotherm model shows better efficiency of 476.1, 293.3, and 162.6 mg g^{-1} for the removal of Pb(II), Cd(II), and Ni(II) respectively with unmodified orange peel; further, the modified orange peel enhances the efficiency by several folds. The kinetic and thermodynamic data revealed that the biosorption process was pseudo-second-order kinetics, and the negative value of free energy change indicates the biosorption process is spontaneous. The potential of banana peel as biosorbent material was tested for the removal of Pb(II), Cu(II), Zn(II), and Ni(II) from aqueous media [35]. The value of the separation factor using the Langmuir and Freundlich adsorption isotherm models was between zero and one indicating the positive biosorption process for the removal of Pb(II), Cu(II), Zn(II), and Ni(II) using banana peel as biosorbent material. The lower experimental solution was 25 mg L^{-1}, and heavy metal removal efficiency was 95%, 87%, 85%, 82% for Pb(II), Cu(II), Zn(II), and Ni(II) respectively.

Some fruit parts of plants having nutshells such as peanut shells and pecan nutshells can be also used as a biosorbent material. Nutshells are common plant waste biomass which are available as low-cost and environmentally friendly biosorbent material. Pecan nutshell is examined for the effective removal of Cr(III), Fe(III), and Zn(II) from an aqueous solution [36]. The nutshell has a carboxylic (–COOH) group which turns a negative ion in acidic media (pH 4–6). The cation of metal effectively binds with this –COO– group and HNO_3 was used to regenerate the biosorbent. The maximum removal efficiency of Cr(III), Fe(III), and Zn(II) from an aqueous solution was 93.01, 76.59, and 107.9 mg g^{-1} respectively using pecan nutshell as biosorbent material. Another study on peanut shell biosorbent material was done for the removal of Cu(II) and Cr(III) from an aqueous solution [37]. They used four kinetic as well as four adsorption isotherm models to find the excellent condition for the removal of these two heavy metals from an aqueous solution. The equilibrium isotherm of the biosorption process using peanut shell as a biosorbent material shows high potential for removal of Cu(II) and Cr(III) with the value of 25 and 28 mg g^{-1} respectively.

An indigenous biosorbent material gains more attention for water treatment like the removal of turbidity and other impurities. *Moringa oleifera* is an excellent indigenous plant biosorbent which can be easily cultivated in all climatic conditions. The seed of *M. oleifera* can be used for the removal of heavy metals due to its coagulant/flocculant property. An attempt to investigate the biosorbent property of *M. oleifera* seed to remove different heavy metals was made with comparison with amine-based ligand [38]. The heavy metal uptake by the seed of *M. oleifera* was found as Pb(II) > Cu(II) > Cd(II) > Ni(II) > Mn(II) and Zn(II) > Cu(II) > Ni(II). The Fourier-transform infrared spectroscopy (FTIR) was employed to characterize the biosorption process. FTIR revealed that the seed of *M. oleifera* contains amino acid, hydroxyl, and carboxylic groups, and amino acid-metal interaction was observed which clarifies the interaction and binding of heavy metals with sorbent material. Their finding suggested that the seed of *M. oleifera* is a low-cost, eco-friendly, and domestic option for the removal of heavy metals from water samples.

There were some pieces of literature available that deal with the importance of grass and weed as biosorbent for the removal of heavy metals from aqueous solution [39, 40]. Weeds, such as *Eichhornia crassipes*, locally named water hyacinth, is a harmful weed to the water ecosystem. Additionally, these weed affects the irrigation and power generation process. The research activity related to the biological control of unconditional growth and application of *E. crassipes* has increased [39]. Cellulose was extracted from *E. crassipes* to prepare biosorbent for the removal of Cu(II) and Zn(II) ions from an aqueous solution. Langmuir, Freundlich, Temkin, and D-R equations were used to analyze the adsorption equilibrium. The sorption capacity for the removal of Cu(II) and Zn(II) from the water sample was 99 and 83 mg g^{-1} respectively. This higher efficiency shows the potential of *E. crassipes* as a biosorbent for the removal of heavy metals. Cellulose and hemicellulose are low-cost and eco-friendly options to use as biosorbent materials widely present in grasses. They are low molecular weight branched polymers with a high degree of polymerization. This plant-based material can be an excellent absorbent hydrogel for the efficient removal of heavy metals. Hemicellulose-chitosan-based absorbent hydrogel showed efficient removal of Pb(II), Cu(II), and Ni(II) from aqueous solution [40]. The source for hemicellulose was switchgrass and coastal Bermuda grass showed a high degree of polymerization. The Hemicellulose-chitosan has second-order kinetic with the efficiency of 2.90, 0.95, and 1.37 mg g^{-1} of Pb(II), Cu(II), and Ni(II) respectively in acidic media.

Some other plants like flowers and agricultural sources such as husks can also be used as biosorbents for the removal of heavy metals from aqueous media [41]. The plant-based biosorbent materials have been commonly used in the removal of toxic heavy metals from aqueous solutions. The removal of heavy metals from aqueous solution using plant-based biosorbent is rapid, cheap, effective, eco-friendly, wide range applicability, etc. The summarized form of plant-based biosorbent material for the removal of heavy metals from aqueous solution is shown in Table 12.1.

TABLE 12.1
Plant-based biosorbent materials for the removal of heavy metals from the aqueous solutions

Plant-based biosorbent	Part	Heavy metals	Biosorption capacity (mg g^{-1})	Reference
E. urograndis	Bark	Cu(II)	227	[30]
		Cd(II)	89	
		Ni(II)	13	
Coffea arabica	Husk	Cu(II)	7	[41]
		Cd(II)	7	
		Zn(II)	6	
		Cr(VI)	7	
Musa paradisiacal	Peel	Cu(II)	36	[33]
Citrus limonum		Cd(II)	67	
Citrus sinensis		Pb(II)	65	
		Cu(II)	70	
		Cd(II)	12	
		Pb(II)	78	
		Cu(II)	67	
		Cd(II)	29	
		Pb(II)	77	
U. carpinifolia	Leaves	Cu(II)	69	[26]
F. excelsior		Cd(II)	80	
		Pb(II)	201	
		Cu(II)	33	
		Cd(II)	67	
		Pb(II)	172	
P. karka	Leaves and stem	Hg(II)	2	[27]
Cercis siliquastrum	Leaves	Pb(II)	12	[42]
		Cu(II)	9	
		Ni(II)	5	
Carya illinoensis	Nutshell	Cr(III)	93	[43]
		Fe(III)	77	
		Zn(II)	108	
C. sinensis	Peel	Pb(II)	476	[34]
		Cd(II)	293	
		Ni(II)	163	
Carica papaya	Wood	Cu(II)	13	[44]
		Cd(II)	12	
		Zn(II)	9	
Arachis hypogaea	Nutshell	Cu(II)	25	[37]
		Cr(III)	28	
Oryza sativa	By-product	Cd(II)	22	[45]
		Pb(II)	12	
Camellia sinensis	Waste part	Pb(II)	2	[46]
		Fe(III)	79	
		Zn(II)	785	
		Ni(II)	515	
E. crassipes	Leaves	Cu(II)	99	[36]
		Zn(II)	83	

12.3 BACTERIAL BIOSORBENT FOR THE REMOVAL OF HEAVY METALS

There are two types of bacteria present in the environment first one is Gram-positive bacteria and the second one is Gram-negative bacteria. These bacteria are differing from each other based on the thickness of peptidoglycan layers in the cell wall [47]. Gram-positive bacteria have better efficiency to remove heavy metals due to the presence of teichoic and teichuronic acids in the larger cell wall. These two types of acids generate a huge amount of electronegative charge for the interaction and binding of heavy metal cations to the surface of Gram-positive bacteria [48]. Bacteria also have oxygen, nitrogen, and sulfur containing functional groups for binding of heavy metals that can enhance the efficiency of removal of toxic heavy metals from aqueous solutions. The major challenge in bacteria-based biosorbent is their mechanical instability as well as the separation of biosorbent after the biosorption process [49]. This problem can be solved by using the bacterial immobilization technique which enhances the removal capacity of heavy metals using a bacterial-based biosorption process. The immobilized form of bacteria can enhance the efficiency of the removal of heavy metals. Sodium alginate and activated carbon were used to immobilize *Bacillus cereus* for application in the removal of Pb(II), Cd(II), and Hg(II) from an aqueous solution [49]. The maximum efficiency, as well as mechanical strength, was observed for the removal of Pb(II) with an efficiency value of 92% in acidic media. Cyanobacteria can remove toxic contamination by metal-binding proteins in their prokaryotic cell. Two cyanobacteria namely *Nostoc linckia* and *Nostoc rivularis* were studied for the removal of Zn(II) and Cd(II) from sewage water [50]. They found that nearly 65% of both metal ions regimented in *N. linckia* and *N. rivularis* show lower metal-binding efficiency. *Bacillus laterosporus* or *Bacillus licheniformis* was isolated from metal ion-polluted soil to estimate the biosorption capability of Cd(II) and Cr(VI) from an aqueous solution [51]. The biosorption process was well studied using the Langmuir adsorption isotherm model and different factor that affect the efficiency of biosorption was optimized. The maximum biosorption efficiency for the removal of Cd(II) and Cr(VI) was 142 and 71 mg g^{-1} in aqueous media. *Streptomyces rimosus* is another important Gram-positive bacteria biomass that can be used as a biosorbent for the removal of heavy metal ions. This species does not need large modifications and has low cost as well as high efficiency as a biosorbent. The FTIR study shows that *S. rimosus* has active groups such as amide, $-NH_2$, C–O, –OH, and –COOH on its surface that contribute to the removal of heavy metals [52]. In this study, the FTIR data revealed that these active functional groups of *S. rimosus* interacted with Ni(II) and Pb(II) in aqueous media. Similarly, the carboxylic hydroxyl and amino group of *Aeromonas hydrophila* contributed to the removal of Pb(II) from the aqueous solution [53]. Another FTIR-based study of the removal of uranium by using *Bacillus amyloliquefaciens* as a biosorbent material was done [54]. A new antisymmetric vibration peak of $[O=U=O]^{2+}$ was observed in FTIR spectra that indicate the binding of the functional group with uranium. Cr(VI) was tested with *Rhodococcus* sp., *Burkholderia cepacia*, *Corynebacterium kutscheri*, and *Pseudomonas aeruginosa* [55]. The Langmuir adsorption isotherm was fitted as compared to Freundlich, and the order of affinity of bacterial-based biosorbent

TABLE 12.2
Bacteria-based biosorbent materials for the removal of heavy metals from the aqueous solutions

Bactria	Heavy metal	Biosorption efficiency (mg g^{-1})	References
B. laterosporus or	Cd(II)	143	[51]
B. licheniformis	Cr(VI)	71	
S. rimosus	Cd(II)	65	[52]
	Cr(III)	29	
	Fe(III)	125	
	Ni (II)	33	
	Pb(II)	137	
	Zn(II)	80	
B. amyloliquefaciens	U(VI)	179	[54]
Rhodococcus sp.	Cr(VI)	107	[55]
B. cereus	Pb(II)	22	[57]
Bacillus pumilus		28	
Staphylococcus xylosus	Cd(II)	250	[58]
Pseudomonas sp.	Cr(VI)	143	
	Cd(II)	278	
	Cr(VI)	95	
B. cereus	Cr(VI)	86	[59]
Bacillus subtilis	Cd(II)	210	[60]
	Hg(II)	332	
	Pb(II)	421	
Ochrobactrum anthropi	Cr(VI)	86	[61]
	Cd(II)	37	
	Cu(II)	32	

towards interaction with Cr(IV) was *Rhodococcus* sp. > *B. cepacia* > *C. kutscheri* > *P. aeruginosa*. The maximum metal uptake was 107 mg g^{-1} by *Rhodococcus sp* was recorded in an aqueous solution. Fixed and suspended forms of *Enterobacter ludwigii*, *Zoogloea ramigera*, and *Comamonas testosteroni* were demonstrated for adsorption and removal of Cu(II) and Pb(II) [56]. This study indicates that the fixed form of bacteria is more significant in the removal of heavy metals as compared to the suspended form of bacteria. The efficiency was 84% and 97% for Pb(II) and Cu(II) respectively.

The management of contaminated water is a major environmental challenge; in this circumstance, bacterium-based biosorption can be an emerged and alternative method for the removal of heavy metals. The high surface-to-volume ratio of bacteria can provide a wider surface for the interaction of metal from aqueous solutions. This significant property interacts attention of researchers to study bacterial bioaccumulation of heavy metals to find out the potential of bacteria for the treatment of metal ion-contaminated water streams. Table 12.2 shows the summarized form of bacterial-based biosorbent material for the removal of heavy metals from an aqueous solution.

12.4 ALGAL BIOSORBENT FOR THE REMOVAL OF HEAVY METALS

Algae can be classified into two types; the first is micro-algae and the second one is macro-algae also called seaweed. Micro-algae are unicellular photosynthetic plants grown in fresh or salted water. The main groups of micro-algae are diatoms, green algae, golden algae, and blue-green algae which are categorized based on their pigmentation, arrangement, and morphological character [62]. Macro-algae are divided into brown, red, and green algae; they are multi-cellular plant species that can grow in fresh or salted water. Algae can be used as natural biomass for the development of biosorbent material for the removal of heavy metals from an aqueous solution. The absorption efficiency of algal biomass is very high due to the structure of the cell wall of algae consisting of compounds such as chitin, polysaccharides, proteins, and lipids [63]. These compounds have oxygen, nitrogen, sulfur, and phosphorus-containing functional groups that contributed to the binding of heavy metals with algae.

Gracilaria corticata is a red alga have an affinity to remove Zn(II), Cu(II), Pb(II), and Cd(II) from an aqueous solution [64]. The biosorption capacity data were fitted in the Langmuir adsorption isotherm model as compared to the Freundlich adsorption isotherm model. The maximum metal uptake by *G. corticata* as a biosorbent material was 2, 1.3, 2, and 1.2 mmol g^{-1} for the removal of Cu(II), Pb(II), Zn(II), and Cd(II) respectively. The lower metal concentration is a fever condition for biosorption with a higher removal percentage. The concentration of *G. corticata* was directly proportional to the percentage of metal adsorbed. The metal uptake capability by *G. corticata* biosorbent has been well described by several analytical terms such as the ionic size of metal, active functional groups, and the approach of interaction. The morphological characteristics of *G. corticata* biosorbent show porous and mesh construction which provides a large surface area for the binding of the metal ion. The gelatinous colonies of algae named Al-VN were composed of rice paddies for the removal of Cu(II), Cd(II), and Pb(II) from an aqueous solution [65]. The absorption potential was studied by the Langmuir and Freundlich adsorption isotherm models and absorption capacity is independent of temperature. The maximum absorption efficiency of sorbent material for the removal of Cu(II), Cd(II), and Pb(II) was 28, 29, and 77 mg g^{-1} respectively, observed in an aqueous media. Algae found in gelatinous colonies were unicellular with two filaments closely related to *Cyanothece* sp. and two filamentous algae, *Leptolyngbya* sp. and *Phormidium* sp. The FTIR spectra of Al-VN show the presence of carbonyl, carboxyl, sulfide, sulfate, hydroxyl, and amine of protein. Among these groups, carboxyl, hydroxyl, sulfate, and amino are negatively charged and can interact with metal cations for binding. *Sargassum* sp. is a marine alga that is demonstrated for biosorption of Pb(II), Cu(II), and Cd(II) from an aqueous solution [66]. The chemical dynamic data shows the maximum metal uptake potential for the removal of these heavy metals by *Sargassum* sp. as a biosorbent was achieved within 10–60 min with an efficacy of greater than 90%. They correlate the electronegative and stability constant of metal hydroxide to know the order of metal uptake. Consistently, the metal uptake order of algae Pb(II) > Cu(II) > Cd(II) is reported. The maximum metal uptake was 1.2, 1.0, and 0.7 mmol g^{-1} for the removal of Pb(II), Cu(II), and Cd(II) respectively. Red macro algal species *Corallina mediterranea*, *Galaxaura oblongata*, *Jania rubens*, and *Pterocladia capillacea* were tested for the removal of Co(II), Cd(II), Cr(III), and Pb(II) ions from the aqueous solution [67]. The maximum absorption efficiency of *G. oblongata* biosorbent for Cr(III) metal ion was 105 mg g^{-1} in acidic media within 60 min using the Langmuir adsorption isotherm model.

The chief characteristic property of algae to contribute to metal ion binding is their cell wall morphology. The main composition of the cell wall of green algae is the protein which has amino, carboxyl, hydroxyl, and sulfate functional groups that contribute to heavy metal binding in the biosorption process [68]. The cell wall of red algae consists of cellulose and polysaccharides which have sulfate as a major functional group attributed to metal binding [69]. The main composition of the cell wall of brown algae is cellulose, the polymer of some aids which can easily bind with light-weight metal ions [69]. Micro-algae consist mainly of polysaccharides, proteins, and lipids, which have functional groups of negative charge on the surface of the cell wall of algae, contributing to the binding of metal cations through counter ion interactions [70]. These properties of algae make them a cheap, eco-friendly, and rapid biosorbent material for the removal of heavy metals from the aqueous solutions. Table 12.3 summarized the efficiency of some algae for metal ion removal from an aqueous solution.

12.5 FUNGAL BIOSORBENT FOR THE REMOVAL OF HEAVY METALS

Fungi also gain attention as a biosorbent material for the removal of heavy metals from the aqueous solution like other microorganisms. The main characteristic of the microorganism to be an excellent biosorbent material is their morphology of cell wall contents and functional groups [85]. Fungi are easily grown biomass by simply fermentation method as compared to plants and algae. Thus, it has more significance including cheap, rapid, eco-friendly removal of heavy metals [86]. Contents like chitin, cellulose, glucan, chitosans, polyuronides, glycoproteins, lipids, inorganic salts, and pigments are major contents of the cell wall of fungi [86]. FTIR data revealed that the fungi have nitrogen, phosphorous, and oxygen-containing functional groups that are involved in binding with heavy metal ions in the biosorption process [87].

Botrytis cinerea was pretreated by heat, detergent, sodium hydroxide, dimethyl sulfoxide, and acetic acid for the removal of Pb(II) from the aqueous sample [87]. The biosorption efficiency was ranging between 30 and 107 mg g^{-1} at pH 4 and a sorption time of 90 min. The possible mechanism of interaction of the cell wall of *B. cinerea* with Pb(II) was determined by FTIR. The vibrational data shows the presence of amine and hydroxyl functional groups within the surface of *B. cinerea* which may bind with Pb(II). *Lactarius acerrimus* belonging to the Russulaceae family that can easily be grown in woodland and the species was investigated for the removal of Hg(II) from an aquatic media [88]. The biosorption efficiency of removal of Hg(II) was measured by the Freundlich and Langmuir adsorption isotherm models. The Langmuir model was fitted and the maximum removal efficiency was 134.9 mg g^{-1} at pH 5 by *L. acerrimus* biosorbent. The FTIR spectra show the vibrational bands of amino, carboxyl, hydroxyl, thiol, and phosphate

TABLE 12.3
Algae as biosorbent materials for the removal of heavy metals from the aqueous solutions

Algae	Heavy metals	Biosorption efficiency (mg g^{-1})	Reference
Gelatinous colonies (referred to as Al-VN)	Cu(II)	28	[65]
	Cd(II)	29	
	Pb(II)	77	
Phaeodactylum sp.	Cu(II)	1.6	[71]
	Cd(II)	1.2	
	Pb(II)	1.5	
Porphyridium sp.	Cu(II)	0.3	[72]
	Cd(II)	0.4	
	Pb(II)	0.3	
Chlamydomonas reinhardtii	Cr(VI)	21	[73]
Oedogonium hatei	Cr(VI)	35	[74]
Callithamnion corymbosum	Cu(II)	65	[75]
	Co(II)	19	
	Zn(II)	37	
Maugeotia genuflexa	As(III)	57	[76]
Catenella repens	U(VI)	303	[77]
Anabaena sphaerica	Cd(II)	111	[78]
	Pb(II)	122	
C. reinhardtii	Hg(II)	106	[79]
	Cd(II)	78	
	Pb(II)	380	
O. hatei	Ni(II)	44	[80]
Gracilaria verrucosa	Zn(II)	91	[81]
Oedogonium sp.	Pb(II)	145	[82]
Nostoc sp.		93	
Desmodesmus sp.	Cu(II)	33	[83]
Chlorella vulgaris	Hg(II)	42	[84]

functional groups present in the cell wall of *L. acerrimus*. These functional groups are responsible for possible physico-chemical interaction between *L. acerrimus* biosorbent and Hg(II) in aqueous media. The aromatic ring is also present in this fungi which can provide a π-bond interaction system to bind Hg(II) for removal. Aqueous removal of Cd(II), Cu(II), and Pb(II) was studied using *Penicillium chrysogenum* and *Aspergillus ustus* fungus as a biosorbent material [89]. This study results in an excellent metal uptake property for the removal of Cd(II), Cu(II), and Pb(II) from different wastewaters. The maximum sorption efficiency for the removal of Cd(II), Cu(II), and Pb(II) was 180, 180, and 190 mg g^{-1} respectively using *P. chrysogenum*. On the other hand, the metal uptake efficiency was

190, 185, and 185 mg g^{-1} for the removal of Cd(II), Cu(II), and Pb(II) respectively by using *A. ustus* as biosorbent material. *Aspergillus* genera were dominant in the industrial area of the Raipur district of Chhattisgarh state and were studied for bioaccumulation of Pb(II) ion [90]. *Cunninghamella elegans* fungus as a biosorbent was tested for aqueous removal of Pb(II) ion [91]. The maximum removal potential of this biosorbent material was 90% for Pb(II) contamination in the aqueous solution. Another fungus *Penicillium oxalicum* was also examined for the removal of Pb(II) [92]. The tolerance limit was found to be 67 and metal uptake efficiency was 90% by *P. oxalicum* as biosorbent material.

Fungi as biosorbent can be an alternate option for the removal of toxic heavy metal contamination from the aqueous solutions. The morphology and contents of the cell wall of fungi make it an excellent biosorbent material. The summarized form of metal removal efficiency by fungi as biosorbent material is represented in Table 12.4.

TABLE 12.4
Fungi as biosorbent materials for the removal of heavy metals from the aqueous solutions

Fungi	Heavy metals	Biosorption efficiency (mg g^{-1})	Reference
L. acerrimus	Hg(II)	135	[88]
P. chrysogenum	Pb(II)	180	[89]
A. ustus	Cd(II)	180	
	Cu(II)	190	
	Pb(II)	190	
	Cd(II)	185	
	Cu(II)	185	
Phanerochaete chrysosporium	Cd(II)	28	[93]
	Pb(II)	46	
	Cu(II)	27	
Aspergillus sp.	Cu(II)	94	[94]
	Pb(II)	172	
Lactarius salmonicolor	Ni(II)	114	[95]
L. salmonicolor	Pb(II)	36	[96]
	N(II)	11	
	Cd(II)	8	
	Zn(II)	8	
Penicillium purpurogenum	As(III)	36	[97]
	Hg(II)	70	
	Cd(II)	110	
	Pb(II)	253	
Penicillium citrinum	U(VI)	127	[98]

12.6 CONCLUSIONS

The biosorption process is the ability of dead biomass of plants, algae, bacteria, and fungi to remove heavy metals from aqueous media and is measured as a more competitive, efficient, and economically attractive treatment option. While using living biomass, challenges like complexity in separation, mass loss, and difficult handling processes occur. The main purpose of this chapter is to highlight the importance, significance, characteristics, and efficacy in the recovery of heavy metals from aqueous samples. Biosorbent discussed in this chapter is based on plant, agricultural, and microbial-origin biomasses freely available in huge amounts on Earth. The chemically treated biosorbent material is more efficient as compared to untreated biomass for the removal of heavy metal contamination from water sources. Biosorption is a low-cost, natural abundance, renewable, eco-friendly, and efficient method for heavy metal removal from aqueous solutions. Many researchers have already made efforts to identify the mechanisms of metal uptake which is necessary to understand the interaction between metal ions and biosorbents. Studies revealed that biosorbents have several negatively charged functional groups that interact with cations of heavy metals for binding and removal from aqueous media. Further, even though a huge number of research articles are already available on heavy metal removal through the biosorption process, the researchers should collaborate to find out industrial-scale applications.

REFERENCES

[1] M. Jaishankar, T. Tseten, N. Anbalagan, B.B. Mathew, K.N. Beeregowda. Toxicity, mechanism and health effects of some heavy metals. *Interdisciplinary Toxicology*, 7: 60–72, 2014.

[2] A. Patle, R. Kurrey, M.K. Deb, T.K. Patle, D. Sinha, K. Shrivas. Analytical approaches on some selected toxic heavy metals in the environment and their socio-environmental impacts: a meticulous review. *Journal of the Indian Chemical Society*, 99, 100545, 2022.

[3] S. Patel, R. Jamunkar, D. Sinha, T.K. Patle, T. Kant, K. Dewangan, K. Shrivas. Recent development in nanomaterials fabricated paper-based colorimetric and fluorescent sensors: a review. Trends in Environmental Analytical Chemistry, 31, e00136, 2021.

[4] G. R. de Freitas, M.G.C. da Silva, M.G.A. Vieira. Biosorption technology for removal of toxic metals: a review of commercial biosorbents and patents. *Environmental Science and Pollution Research*, 26, 19097–19118, 2019.

[5] A. Barakat, W. Ennaji, S. Krimissa, M. Bouzaid. Heavy metal contamination and ecological-health risk evaluation in peri-urban wastewater-irrigated soils of Beni-Mellal city (Morocco). *International Journal of Environmental Health Research*, 30, 372–387, 2020.

[6] S.K. Kumbhakar, R. Chauhan, V. Singh, S.K. Jadhav, A. Quraishi. Screening of a new candidate tree legume-*Pithecellobium dulce* (Roxb.) Benth., for lead remediation. *Brazilian Journal of Botany*, 45, 929–942, 2022.

[7] S.K. Kumbhakar, R. Chauhan, S.K. Jadhav, A. Quraishi. Lead induced-toxicity in vegetables, its mitigation strategies, and potential health risk assessment: a review. *International Journal of Environmental Science and Technology*, 20, 5773–5798, 2023.

[8] C.F. Carolin, P.S. Kumar, A. Saravanan, G.J. Joshiba, M. Naushad. Efficient techniques for the removal of toxic heavy metals from aquatic environment: a review. *Journal of Environmental Chemical Engineering*, 5, 2782–2799, 2017.

[9] X. Tang, H. Zheng, H. Teng, Y. Sun, J. Guo, W. Xie, Q. Yang, W. Chen. Chemical coagulation process for the removal of heavy metals from water: a review. *Desalination and Water Treatment*, 57, 1733–1748, 2016.

[10] N.A. Qasem, R.H. Mohammed, D.U. Lawal. Removal of heavy metal ions from wastewater: a comprehensive and critical review. *NPJ Clean Water*, 4, 1–15 2021.

[11] K. Atkovska, K. Lisichkov, G. Ruseska, A.T. Dimitrov, A. Grozdanov. Removal of heavy metal ions from wastewater using conventional and nanosorbents: a review. *Journal of Chemical Technology & Metallurgy*, 53, 202–219, 2018.

[12] A.E. Burakov, E.V. Galunin, I.V. Burakova, A.E. Kucherova, S. Agarwal, A.G. Tkachev, V.K. Gupta. Adsorption of heavy metals on conventional and nanostructured materials for wastewater treatment purposes: a review. *Ecotoxicology and Environmental Safety*, 148, 702–712, 2018.

[13] M. Priyadarshanee, S. Das. Biosorption and removal of toxic heavy metals by metal tolerating bacteria for bioremediation of metal contamination: a comprehensive review. *Journal of Environmental Chemical Engineering*, 9,104686, 2021.

[14] A.A. Beni, A. Esmaeili. Biosorption, an efficient method for removing heavy metals from industrial effluents: a review. *Environmental Technology & Innovation*, 17, 100503, 2020.

[15] C. Zamora-Ledezma, D. Negrete-Bolagay, F. Figueroa, E. Zamora-Ledezma, M. Ni, F. Alexis, V.H. Guerrero. Heavy metal water pollution: a fresh look about hazards, novel and conventional remediation methods. *Environmental Technology & Innovation*, 22, 101504, 2021.

[16] D. Park, Y.S. Yun, J.M. Park. The past, present, and future trends of biosorption. *Biotechnology and Bioprocess Engineering*, 15, 86–102, 2010.

[17] A. Ali Redha. Removal of heavy metals from aqueous media by biosorption. *Arab Journal of Basic and Applied Sciences*, 27, 183–193, 2020.

[18] M. Fomina, G.M. Gadd. Biosorption: current perspectives on concept, definition and application. *Bioresource Technology*, 160, 3–14, 2014.

[19] J.C. Zheng, H.M. Feng, M.H.W. Lam, P.K.S. Lam, Y.W. Ding, H.Q. Yu. Removal of Cu (II) in aqueous media by biosorption using water hyacinth roots as a biosorbent material. *Journal of Hazardous Materials*, 171, 780–785, 2009.

[20] K. Chojnacka, M. Mikulewicz. Green analytical methods of metals determination in biosorption studies. *TrAC Trends in Analytical Chemistry*, 116, 254–265, 2019.

[21] R. Md Salim, J. Asik, M.S. Sarjadi. Chemical functional groups of extractives, cellulose and lignin extracted from native *Leucaena leucocephala* bark. *Wood Science and Technology*. 55, 295–313, 2021.

[22] F. Noli, E. Kapashi, M. Kapnisti. Biosorption of uranium and cadmium using sorbents based on Aloe vera wastes. *Journal of Environmental Chemical Engineering*, 7, 102985, 2019.

[23] S.O. Prasher, M. Beaugeard, J. Hawari, P. Bera, R.M. Patel, S.H. Kim. Biosorption of heavy metals by red algae (*Palmaria palmata*). *Environmental Technology*, 25, 1097–1106, 2004.

[24] S. Rangabhashiyam, R. Jayabalan, M. Asok Rajkumar, P. Balasubramanian. Elimination of toxic heavy metals from aqueous systems using potential biosorbents: a review. *Green Buildings and Sustainable Engineering*, 291–311, 2019.

[25] C.K. Jain, D.S. Malik, A.K. Yadav. Applicability of plant based biosorbents in the removal of heavy metals: a review. *Environmental Processes*, 3, 495–523, 2016.

[26] M.R. Sangi, A. Shahmoradi, J. Zolgharnein, G.H. Azimi, M. Ghorbandoost. Removal and recovery of heavy metals from aqueous solution using *Ulmus carpinifolia* and *Fraxinus excelsior* tree leaves. *Journal of Hazardous Materials*, 155, 513–522, 2008.

[27] M.H. Raza, A. Sadiq, U. Farooq, M. Athar, T. Hussain, A. Mujahid, M. Salman. *Phragmites karka* as a biosorbent for the removal of mercury metal ions from aqueous solution: effect of modification. *Journal of Chemistry*, 293054, 2015. https://doi.org/10.1155/2015/293054
[28] J. Khabibi, W. Syafii, R.K. Sari. Reducing hazardous heavy metal ions using mangium bark waste. *Environmental Science and Pollution Research*, 23, 16631–16640, 2016.
[29] A. Şen, H. Pereira, M.A. Olivella, I. Villaescusa. Heavy metals removal in aqueous environments using bark as a biosorbent. *International Journal of Environmental Science and Technology*, 12, 391–404, 2015.
[30] E.C. de Souza, A.S. Pimenta, A.J.F. da Silva, P.F.P. do Nascimento, J.O. Ighalo. Oxidized eucalyptus charcoal: a renewable biosorbent for removing heavy metals from aqueous solutions. *Biomass Conversion and Biorefinery*, 13, 4105–4119, 2023.
[31] K.M. Al-Qahtani. Water purification using different waste fruit cortexes for the removal of heavy metals. *Journal of Taibah University for Science*, 10, 700–708, 2016.
[32] S. Ganesan. Waste fruit cortexes for the removal of heavy metals from water. In Inamuddin, Mohd Imran Ahamed, Eric Lichtfouse, Abdullah M. Asiri (Eds.), *Green Adsorbents to Remove Metals, Dyes and Boron from Polluted Water*, 323–350, 2021. Springer, Cham.
[33] K. Kelly-Vargas, M. Cerro-Lopez, S. Reyna-Tellez, E.R. Bandala, J.L. Sanchez-Salas. Biosorption of heavy metals in polluted water, using different waste fruit cortex. *Physics and Chemistry of the Earth, Parts A/B/C*, 37, 26–29, 2012.
[34] N. Feng, X. Guo, S. Liang, Y. Zhu, J. Liu. Biosorption of heavy metals from aqueous solutions by chemically modified orange peel. *Journal of Hazardous Materials*, 185, 49–54, 2011.
[35] M.A. Ashraf, A. Wajid, K. Mahmood, M.J. Maah, I. Yusoff. Low cost biosorbent banana peel (*Musa sapientum*) for the removal of heavy metals. *Scientific Research and Essays*, 6, 4055–4064, 2011.
[36] J.C. Vaghetti, E.C. Lima, B. Royer, N.F. Cardoso, B. Martins, T. Calvete. Pecan nutshell as biosorbent to remove toxic metals from aqueous solution. *Separation Science and Technology*, 44, 615–644, 2009.
[37] A. Witek-Krowiak, R.G. Szafran, S. Modelski. Biosorption of heavy metals from aqueous solutions onto peanut shell as a low-cost biosorbent. *Desalination*, 265, 126–134, 2011.
[38] V. Obuseng, F. Nareetsile, H.M. Kwaambwa. A study of the removal of heavy metals from aqueous solutions by *Moringa oleifera* seeds and amine-based ligand 1, 4-bis [N, N-bis (2-picoyl) amino]butane. *Analytica Chimica Acta*, 730, 87–92, 2012.
[39] A. Buasri, N. Chaiyut, K. Tapang, S. Jaroensin, S. Panphrom. Biosorption of heavy metals from aqueous solutions using water hyacinth as a low cost biosorbent. *Civil and Environmental Research*, 2, 17–25, 2012.
[40] A. Ayoub, R.A. Venditti, J.J. Pawlak, A. Salam, M.A. Hubbe. Novel hemicelluloses-chitosan biosorbent for water desalination and heavy metal removal. *ACS Sustainable Chemistry & Engineering*, 1, 1102–1109, 2013.
[41] W.E. Oliveira, A.S. Franca, L.S. Oliveira, S.D. Rocha. Untreated coffee husks as biosorbents for the removal of heavy metals from aqueous solutions. *Journal of Hazardous Materials*, 152, 1073–1081, 2008.
[42] L. P. Salehi, B. Asghari, F. Mohammadi. Removal of heavy metals from aqueous solutions by *Cercis siliquastrum*. *Journal of the Iranian Chemical Society*, 5, 80–86, 2008.
[43] J.C. Vaghetti, E.C. Lima, B. Royer, N.F. Cardoso, B. Martins, T. Calvete. Pecan nutshell as biosorbent to remove toxic metals from aqueous solution. *Separation Science and Technology*, 44, 615–644, 2009.
[44] A. Saeed, M.W. Akhter, M. Iqbal. Removal and recovery of heavy metals from aqueous solution using papaya wood as a new biosorbent. *Separation and Purification Technology*, 45, 25–31, 2005.

[45] C.R.T. Tarley, M.A.Z. Arruda. Biosorption of heavy metals using rice milling by-products. Characterisation and application for removal of metals from aqueous effluents. *Chemosphere*, 54, 987–995, 2004.
[46] S.S. Ahluwalia, D. Goyal. Removal of heavy metals by waste tea leaves from aqueous solution. *Engineering in Life Sciences*, 5, 158–162, 2005.
[47] A. Zyoud, R. Alkowni, O. Yousef, M. Salman, S. Hamdan, M. H. Helal, H. S. Hilal. Solar light-driven complete mineralization of aqueous Gram-positive and Gram-negative bacteria with ZnO photocatalyst. *Solar Energy*, 180, 351–259, 2019.
[48] O. Abdi, M. Kazemi. A review study of biosorption of heavy metals and comparison between different biosorbents. *Journal of Materials and Environmental Science*, 6, 1386–1399, 2015.
[49] K. Todorova, Z. Velkova, M. Stoytcheva, G. Kirova, S. Kostadinova, V. Gochev. Novel composite biosorbent from *Bacillus cereus* for heavy metals removal from aqueous solutions. *Biotechnology & Biotechnological Equipment*, 33, 730–738, 2019.
[50] A.E. El-Enany, A.A. Issa. Cyanobacteria as a biosorbent of heavy metals in sewage water. *Environmental Toxicology and Pharmacology*, 8, 95–101, 2000.
[51] A.I. Zouboulis, M.X. Loukidou, K.A. Matis. Biosorption of toxic metals from aqueous solutions by bacteria strains isolated from metal-polluted soils. *Process Biochemistry*, 39, 909–916, 2004.
[52] M.N. Sahmoune. Performance of *Streptomyces rimosus* biomass in biosorption of heavy metals from aqueous solutions. *Microchemical Journal*, 141, 87–95, 2018.
[53] V. Javanbakht, S.A. Alavi, H. Zilouei. Mechanisms of heavy metal removal using microorganisms as biosorbent. *Water Science and Technology*, 69, 1775–1787, 2014.
[54] L. Liu, J. Liu, X. Liu, C. Dai, Z. Zhang, W. Song, Y. Chu. Kinetic and equilibrium of U (VI) biosorption onto the resistant bacterium *Bacillus amyloliquefaciens*. *Journal of Environmental Radioactivity*, 203, 117–124, 2019.
[55] G.O. Oyetibo, M.O. Ilori, O.S. Obayori, O.O. Amund. Chromium (VI) biosorption properties of multiple resistant bacteria isolated from industrial sewerage. *Environmental Monitoring and Assessment*, 185, 6809–6818, 2013.
[56] R. Black, M. Sartaj, A. Mohammadian, H.A. Qiblawey. Biosorption of Pb and Cu using fixed and suspended bacteria. *Journal of Environmental Chemical Engineering* 2, 163–1671, 2014.
[57] F. Çolak, N. Atar, D. Yazıcıoğlu, A. Olgun. Biosorption of lead from aqueous solutions by *Bacillus* strains possessing heavy-metal resistance. *Chemical Engineering Journal*, 173, 422–428, 2011.
[58] M. Ziagova, G. Dimitriadis, D. Aslanidou, X. Papaioannou, E.L. Tzannetaki, M. Liakopoulou-Kyriakides. Comparative study of Cd (II) and Cr (VI) biosorption on *Staphylococcus xylosus* and *Pseudomonas* sp. in single and binary mixtures. *Bioresource Technology*, 98, 2859–2865, 2007.
[59] P.F. Nguema, Z. Luo, J. Lian. The biosorption of Cr (VI) ions by dried biomass obtained from a chromium-resistant bacterium. *Frontiers of Chemical Science and Engineering*, 8, 454–464, 2014.
[60] T. Wang, H. Sun. Biosorption of heavy metals from aqueous solution by UV-mutant *Bacillus subtilis*. *Environmental Science and Pollution Research*, 20, 7450–7463, 2013.
[61] G. Ozdemir, T. Ozturk, N. Ceyhan, R. Isler, T. Cosar. Heavy metal biosorption by biomass of *Ochrobactrum anthropi* producing exopolysaccharide in activated sludge. *Bioresource Technology*, 90, 71–74, 2003.
[62] I. Anastopoulos, G. Z. Kyzas. Progress in batch biosorption of heavy metals onto algae. *Journal of Molecular Liquids*, 209, 77–86. 2015.
[63] T.A. Davis, B. Volesky, A. Mucci. A review of the biochemistry of heavy metal biosorption by brown algae. *Water Research*, 37, 4311–4330, 2003.

[64] A. Teimouri, S. Eslamian, A. Shabankare. Removal of heavy metals from aqueous solution by red alga *Gracilaria corticata* as a new biosorbent. *Trends in Life Science*, 5, 236–243, 2016.
[65] H.T. Tran, N.D. Vu, M. Matsukawa, M. Okajima, T. Kaneko, K. Ohki, S. Yoshikawa. Heavy metal biosorption from aqueous solutions by algae inhabiting rice paddies in Vietnam. *Journal of Environmental Chemical Engineering*, 4, 2529–2535, 2016.
[66] P.X. Sheng, Y.P. Ting, J.P. Chen. Biosorption of heavy metal ions (Pb, Cu, and Cd) from aqueous solutions by the marine alga *Sargassum* sp. in single-and multiple-metal systems. *Industrial & Engineering Chemistry Research*, 46, 2438–2444, 2007.
[67] W.M. Ibrahim. Biosorption of heavy metal ions from aqueous solution by red macroalgae. *Journal of Hazardous Materials*, 192, 1827–1835, 2011.
[68] D. Ivanova, J. Kadukova, J. Kavulicova, H. Horvathova. Determination of the functional groups in algae *Parachlorella kessleri* by potentiometric titrations. *Nova Biotechnologica et Chimica*, 11, 93–99, 2012.
[69] J. He, J.P. Chen. A comprehensive review on biosorption of heavy metals by algal biomass: materials, performances, chemistry, and modeling simulation tools. *Bioresource Technology*, 160, 67–78, 2014.
[70] O. Spain, M. Plöhn, C. Funk. The cell wall of green microalgae and its role in heavy metal removal. *Physiologia Plantarum*, 173, 526–535, 2021.
[71] R. Jalali, H. Ghafourian, Y. Asef, S. J. Davarpanah, S. Sepehr. Removal and recovery of lead using nonliving biomass of marine algae. *Journal of Hazardous Material*, 92, 253–262, 2002.
[72] D. Schmitt, A. Mu¨ller, Z. Cso¨go¨r, F. H. Frimmel, C. Posten. The adsorption kinetics of metal ions onto different microalgae and siliceous earth. *Water Research*, 35, 779–785, 2001.
[73] M.Y. Arıca, İ. Tüzün, E. Yalçın, Ö. İnce, G. Bayramoğlu. Utilisation of native, heat and acid-treated microalgae *Chlamydomonas reinhardtii* preparations for biosorption of Cr (VI) ions. *Process Biochemistry*, 40, 2351–2358, 2005.
[74] V.K. Gupta, A. Rastogi. Biosorption of hexavalent chromium by raw and acid-treated green alga *Oedogonium hatei* from aqueous solutions. *Journal of Hazardous Materials*, 163, 396–402, 2009.
[75] A.R. Lucaci, D. Bulgariu, I. Ahmad, L. Bulgariu. Equilibrium and kinetics studies of metal ions biosorption on alginate extracted from marine red algae biomass (*Callithamnion corymbosum* sp.). *Polymers*, 12, 1–16, 2020.
[76] A. Sarı, Ö.D. Uluozlü, M. Tüzen. Equilibrium, thermodynamic and kinetic investigations on biosorption of arsenic from aqueous solution by algae (*Maugeotia genuflexa*) biomass. *Chemical Engineering Journal*, 167, 155–161, 2011.
[77] S.V. Bhat, J.S. Melo, B.B. Chaugule, S.F. D'souza. Biosorption characteristics of uranium (VI) from aqueous medium onto *Catenella repens*, a red alga. *Journal of Hazardous Materials*, 158, 628–635, 2008.
[78] A.M. Abdel-Aty, N.S. Ammar, H.H.A. Ghafar, R.K. Ali. Biosorption of cadmium and lead from aqueous solution by fresh water alga *Anabaena sphaerica* biomass. *Journal of Advanced Research*, 4, 367–374, 2013.
[79] G. Bayramoğlu, I. Tuzun, G. Celik, M. Yilmaz, M.Y. Arica. Biosorption of mercury (II), cadmium (II) and lead (II) ions from aqueous system by microalgae *Chlamydomonas reinhardtii* immobilized in alginate beads. *International Journal of Mineral Processing*, 81, 35–43, 2006.
[80] V.K. Gupta, A. Rastogi, A. Nayak. Biosorption of nickel onto treated alga (*Oedogonium hatei*): application of isotherm and kinetic models. *Journal of Colloid and Interface Science*, 342, 533–539, 2010.
[81] Y. Hannachi, A. Dekhila, T. Boubakera. Biosorption potential of the red alga, *Gracilaria verrucosa* for the removal of Zn^{2+} ions from aqueous media: equilibrium,

kinetic and thermodynamic studies. *International Journal of Current Engineering and Technology*, 3, 2277–4106, 2013.

[82] V.K. Gupta, A. Rastogi. Biosorption of lead (II) from aqueous solutions by non-living algal biomass *Oedogonium* sp. and *Nostoc* sp.-a comparative study. *Colloids and Surfaces B: Biointerfaces*, 64, 170–178, 2008.

[83] L. Rugnini, G. Costa, R. Congestri, L. Bruno. Testing of two different strains of green microalgae for Cu and Ni removal from aqueous media. *Science of the Total Environment*, 601, 959–967, 2017.

[84] M. Kumar, A.K. Singh, M. Sikandar. Biosorption of Hg (II) from aqueous solution using algal biomass: kinetics and isotherm studies. *Heliyon*, 6, e03321, 2020.

[85] J.R.M. Benila Smily, P.A. Sumithra.Optimization of chromium biosorption by fungal adsorbent, *Trichoderma* sp. BSCR02 and its desorption studies. *HAYATI Journal of Biosciences*, 24, 65–71, 2017.

[86] R. Dhankhar, A. Hooda. Fungal biosorption-an alternative to meet the challenges of heavy metal pollution in aqueous solutions. *Environmental Technology*, 32, 467–491, 2011.

[87] T. Akar, S. Tunali, I. Kiran. *Botrytis cinerea* as a new fungal biosorbent for removal of Pb (II) from aqueous solutions. *Biochemical Engineering Journal*, 25, 227–235, 2005.

[88] M. Tuzen, A. Sarı, I. Turkekul. Influential bio-removal of mercury using *Lactarius acerrimus* macrofungus as novel low-cost biosorbent from aqueous solution: isotherm modeling, kinetic and thermodynamic investigations. *Materials Chemistry and Physics*, 249, 123–168, 2020.

[89] M.J. Melgar, J. Alonso, M.A. García. Removal of toxic metals from aqueous solutions by fungal biomass of *Agaricus macrosporus*. *Science of the Total Environment*, 385, 12–19, 2007.

[90] P.K. Mahish, K.L. Tiwari, S.K. Jadhav. Biodiversity of fungi from lead contaminated industrial waste water and tolerance of lead metal ion by dominant fungi. *Research Journal of Environmental Sciences*, 9, 159–168, 2015.

[91] S.K. Jadhav, P.K. Mahish. Biosorption of Lead from Aqueous Solution using a new fungal strain, *Cunninghamella elegans* TUFC20022. *Research Journal of Biotechnology*, 15, 35–42, 2020.

[92] P. K. Mahish, K.L. Tiwari, S.K. Jadhav. Biosorption of lead by biomass of resistant *Penicillium oxalicum* isolated from industrial effluent. *Journal of Applied Sciences*, 18, 41–47, 2018.

[93] R. Say, A. Denizli, M.Y. Arıca. Biosorption of cadmium (II), lead (II) and copper (II) with the filamentous fungus *Phanerochaete chrysosporium*. *Bioresource Technology*, 76, 67–70, 2001.

[94] S. Iram, R. Shabbir, H. Zafar, M. Javaid. Biosorption and bioaccumulation of copper and lead by heavy metal-resistant fungal isolates. *Arabian Journal for Science and Engineering*, 40, 1867–1873, 2015.

[95] T. Akar, S. Celik, A. Gorgulu Ari, S. Tunali Akar. Nickel removal characteristics of an immobilized macro fungus: equilibrium, kinetic and mechanism analysis of the biosorption. *Journal of Chemical Technology & Biotechnology*, 88, 680–689, 2013.

[96] G. Yan, T. Viraraghavan. Heavy-metal removal from aqueous solution by fungus *Mucor rouxii*. *Water Research*, 37, 4486–4496, 2003.

[97] R. Say, N. Yılmaz, A. Denizli. Biosorption of cadmium, lead, mercury, and arsenic ions by the fungus *Penicillium purpurogenum*. *Separation Science and Technology*, 38, 2039–2053, 2003.

[98] C. Pang, Y.H. Liu, X.H. Cao, M. Li, G.L. Huang, R. Hua, C.X. Wang, Y.T. Liu, X.F. An. Biosorption of uranium (VI) from aqueous solution by dead fungal biomass of *Penicillium citrinum*. *Chemical Engineering Journal*, 170, 1–6. 2011.

13 Application of Biosorbents in Dye Removal

Veena Thakur and Tikendra Kumar Verma

13.1 INTRODUCTION

For many years, handling and treating contaminated aqua reservoirs has been a key source of scientific concern. Numerous sectors use a lot of water and organic based chemicals, including textile, paper and paper and pulp, printing, iron-steel, coke, petroleum, insecticide, paint, solvent and pharmaceutics. Due to severe regulations imposed on the organic content of industrial effluents, industry and scientists are currently working to develop effective solutions for wastewater remediation. Pollution emissions into the environment have increased due to the recent rapid industrialization (Nigam et al., 2000). Water contamination results due to the disposal of industrial, municipal and other wastes into water bodies (Ahalya et al., 2003). Industries pollute water and soil by discharging untreated or insufficiently treated wastewater into the environment. Most manufacturing processes for synthetic colours are found in the pharmaceutical, textile, food, paper and pulp, and cosmetic industries. A significant amount of water is used during the fixing, dying and washing stages of the textile industry's dyeing process. The dyes are soluble organic substances more precisely those that fall under the categories of direct, reactive, acidic and basic. Due to their great solubility in water, dyes are more difficult to remove using standard methods (Ramachandra et al., 2005).

The effluents from industries that involve both dye production and dye use are very colourful. Such effluents also contain salts, dissolved and suspended particles, acids, alkalis, and other hazardous substances, in addition to colours (Vijayaraghavan and Yeoung-Sang, 2008). The formation of the desired aquatic life required for self-purification is inhibited by colour acquired by receiving water bodies like rivers or lake because it reduces sunlight penetration and, as a result, photosynthetic activity. Utilities located downstream to coloured water basins have substantial water treatment costs (Onal, 2006). Metal ions and colour-producing substances can combine to create very poisonous effect for fish and other aquatic life. Population growth at a rapid rate, urbanization, and industrialization increased the need for fresh water resources. By 2030, it is predicted that there will be a global water shortfall of about 40%. Water scarcity is already a major problem in India. Although only 4% of the world's fresh water resources are available, 16% of the world's population still needs to eat. It is obvious that effective fresh water use, reuse, recycling, and wastewater

DOI: 10.1201/9781003366058-13

reclamation would make things better (Manna et al., 2017). Major Indian cities produce an estimated 38,354 million litres of residential sewage each day. However, only 11,786 million litres of sewage per day receive appropriate treatment. Additionally, only 60% of the industrial wastewater produced by large-scale enterprises is reportedly handled. According to these figures, a significant amount of home and industrial wastewater is dumped into the environment without being properly treated. These activities increased the amount of dangerous substances in water that could be harmful to living things health overtime (Kaur et al., 2012) (Figure 13.1).

13.1.1 Dye

"It is defined as the compound which containing chromophore and auxochrome groups called dye. Chomophore group is responsible for dye colour due to their saturation. Auxochrome group is responsible for dye fibre reaction."

13.2 TYPES OF DYES

13.2.1 Natural Dyes

These are the dye compounds or pigments that are obtained from natural sources like minerals, invertebrates or plants. Majority of dyes are plant based i.e. they are derived from various plant parts like roots, wood, bark, leaves and flowers. There are other sources also from which the natural dyes can be obtained like fungi. They were the primary source of colours in the past. Man has utilized colouring agents for tens of thousands of years. This method has been used to modify everything from homes to pottery, food, and leather. These dyes are less expensive and easier to obtain Carmine, orcein and haematoxylin are some natural dyes that are still used today. Even today, some of the most popular dyes we use come from natural sources. Negative charges are common in natural colours. Natural dyes that are positively charged do exist, although they are uncommon. In other words, the anion is typically the coloured portion of the uncommon. In other words, the anion is typically the coloured portion of the molecule. The entire molecule is charged, despite the fact

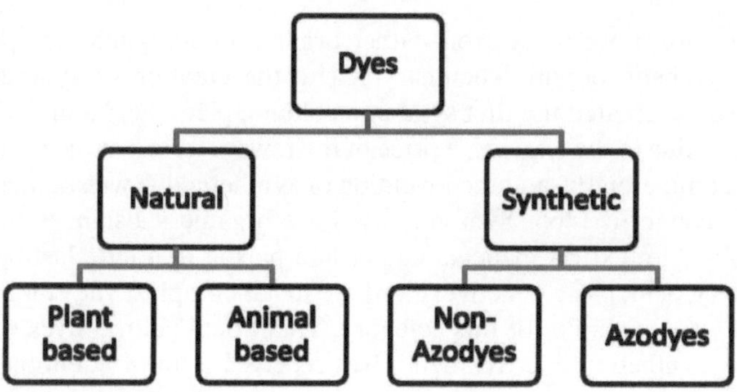

FIGURE 13.1 Classifications of dye

TABLE 13.1
Some natural dyes, their sources and examples

S. No.	Types of dye	Sources	Name of dye
1.	Natural dyes (animal derived)	Cochineal (insect)	Carmine
		Cow urine	
		Lac insect	Indian yellow
		Murex snail	Red, violet
		Octopus/cuttle fish	Purple, indigo blue
2.	Natural dyes (plant derived)	Catechu or cutch	Sepia brown
		Gamboge tree resin	Brown
		Chestnut hulls	
		Himalayan Rhubarb root	Dark mustard yellow
		Indigofera leaves	Peach to brown
		Kamala seed pods	Bronze, yellow
		Maddar root	Blue
		Mangosteen peel	Yellow
		Pomegranate rind	Red, pink, orange
		Teak leaf	Green, brown, dark brown, purple, crimson
		Weld herb	Yellow
		Juglans nigra or black walnut hulls	Crimson to maroon
		Rhus typhina or Staghorn Sumac tree	Yellow
			Brown, black, source of tannin
			Brown, sources of tannin

that the molecular charge is frequently indicated on a particular atom in structural equations. The use of a mordant is necessary for many natural dyes, but not all of them (Benkhaya et al., 2020) (Table 13.1).

13.2.2 Synthetic Dyes

Synthetic dyes are those made from either organic or inorganic compounds. The invention of synthetic organic chemistry led to the creation of synthetic colours. Woulfe and Runge created the first synthetic colours, picric acid and aurine, respectively. However, due to the expensive price of the raw materials at the time, they were not prepared commercially, and the invention of synthetic dye was known as Perkins Mauve or Mauveine in 1865. Synthetic dyes are organic substances that are used to provide colour to textiles, papers, leather and plastic in a long-lasting or permanent way. Direct, acid, basic, reactive, mordant, metal complex, vat, sulphur, disperse and other dyes are examples of this category. Traditional natural dyes were quickly superseded by synthetic dyes. Synthetic dyes received a lot of attention by 1900 as a growing demand and were eventually used in place of natural dyes for a number

TABLE 13.2
Some synthetic dyes and their examples

S. No.	Name of the synthetic dye	Examples
1.	Acid dyes	Acid yellow 36
2.	Basic or cationic dyes	Basic brown 1
3.	Direct dyes	Direct orange 36
4.	Azoic dyes	Bluish red azoic dye
5.	Nitro dyes	Maritus yellow 3
6.	Disperse dyes	Disperse red 3
7.	Vat dyes	Vat blue 4
8.	Mordant dyes	Mordant red 11
9.	Reactive dyes	Reactive blue 5
11.	Solvent dyes	Solvent yellow 32
12.	Sulphur dyes	Sulphur Red 7

of reasons, including their ease of availability, greater colour stability, resistance to light, pH changes, oxygen and other factors as well as their low cost of production (Table 13.2).

13.3 EFFECTS OF DYE IN THE ENVIRONMENT

For the survival of the planets life and the advancement of humanity, water is a crucial resource. One of the anthropogenic activities that pollutes and consumes the most water is the textile industry. Even at the low concentrations, dyes have negative environmental effects. Textile dyes seriously degrade water bodies visual appeal, raise Biological oxygen demand (BOD) and Chemical oxygen demand (COD) levels, hinder photosynthesis, stunt plant growth, infiltrate the food chain, give recalcitrant and bioaccumulation and may be poisonous, mutagenic and carcinogenic. By reducing light, colour acquired from receiving water bodies like rivers or lakes stunts the development of the aquatic life that is needed for self-purification. Decrease in photosynthetic activity as a result of less sunlight penetration. Utilities located downstream to coloured water bodies have substantial water treatment costs. Metal ions and colour-producing substances can combine to create very harmful effect for fishes and other aquatic life. Most of the technologies currently used for physicochemical mechanisms explain the removal of dyes. Physicochemical methods like dilution, coagulation, flocculation, chemical precipitation, oxidation, ion-ultra filteration and reverse osmosis are some processes used for the removal of dyes. The other processes are dependent upon the biodegradability of the dyes. Despite this, the enzymatic transformation or mineralization of these pollutants by biomasses of plants, bacteria, extremophiles and fungus is known as the bioremediation of textile dyes. Adsorption process can also be applied for the dye removal techniques (Table 13.3).

TABLE 13.3
Techniques used for treatment of wastewater for dye removal

Various techniques used for the removal of dyes from effluent:		
Physicochemical techniques	**Biological techniques**	**Advanced oxidation processes**
• Adsorption • Biosorption • Coagulation • Ion-exchange • Membrane filtration	• Aerobic treatment • Anaerobic treatment • Sequential aerobic–anaerobic treatment • Phytoremediation • Microbial remediation	• Phytolysis • Ozonation • Electrochemical oxidation • Sonolysis • Fenton Process

13.4 MECHANISM OF BIOSORPTION

Biosorption can be defined as the ability of biological materials to accumulate metal ions on the cell wall and extracellular components to absorb metals through metabolically mediated or physicochemical pathways of uptake by live or dead biomass. Biosorption is a type of adsorption in which the sorbent is a biological matrix (Michalak et al., 2013).

The biosorption process is divided into two phases. One phase is solid (biomass/sorbent/biosorbent/biological material) and the other is liquid (solvent, generally water) with a dissolved species to be sorbed (sorbate/metal ion). Because of the biosorbent's attraction for the sorbate, the sorbate is coupled with biosorbent by multiple processes, and this process continues until an equilibrium is reached between the amount of solid-bound sorbate species and its fraction remaining in solution. The sorbate's distribution between the solid and liquid phases is determined by the biosorbent's affinity for it (Ahalya et al., 2003).

In biosorption, the ions from aqueous solutions bind to functional groups which are present on the surface of biomass in a fast and reversible process. This mechanism is unaffected by cellular metabolism (Davis et al., 2003). It is a complicated process that is influenced by a variety of elements including cell physiology, physicochemical factors such as pH, temperature, contact time, ionic strength and metal concentration, metal ion chemistry and microbe cell wall composition (Gabr et al., 2008).

Biosorbents have a complex structure with multiple functional or chemical groups that bind and sequester heavy metals. Phosphodiester, phosphate, phenolic, thioether, sulfhydryl, sulfonate, imidazole, imine, hydroxyl, carbonyl, carboxyl, amine and amide groups are examples of functional groups (Park et al., 2010).

Based on cellular metabolism, the biosorption mechanism can be divided into metabolism-dependent and non-metabolism-dependent categories, while extracellular accumulation and precipitation, cell surface sorption and precipitation, and intracellular accumulation are categorized according to where the sorbate species are found (Alluri et al., 2007).

According to a variety of literary works, the living microbes may eliminate absorbent pollutants like heavy metals and dyes while dead biomass also has the ability to adsorb these pollutants (Gadd, 1990). Living cells necessitate a continuous flow of nutrients, are more susceptible to changes in pH and temperature, and require a sophisticated bioreactor system to sustain microbial culture.Therefore, non-living biomass has greater potential for large-scale applications than living microorganisms (Das et al., 2007).

The only mechanism for non-living biomass is metal binding to cell walls and external surfaces without regard to metabolism. In metabolism-independent absorption, adsorption techniques like ionic, chemical and physical adsorption are employed. Interactions include ion exchange, complexation, physical force adsorption, precipitation and trapping in interior spaces, which are the main mediators of metal or dye biosorption (Sud et al., 2008).

13.5 BIOSORPTION MODELLING OF EQUILIBRIUM ISOTHERMS

The equilibrium interactions between a solid phase (sorbent) and a liquid phase (solvent typically water) containing dissolved species of sorbate were defined by biosorption (adsorption) isotherms (e.g. metal ions). In general, the ratio of the adsorbed amount of metal ions to the remaining amount in the solution at constant temperature is expressed by the adsorption isotherm (Ahemad and Kibret, 2013).

Adsorption isotherms are shown as a plot of the residual concentration of sorbate in the solution at equilibrium state, C_{eq}, and the quantity of sorbate bound by the per unit weight of biosorbent (adsorbent), q_{eq}. One or more isotherms can be used to characterize the connection between metal ions and the biomass in the biosorption process, such as Langmuir model, Freundlich model, Temkin model, Dubinin–Radushkevich (D–R) model, Redlich–Peterson model and Brunauer–Emmett–Teller (BET) model.

The kinetic equations of these isotherms are as follows (Febrianto et al., 2009; He and Chen, 2014; Verma, 2017):

1. Langmuir isotherm: $q_e = q_{\max}\dfrac{K_L C_e}{1+K_L C_e}$

   where q_e is the amount of solute adsorbed at equilibrium condition, q_{max} is the saturated monolayer adsorption capacity, K_L is the sorption equilibrium constant and C_e is the equilibrium concentration of ions.

2. Freundlich isotherm: $q_e = K_F C_e^{1/n}$

   where q_e is the amount of solute adsorbed at equilibrium condition, K_F is the characteristic constant related to the adsorption, C_e is the equilibrium concentration of ions and n is the characteristic constant related to adsorption intensity or degree of favourability of adsorption capacity.

3. Temkin isotherm: $q_e = \dfrac{RT}{b}\ln(aC_e)$

   where q_e is the amount of solute adsorbed at equilibrium condition, R is the gas constant (0.0083 kJ (mol K)$^{-1}$), T is the absolute temperature (K), a is the Temkin isotherm constant (L g^{-1}) and C_e is the equilibrium concentration of ions.

4. D–R isotherm: $q_e = q_{\max} \exp\left(-\left(\frac{RT\ln(C_e/C_S)}{\beta E_o}\right)^2\right)$

where q_e is the amount of solute adsorbed at equilibrium condition, q_{max} is the saturated monolayer adsorption capacity, R is the gas constant (0.0083 kJ (mol K)$^{-1}$), T is the absolute temperature (K), C_e is the equilibrium concentration of ions, C_S is the sorbate solubility at a given temperature, E_o is the solid characteristic energy towards a reference compound and β is the a constant (proportional to the liquid molar volume).

5. Redlich–Peterson isotherm: $qe = \frac{K_{RP}C_e}{1 + a_{RP}C_e^{\beta}}$

where q_e is the amount of solute adsorbed at equilibrium condition, K_{RP}, a_{RP} and β are the Redlich–Paterson's parameters andC_e is the equilibrium concentration of ions.

6. BET isotherm: $q_e = q_{\max} \frac{BC_e}{(C_e - C_S)[(1 + B - 1)(C_e - C_S)]}$

where q_e is the amount of solute adsorbed at equilibrium condition, q_{max} is the saturated monolayer adsorption capacity, B is the a constant related to energy of adsorption, C_e is the equilibrium concentration of ions andC_S is the sorbate solubility at a given temperature.

The maximal absorption capacity in terms of metal removal is provided by these biosorption isotherms models, which are mathematical depictions of biosorption equilibrium. These models allow us to determine the maximal metal-removing capability of each species of biosorbent and compare their removal rates. Utilizing a comprehensive single sorbate sorption isotherm curve, the biosorption performance must be compared under identical environmental circumstances, including pH, temperature, ionic strength etc. (Volesky, 1994). The two most popular isotherms models for describing the biosorption process are the Langmuir and Freundlich models.

13.6 REMOVAL OF DYE BY BIOSORPTION

Adsorption is rapidly growing to the forefront of treatment technologies as a means of handling aqueous effluent. The possibility of low-cost regeneration, the availability of well-known process equipment, sludge-free operation and sorbate recovery are a few benefits of the adsorption process. The physicochemical process of biosorption makes use of the mechanisms of ion exchange, surface complexation, adsorption and precipitation processes. It is a natural mechanism unrelated to microbial metabolism. Biosorbents are used in biotechnology to extract organic and inorganic components from the solution. A crucial process in protecting the environment is biosorption. The passive adsorption of harmful chemicals by inert, dead or biologically produced materials is known as biosorption. The mechanisms responsible for the absorption of pollutant differ depending on the type of applied biomass. Biosorption is the result of many metabolic processes independent of the cell membrane.

13.6.1 Fungal Biosorbents for the Removal of Dyes

Many researchers have reported different types of fungal biomasses for dye removal from industrial wastewater under several conditions. Takey (2014) using dead biomass of fungus *Aspergillus flavus* for biosorption of three toxic components such as methyl orange, chromuium and lead of textile industry effluent and found to be 53.62%, 72.18% and 76.12% of removal, respectively, under different parameter conditions.

Chew and Ting (2015) studied dye-removal potential of common isolate *Trichoderma asperellum* on four triphenylmethane dyes. The four dyes crystal violet (CV), methyl violet (MV), cotton blue (CB) and malachite green (MG) showed the absorption of 12.97, 12.54, 14.34 and 11.44 mg g–1 by free cells, respectively, and 60.64, 50.29, 49.91 and 16.61 mg g–1 by alginate-immobilized forms, respectively. The biosorption of brilliant green and methylene blue (MB) dyes in binary mixture using *Saccharomyces cerevisiae* is studied by Ghaedi et al. (2013). Kabbout and Taha (2014) showed the ability of dead fungal biomass of *Aspergillus fumigatu* for the biosorption of MB in industrial coloured effluents and optimized different parameter conditions for better adsorption. The maximum percentage removal of MB is 93.43% in this study.

Khelifi et al. (2015) showed nonviable *Aspergillus alliaceus* to remove the textile indigo dye from aqueous solution with different parameter. In batch experiment, they founded 99% of dye removed at pH 4 and 240 min. of incubation. The *Aspergillus niger* strain was tested for azo dye under different environmental conditions and by Mahmoud et al. (2017). The highest capacity of the removal of azo dye was found at pH 9, agitation speed 250, biosorbent dose 3.5 g and 120 min of contact time.

Ghariani et al. (2019) were used heat treated-lyophilized fungal biomass of *Neonectria radicicola* for the removal of the three acid dyes, Acid Orange 51 (AO 51), Reactive Red 75 (RR 75) and Direct Blue 86 (DB 86), from aqueous solutions. Under different environmental conditions, the maximum adsorption was found as 120.6, 75.37 and 200.5 L g^{-1} for AO 51, RR 75 and DB 86, respectively.

The capacity of inactive fungal biomass of *Phoma* sp. was evaluated by Drumm et al. (2019). In batch experiment, the maximum value of biosorption capacity was 63.58 mg g^{-1} under different parameter for the removal of Acid Red 18 (AR 18) dye. Pecková et al. (2020) reported three different white rot fungi namely *Pycnoporus cinnabarinus*, *Pleurotus ostreatus* and *Trametes hirsute* for the mono azo dye Allura Red AC (AR) removal from the aqueous solutions. The maximum sorption capacity of 118.3±9.9 mg g^{-1} was found by heat treated-dried biomass of *P. ostreatus* (Table 13.4).

13.6.2 Algal Biosorbents for the Removal of Dyes

Ozer et al. (2006) studied the removal of AR 274 dye in a batch mode as a function of the initial pH at 3.0, temperature at 30°C and algae concentrations at 0.5 g L^{-1} by *Spirogyra rhizopus*, and in this optimal conditions, up to 99% removal of dye was obtained. Daneshvar et al. (2007) reported a green microalgae *Cosmarium* species for the removal of MG, and they also obtained *Cosmarium* species removed

TABLE 13.4
Some fungal species used as biosorbent for the removal of dyes from wastewater

S. No.	Fungal biosorbent	Dye	References
1.	*Agaricus bisporus*	RR 2	Akar and Divriklioglu (2010)
2.	*Aspergillus lentulus*	Acid navy blue, orange High fast (HF), fast red A, acid sulphone blue and acid magenta	Kaushik and Malik (2010)
3.	*Aspergillus versicolor*	Reactive Black 5	Huang et al. (2016)
4.	*Aspergillus iizukae*	Remazol brilliant blue R dye	Noman et al. (2020)
5.	*Ceriporia lacerata*	CV	Lin et al. (2011)
6.	*Coriolopsis* sp.	CV, MV, CB, MG	Chen and Ting (2015)
7.	*Cunninghamella elegans*	Acid Blue 62, AR 266 and Acid Yellow 49	Russo et al. (2010)
8.	*Mucor circinelloides*	Congo red	Azin and Moghimi (2018)
9.	*Nigrospora sp.*	Procion red H-E7B dye	Tonato et al. (2019)
10.	*Penicillium restrictum*	Reactive yellow 145	Caner et al. (2011)

maximum up to 89.1% in their experimental results. Kousha et al. (2012) studied the effects of different individual parameters such as biomass dosage, dye concentration and pH of initial concentration by three brown macroalgae on decontamination of Acid Black 1 (AB1) dye and the maximum removal efficiency of 99.27%, 98.12% and 97.62% for *Nizamuddin zanardini, Sargassum glaucescens* and *Stoechospermum marginatum*, respectively, was observed.

Khataee et al. (2013) reported that green alga *Spirogyra* sp. was efficient biosorbent for the removal of AO 7, Basic Red 46 (BR46) and Basic Blue 3 (BB3) dyes, and they examined that the biosorption capacity were 13.2, 12.2 and 6.2 mg g^{-1} for BR46, BB3 and AO7, respectively. Vijayaraghavan et al. (2016) examined the biosorption of MB from aqueous solutions by *Gracilaria corticata* (red seaweed) and they found highest biosorption capacity of 28.9 mg g^{-1} of biomass was observed at pH 8. Gul et al. (2019) investigated the dye biosorption capacities of cyanobacteria (*Phormidium animale*) and microalgae (*Scenedesmus* sp.) to remove Acid Red P-2BX (ARP-2BX) dye. Their results showed that dried *P. animale* had a maximum dye removal efficiency of 99.70 ± 0.27%.

Bonyadi et al. (2022) studied on blue green algae *Spirulina platensis* for the removal of MG dye and they reported the maximum dye removal 94.12% under dye concentration 100.54 mg L^{-1}, pH 7.57, contact time 52.43 min and biosorbent dose 0.98 g L^{-1} was obtained. Nielsen et al. (2022) examined the different *Sargassum* species (*S. muticum, S. fluitans, S. natans* I and *S. natans* VIII) for the removal of MB up to 93% and they also concluded that *Sargassum* biomass is suitable for cationic dyes removal but without chemical modification it may not be suitable for anionic substances (Table 13.5).

TABLE 13.5
Some algal species used as biosorbent for the removal of dyes from wastewater

S. No.	Algal biosorbent	Dye	References
1.	*Chlorella pyrenoidosa*	Direct Red-31	Sinha et al. (2016)
2.	*C. pyrenoidosa*	Rhodamine B dye	da Rosa et al. (2018)
3.	*Chlorella vulgaris*	Supranol Red 3BW, Lanaset Red 2 GA and Levafix Navy Blue EBNA (Epstein–Barr virus nuclear antigen 1)	Lim et al. (2010)
4.	*Enteromorpha prolifera*	Direct Fast Scarlet 4BS	Sun et al. (2019)
5.	*Fucus vesiculosus*	MB	Lebron et al. (2019)
6.	*Laminaria japonica*	MB	Shao et al. (2017)
7.	*Sargassum dentifolium*	MB	Moghazy et al. (2019)
8.	*S. marginatum*	Acid Blue 25, AO 7 and AB1	Daneshvar et al. (2012)
9.	*Ulothrix* sp.	MB	Dogar et al. (2010)
10.	*Ulva fasciata*	MB	Moghazy et al.(2019)

13.6.3 Plant and Agriculture Waste Biosorbents for the Removal of Dyes

Both unprocessed agriculture solid wastes and forestry industry byproducts like sawdust and bark have been used as biosorbents. Due to their physicochemical properties and inexpensive cost, these materials are readily available and may have potential as sorbents (Kurniawan, 2006). Many plants and agricultural wastes have been used as inexpensive adsorbents to remove different dyes from aqueous solutions. Some of the biomass may be used in a wastewater treatment process at an industrial size, whereas other biomass may only be used at a laboratory scale. The greatest barrier to adopting their alternative adsorbents for the industrial-scale treatment of wastewater containing dyes is the low adsorption capacity of agricultural and plant waste material (Kamsonlian et al., 2012). To improve the adsorption property of the agricultural wastes, various techniques such as physical and chemical treatments are used. Activated carbons or activated biochar are porous materials that are created when lignocellulosic materials are thermally modified. This activated carbon was typically combined with other materials to create novel composite materials with high adsorption performance or capacity in order to increase its adsorption capacity (Tran et al., 2017) (Table 13.6).

13.6.4 Bacterial Biosorbents for the Removal of Dyes

Numerous microorganisms have been identified as promising biosorbents for the removal of dyes, metals, aromatic compounds, insecticides and other contaminants. They are non-hazardous, regenerable and have a large capacity for pollution uptake. They also have low operating costs. Many metals, dye particles and other

TABLE 13.6
Some agricultural wastes and plant based-biosorbent used for the removal of dyes from wastewater

S. No.	Adsorbent	Dye	References
1.	Banana peel	Methyl orange BB 9 Basic violet 10	Annadurai et al. (2002)
2.	Raw date pits	BB 9	Banat et al. (2003)
3.	Fly ash	BB 9	Janos et al. (20032004)
5.	Bagasse	BR 22	Juang and Tseng (2002)
6.	Corncob	BR 22	Juang and Tseng (2002)
7.	Sugarcane bagasse	AO 10	Gao et al. (2003)
8.	Sugarcane bagasse	Methyl red	Saad et al. (2010)
9.	Groundnut shell biochar	BR 09	Praveen et al. (2021)
10.	Coconut shell biochar	BR 09	Praveen et al. (2021)
11.	Rice husk biochar	BR 09	Praveen et al. (2021)

TABLE 13.7
Some bacterial species used as biosorbents for dye removal from wastewater

S. no	Bacteria	Dye	Reference
1.	*Corynebacterium glutamicum*	Reactive black RR	Vijayaraghavan and Yun (2007a) Won et al. (2005)
2.	*Streptomyces rimosus*	MB	Nacèra and Aicha (2006)
3.	Bacterial consortium (mixture of *Ochrobactrum sp., Salmonella enterica* and *Pseudomonas aeruginosa*)	Reactive Black-B	Kilic et al. (2007)
4.	*Pseudomonas putida* (NCIMB 9776)	Tectilon Blue	Walker and Weatherley (2000)
5.	Streptomyces BW130	Azo-copper Red 171	Zhou and Zimmermann (1993)
6.	*Pseudomonas luteola*	Red G	Hu (1994)
7.	Pseudomonas cepacia 13NA (immobilized)	p-Aminoazobenzene	Ogawa and Yatome (1990)
8.	Mixed anaerobic culture	Diazo-linked chromophores	Knapp and Newby (1999)
9.	*S. rimosus*	MB	Nacera and Aicha (2006)

contaminants can be used by pure and mixed cultures of bacteria as a source of carbon or nitrogen. Researchers claimed that the components responsible for the bacterial cell wall's ability to bind metals were the anionic functional groups found in the peptidoglycan, phospholipids and lipopolysaccharides of gram-negative bacteria as well as the peptidoglycan, teichoic acids, of gram-positive bacteria (Beveridge,

1999). Additionally, extracellular polysaccharides can bind ions. However, they are easily separated from one another by mechanical friction or chemical cleaning. The ions or solutes in the case of bacterial biosorption can either be adhered to the surface of the bacterial cell or deposited with in the cell wall structure. Numerous dyes molecules, which separate into cations in liquids, are drawn to negatively charged groups, particularly carboxyl groups, and these charged molecules get attached to the bacterial cell wall by hydrogen bonding and electrostatic interactions (Vijayaraghavan and Yun, 2007a, b). Few of the bacteria are listed below which have been used for as biosorbents for the removal of dye from the wastewater (Table 13.7).

13.7 CONCLUSION

A comprehensive overview of the removal and degradation of dyes from wastewater utilizing traditional removal techniques and biomass-based adsorbents was provided in the chapter with a focus on the various bioadsorbents such as biomass from fungi, bacteria, algae and agricultural wastes. This chapter has also covered the mechanisms, kinetics and models of these types of biosorption processes. It is clear that a biomass-based colour removal technique based on bacteria, fungi and algae could be a promising and environmentally benign alternative to the process currently in use.

REFERENCES

Ahalya, N., Ramachandra, T. V., Kanamadi, R. D., 2003. Biosorption of heavy metals. *Res. J. Chem. Environ.* 7 (4), 71–79.

Ahemad, M., Kibret, M., 2013. Recent trends in microbial biosorption ofheavy metals: A review. *Biochem. Mol. Biol.* 1(1), 19–26.

Akar, T., Divriklioglu, M., 2010. Biosorption applications of modified fungal biomass for decolorization of Reactive Red 2 contaminated solutions: batch and dynamic flow mode studies. *Bioresour. Technol.* 101, 7271–7277.

Alluri, H. K., Ronda, S.R., Settalluri, V. S., Bondili, J. S., Suryanarayana, V., Venkateshwar, P., 2007. Biosorption: an eco-friendly alternative for heavy metal removal. *Afr. J. Biotechnol.* 6(25), 2924–2931.

Annadurai, G., Juang, R. S., Lee, D. J., 2002. Use of cellulose-based wastes for adsorption of dyes from aqueous solutions. *J. Hazard Mater.: B* 92, 263–274.

Azin, E., Moghimi, H., 2018. Efficient mycosorption of anionic azo dyes by Mucor circinelloides: surface functional groups and removal mechanism study. *J. Environ. Chem. Eng.* 6, 4114–4123.

Banat, F., Al-Asheh, S., Al-Makhadmeh, L., 2003. Evaluation of the use of raw activated date pits as potential adsorbents for dye containing waters. *J. Chem. Technol. Biotech.* 54, 192–196.

Benkhaya, S., Mrabet, S., Elharfi, A., 2020. A review on classifications, recent synthesis and applications of textile dyes. *Inorg. Chem. Commun.* 115, 107891.

Beveridge, T.J., 1999. Structures of Gram-negative cell walls and their derived membrane vesicles. *J. Bacteriol.* 181, 4725–4733.

Bonyadi, Z., Nasoudari, E., Ameri, M., et al., 2022. Biosorption of malachite green dye over *Spirulina platensis* mass: process modeling, factors optimization, kinetic, and isotherm studies. *Appl. Water Sci.* 12, 167.

Caner, M., Kiran, I., Ilhan, S., Pinarbasi, A., Iscen, C.F., 2011. Biosorption of reactive yellow 145 dye by dried penicillum restrictum: isotherm, kinetic, and thermodynamic studies. *Separ. Sci. Technol.* 46, 2283–2290.

Chen, S.H., Ting, A.S.Y., 2015. Biodecolorization and biodegradation potential of recalcitrant triphenylmethane dyes by *Coriolopsis* sp. isolated from compost. *J. Environ. Manage.* 150, 274–280.

Chew, S. Y., Ting, A. S. Y., 2015. Common filamentous *Trichoderma asperellum* for effective removal of triphenylmethane dyes. *Desalin. Water Treat.* 57(29), 13534–13539.

da Rosa, A.L.D., Carissimi, E., Dotto, G.L., Sander, H., Feris, L.A., 2018. Biosorption of rhodamine B dye from dyeing stones effluents using the green microalgae *Chlorella pyrenoidosa*. *J. Clean. Prod.* 198, 1302–1310.

Daneshvar, E., Kousha, M., Sohrabi, M.S., Khataee, A., Converti, A., 2012. Biosorption of three acid dyes by the brown macroalga Stoechospermum marginatum: isotherm, kinetic and thermodynamic studies. *Chem. Eng. J.* 195–196, 297–306.

Daneshvar, N., Ayazloo, M., Khataee, A.R., Pourhassan, M., 2007. Biological decolorization of dye solution containing Malachite Green by microalgae *Cosmarium* sp. *Bioresour. Technol.* 98(6), 1176–1182.

Das, N., Charumathi, D., Vimala, R., 2007. Effect of pretreatment on Cd^{2+} biosorption by mycelial biomass of Pleurotus florida. *Afr. J. Biotechnol.* 6(22), 2555–2558.

Davis, T., et al., 2003. A review of the biochemistry of heavy metal biosorption by brown algae. *Elsevier*, 37, 4311–4330. https://doi.org/10.1016/S0043–1354(03)00293-8.

Dogar, C., Gurses, A., Acıkyıldız, M., Ozkan, E., 2010. Thermodynamics and kinetic studies of biosorption of a basic dye from aqueous solution using green algae *Ulothrix* sp. *Colloids Surf. B Biointerfaces* 76, 279–285.

Drumm, F.C., Grassi, P., Georgin, J., Tonato, D., Pfingsten Franco, D.S., Chaves Neto, J.R., Mazutti, M.A., Jahn, S.L., Dotto, G.L., 2019 Nov. Potentiality of the *Phoma* sp. inactive fungal biomass, a waste from the bioherbicide production, for the treatment of colored effluents. *Chemosphere* 235, 596–605.

Febrianto, J., Kosasih, A.N., Sunarso, J., Ju, Y.H., Indraswati, N., Ismadji, S., 2009 quilibrium and kinetic studies in adsorptionof heavy metals using biosorbent: a summaryof recent studies. *J. Hazard. Mater.* 162, 616–645.

Gabr, R.M., Hassan, S.H.A., Shoreit, A.A.M., 2008. Biosorption of lead andnickel by living and non-living cells of *Pseudomonas aeruginosa* ASU 6a. *Int. Biodeter. Biodegrad.* 62, 195–203.

Gadd, G.M., 1990. Heavy metal accumulation by bacteria and other microorganisms. *Experientia* 46, 834–840.

Gao, Y., Yang, S., Fu, W., Qi, J., Li, R., Wang, Z., Xu, H., 2003. Adsorption of malachite green on micro- and mesoporous rice husk based active carbon. *Dyes Pig.* 56, 219–229.

Ghaedi, M., Hajati, S., Barazesh, B., Karimi, F., Ghezelbash, G., 2013. Saccharomyces cerevisiae for the biosorption of basic dyes from binary component systems and the high order derivative spectrophotometric method for simultaneous analysis of Brilliant green and Methylene blue. *J. Ind. Eng. Chem.* 19(1), 227–233.

Ghariani, B., Hadrich, B., Louati, I., et al., 2019. Porous heat-treated fungal biomass: preparation, characterization and application for removal of textile dyes from aqueous solutions. *J. Porous Mater.* 26, 1475–1488.

Gul, U.D., Tastan B.E., Bayazıt, G., 2019. Assessment of algal biomasses having different cell structures for biosorption properties of acid red P-2BX dye. *S. Afr. J. Bot.* 127, 147–152.

He, J., Chen, J.P., 2014. A comprehensive review on biosorption of heavy metals by algal biomass: materials, performances, chemistry and modeling simulation tools. *Bioresour. Technol.* 160, 67–78.

Hu, T.L., 1994. Decolorization of reactive azo dyes by transformation with Pseudomonas luteola. *Bioresour. Technol.* 49, 47–51.

Huang, J., Liu, D., Lu, J., Wang, H., Wei, X., Liu, J., 2016. Biosorption of reactive black 5 by modified Aspergillus versicolor biomass: kinetics, capacity and mechanism studies. *Colloid Surf. Phys. Eng. Aspect.* 492, 242–248.

Janos, P., Buchtová, H., Rýznarová, M., 2004. Sorption of dyes from aqueous solutions onto fly ash. *Water Res.* 37(20), 4938–4944. https://doi.org/10.1016/j.watres.2003.08.011

Juang, R.S., Wu, F.C., Tseng, R.L., 2002. Characterization and use of activated carbon prepared from bagasse for liquid-phase adsorption. *J. Colloid Surf. 17: Physicochem. Eng. Aspect* 201 191–199.

Kabbout, R., Taha, S., 2014. Biodecolorization of textile dye effluent by biosorption on fungal biomass materials. *Phys. Proc.* 55, 437–444.

Kamsonlian, S., Suresh, S., Ramanaiah, V., Majumder, C., Chand, S., Kumar, A., 2012. Biosorptive behaviour of mango leaf powder and rice husk for arsenic (III) from aqueous solutions. *Int. J. Environ. Sci. Technol.* 9, 565–578.

Kaur, R., Wani, S.P., Singh, A.K., Lal, K., 2012. Wastewater production, treatment and use in India. *National report presented at the 2nd regional workshop on safe use of wastewater in agriculture*, May 16–18, 2012, New Delhi, India.

Kaushik, P., Malik, A., 2010. Alkali, thermo and halo tolerant fungal isolate for the removal of textile dyes. *Colloids Surf. B: Biointerfaces* 81, 321–328.

Khataee, A.R., Vafaei, F., Jannatkhah, M., 2013. Biosorption of three textile dyes from contaminated water by filamentous green algal Spirogyra sp.: kinetic, isotherm and thermodynamic studies. *Int. Biodeter. Biodegrad.* 83, 33–40.

Khelifi, E., Touhami, Y., Bouallagui, H., Hamdi, M., 2015. Biosorption of indigo from aqueous solution by dead fungal biomass *Aspergillus alliaceus*. *Desalin. Water Treat.* 53(4), 976–984.

Kilic, N.K., Nielson, J.L., Yuce, M., Donmez, G., 2007. Characterization of a simple bacterial consortium for effective treatment of wastewaters with reactive dyes and Cr(VI). *Chemosphere* 67(4), 826–831.

Knapp, J.S., Newby, P.S., 1999. The decolorization of a chemistry industry effluent by white-rot fungi. *Water Res.* 33(2), 575–577

Kousha, M., Daneshvar, E., Dopeikar, H., Taghavi, D., Bhatnagar, A., 2012. Box–Behnken design optimization of Acid Black 1 dye biosorption by different brown macroalgae. *Chem. Eng. J.* 179, 158–168.

Kurniawan, T.A., 2006. Comparison of low-cost adsorbents for treating wastewater laden with heavy metals. *Sci. Total Environ.* 366(3), 407–424.

Lebron, Y.A.R., Moreira, V.R., Santos, L.V.S., 2019. Studies on dye biosorption enhancement by chemically modified Fucus vesiculosus, Spirulina maxima and Chlorella pyrenoidosa algae. *J. Clean. Prod.* 240, 118–197.

Lim, S.L., Chu, W.L., Phang, S.M., 2010. Use of Chlorella vulgaris for bioremediation of textile wastewater. *Bioresour. Technol.* 101, 7314–7322.

Lin, Y., He, X., Han, G., Tian, Q., Hu, W., 2011. Removal of Crystal Violet from aqueous solution using powdered mycelial biomass of Ceriporia lacerata P2. *J. Environ. Sci.* 23(12), 2055–2062.

Mahmoud, M.S., Mostafa, M.K, Mohamed, S.A., Sobhy, N.A., Mahmoud, N., 2017. Bioremediation of red azo dye from aqueous solutions by *Aspergillus niger* strain isolated from textile wastewater. *J. Environ. Chem. Eng.* 5(1), 547–554.

Manna, S., Roy, D., Saha, P., Gopakumar, D., Thomas, S., 2017. Rapid methylene blue adsorption using modified lignocellulosic materials. *Process Saf. Environ. Prot.* 107, 346–356. https://doi.org/10.1016/j.psep.2017.03.008

Michalak, I., Chojnacka, K., Witek-Krowiak, A., 2013. State of the art for the biosorption process – a review. *Appl. Biochem. Biotechnol.* 170(6), 1389–1416.

Moghazy, R.M., Labena, A., Husien, S., 2019. Eco-friendly complementary biosorption process of methylene blue using micro-sized dried biosorbents of two macro-algal species (Ulva fasciata and Sargassum dentifolium): full factorial design, equilibrium, and kinetic studies. *Int. J. Biol. Macromol.* 134, 330–343.

Nacera, Y., Aicha, B., 2006. Equilibrium and kinetc modeling of Methylene Blue biosorption by pretreated dead Streptomyces rimosus: effect of temperature. *Chem. Eng. J.* 119, 121–125.

Nielsen, B.V., Maneein, S., Anghan, J.D., Anghan, R.M., Al Farid, M.M., Milledge, J.J., 2022. Biosorption potential of sargassum for removal of aqueous dye solutions. *Appl. Sci.* 12, 4173.

Nigam, P., Armour, G., Banat, I.M., Singht, D., Marchant, R., 2000. Physical removal of textile dyes from effluents and solid state fermentation of dye-adsorbed agricultural residues. *Bioresour. Technol.* 72, 219–226.

Noman, E., Al-Gheethi, A.A., Talip, B., Mohamed, R., Kassim, A.H., 2020. Mycoremediation of Remazol Brilliant Blue R in greywater by a novel local strain of *Aspergillus iizukae* 605EAN: optimisation and mechanism study. *Int. J. Environ. Anal. Chem.* 100(14), 1650–1668.

Ogawa, T., Yatome, C., 1990. Biodegradation of azo dyes in multistage rotating biological contactor immobilized by assimilating bacteria. *Bull. Environ. Contam. Toxicol.* 44, 561–566.

Onal, M., 2006. Determination of chemical formula of a smectite. *Commun. Fac. Sci.* 52(2), 1–6.

Ozer, A., Akkaya, G., Turabik, M., 2006. The removal of Acid Red 274 from wastewater: combined biosorption and biocoagulation with *Spirogyra rhizopus*. *Dyes Pig.* 71(2), 83–89.

Park, D., Yun, Y.-S., Park, J.-M., 2010. The past, present, and future trends of biosorption. *Springer*, 15(1), 86–102. https://doi.org/10.1007/s12257-009-0199-4.

Pecková, V., Legerska, B., Chmelová, D., Horník, M., Ondrejovič, M., 2020. Comparison of efficiency for monoazo dye removal by different species of white-rot fungi. *Int. J. Environ. Sci. Technol.* 18. https://doi.org/10.1007/s13762-020-02806-w.

Praveen, S., Gokulan, R., Pushpa, T.B., Jegan, J., 2021. Techno-economic feasibility of biochar as biosorbent for basic dye sequestration. *J. Indian Chem. Soc.* 98 Article 100107.

Ramachandra, T., Ahalya, N., Kanamadi, R., 2005. Biosorption: techniques and mechanisms. *CES Technical Report 110*. Centre for Ecological Sciences, Indian Institute of Science Bangalore.

Russo, M.E., Di Natale, F., Prigione, V., Tigini, V., Marzocchella, A., Varese, G.C., 2010. Adsorption of acid dyes on fungal biomass: equilibrium and kinetics characterization. *Chem. Eng. J.* 162, 537–545.

Saad, S.A., Md. Isa, K. Bahari, R., 2010. Chemically modified sugarcane bagasse as a potentially low-cost biosorbent for dye removal. *Desalination* 264, 123–128.

Shao, H., Li, Y., Zheng, L., Chen, T., Liu, J., 2017. Removal of methylene blue by chemically modified defatted brown algae Laminaria japonica. *J. Taiwan Inst. Chem. Eng.* 80, 525–532.

Sinha, S., Singh, R., Chaurasia, A.K., Nigam, S., 2016. Self-sustainable Chlorella pyrenoidosa strain NCIM2738 based photo bioreactor for removal of Direct Red-31 dye along with other industrial pollutants to improve the water-quality. *J. Hazard Mater.* 306, 386–394.

Sud, D., Mahajan, G., Kaur, M.P., 2008 Sep. Agricultural waste material as potential adsorbent for sequestering heavy metal ions from aqueous solutions – a review. *Bioresour. Technol.* 99(14), 6017–6027. https://doi.org/10.1016/j.biortech.2007.11.064. Epub 2008 Feb 14. PMID: 18280151.

Sun, W., Sun, W., Wang, Y., 2019. Biosorption of Direct Fast Scarlet 4BS from aqueous solution using the green-tide-causing marine algae Enteromorpha prolifera. *Spectrochim. Acta Mol. Biomol. Spectrosc.* 223, 117347.

Takey, M., 2014. Bioremediation of xenobiotics: use of dead fungal biomass as biosorbent. *Int. J. Res. Eng. Technol.* 03, 565–570. https://doi.org/10.15623/ijret.2014.0301094.

Tonato, D., Drumm, F.C., Grassi, P., Georgin, J., Gerhardt, A.E., Dotto, G.L., Mazutti, M.A., 2019. Residual biomass of Nigrospora sp. from process of the microbial oil extraction for the biosorption of procion red H–E7B dye. *J. Water Process Eng.* 31, 100818.

Tran, H.N., You, S.-J., Chao, H.-P., 2017. Activated carbons from golden shower upon different chemical activation methods: synthesis and characterizations. *Adsorp. Sci. Technol.* https://doi.org/10.1177/0263617416684837.

Verma, T.K., 2017. *Screening of Fungal Species for Biosorption of Iron in Industrial Effluents.* http://hdl.handle.net/10603/220831.

Vijayaraghavan, J., Bhagavathi Pushpa, T., Sardhar Basha, S.J., Jegan, J., 2016. Isotherm, kinetics and mechanistic studies of methylene blue biosorption onto red seaweed *Gracilaria corticata*. *Desalin. Water Treat.* 57:29, 13540–13548.

Vijayaraghavan, K., Yeoung-Sang, Y., 2008. Bacterial biosorbents and biosorption. *Biotechnol. Adv.* 26, 266–291.

Vijayaraghavan, K., Yun, Y.S., 2007a. Utilization of fermentation waste (Corynebacterium glutamicum) for biosorption of Reactive Black 5 from aqueous solution. *J. Hazard. Mater.* 141, 45–52.

Vijayaraghavan, K., Yun, Y.S., 2007b. Chemical modification and immobilization of Corynebacterium glutamicum for biosorption of Reactive Black 5 from aqueous solution. *Ind. Eng. Chem. Res.* 46(2), 608–617.

Volesky, B., 1994. Advances in biosorption of metals: selection of biomass types. *FEMS Microbiol. Rev.* 14, 291–302.

Walker, G.M., Weatherley, L.R., 2000. Biodegradation and biosorption of acid anthraquinone dye. *Environ. Pollut.* 108(2), 219–223.

Won, S.W., Choi, S.B., Chung, B.W., Park, D., Park, J.M., Yun, Y.S., 2005. Biosorptive decolorization of reactive orange 16 using the waste biomass of *Corynebacterium glutamicum*. *Ind. Eng. Chem. Res.* 2004, 43, 24, 7865–7869.

Zhou, W., Zimmermann, W., 1993. Decolorization of industrial effluents containing reactive dyes by actinomycetes. *FEMS Microbiol. Lett.* 107, 157–162.

14 Application of Biosorbents in Separation of Radionuclides

Elyor Berdimurodov, Khasan Berdimuradov, Ilyos Eliboyev, Dakeshwar Kumar Verma, and Omar Dagdag

14.1 INTRODUCTION

14.1.1 Separation of Radionuclides

The radionuclides are dangerous elements for human and animal health. The radionuclides are not stable, causing serious cancer illness. The spread of radionuclides has occurred from the nuclear processes [1]. The degree of radioactivity is an important factor for radioecology monitoring [2–4]. The high level of radioactivity is a crucial factor in the human body. Therefore, the separation of radionuclides from the wastewater, atmosphere and other environmental samples is important task in materials science. For example, the ^{125}Sb, ^{54}Mn, ^{60}Co, ^{155}Eu, ^{154}Eu, ^{137}Cs and ^{134}Cs were most dangerous radionuclides. Their gamma-emitting level is much higher in the wastewater, atmosphere and other environmental samples [5, 6].

The various separation methods for Cm, Am, Pu, Np, U, Th, Tc, Sr and Cs were used in the separation science. Before the separation of radionuclides, various analytical methods were used to determine the amounts of radionuclides [7, 8]. Thermal ionisation mass spectrometry inductively coupled plasma mass spectrometry, mass spectrometry, γ-spectrometry, liquid scintillation, low background beta counting, β-spectrometry and α-spectrometry were the most used methods in the determination of radionuclides. The extraction chromatography, ion exchange, membrane separation, liquid-liquid extraction and separation approaches including precipitation methods were used in the separation of radionuclides [9].

Ion-exchange chromatography, solid-phase extraction and other extraction chromatographic methods can separate the radionuclides related to the ion-exchange processes, in which the metal ions (radionuclides) were exchanged with the corresponding anions on the surface or matrix of ion exchange. As a result, the radionuclides were chemically or physically linked to the surface or matrix of the polymer ion exchanger. For example, ion-selective resins were used to separate the lanthanides and actinides-based radionuclides. The ion exchange is the next important technique for the separation of radionuclides. The ion exchangers are categorised into anion- and cation-exchange resins [10, 11]. The following functional groups are functionalised

 DOI: 10.1201/9781003366058-14

to anion ion exchangers: $-N(CH_3)_2OH$, $-N(CH_3)_3OH$, while the cation exchanger contained the $-COOH$ and $-SO_3H$. The separation of radionuclides is linked to the following factors: solution temperature, pH and other ions (Figure 14.1). The radionuclides were separated from the wastewater using membrane-based separation. For example, reverse osmosis, nanofiltration, ultrafiltration and microfiltration methods

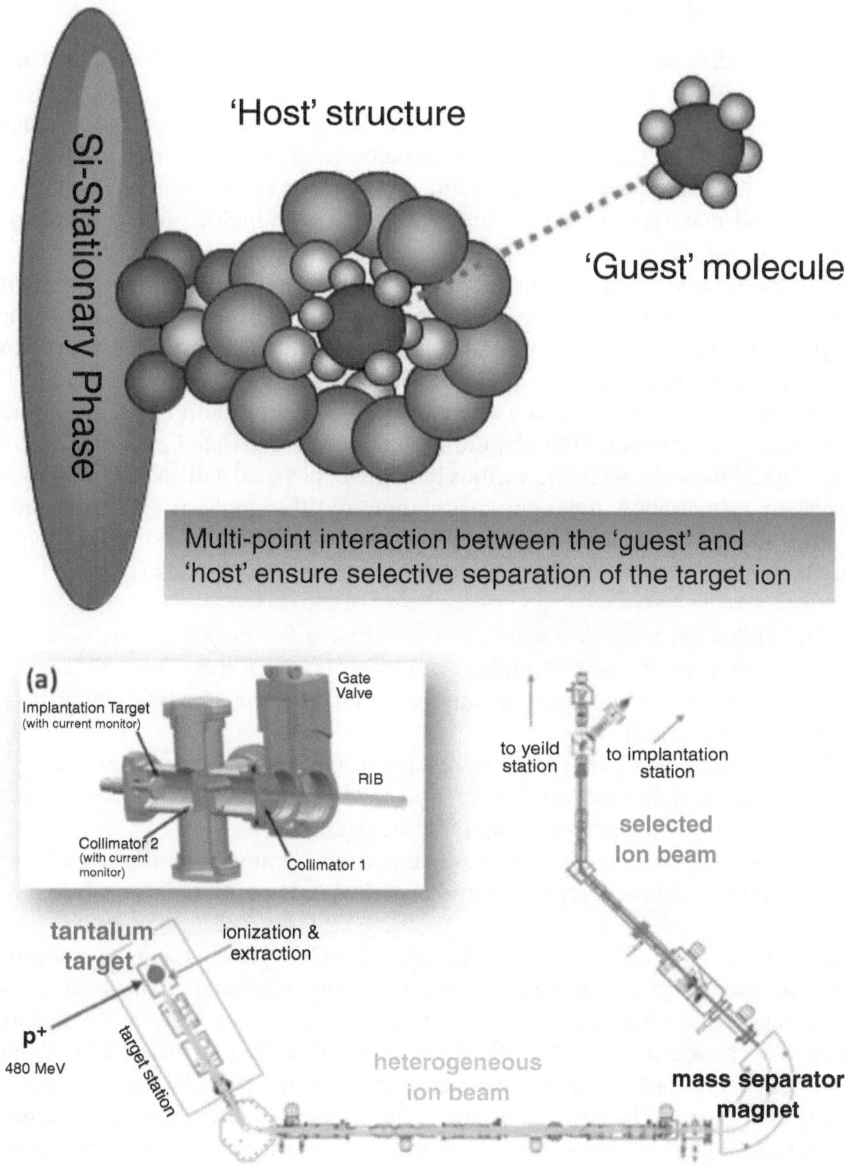

FIGURE 14.1 (a) Separation technique [12], (b) TRIUMF ISOL method for radionuclides separation [13]

are membrane-based methods. The selective membrane must be more porosity material, because the radionuclides are separated related to the size of the pore. Therefore, porosity is an important fact for membrane-based separation [12, 13].

14.2 CHITOSAN-BASED BIOSORBENTS IN THE SEPARATION OF RADIONUCLIDES

The radionuclides were serious pollutants for the environment. In recent times, various methods were used to separate the radionuclides from the environmental samples. Chitosan-based biosorbents were new and green technology in the separation of radionuclides. These types of biosorbents were more efficient and low-cost materials in the separation of radionuclides [14, 15]. The preparation methods of chitosan-based materials were easy and did not require high-degree technologies. Chitosan is a natural product and green polymer. There are many research works related to the separation of radionuclides from wastewater and other environmental samples. The adsorption of radionuclides depends on the following main properties: biosorption mechanisms, isotherm and kinetics models, desorbing agents, pH and contact time, and the influence of the intrinsic nature of metal ions and metal sorption capacities. The cobalt (Co-60), caesium (Cs-137), strontium (Sr-90), polonium (Po), radium (Ra), thorium (Th) and uranium (U) were separated effectively by the chitosan-based biosorbents in the various biological, physical and chemical methods: electrochemical methods, flotation, coagulation and flocculation, membrane filtration, adsorption, ion exchange and chemical precipitation. The radionuclide separation by biological methods is classified into the following categories [1, 16, 17]:

i. The biological processes were used to separate the radionuclides, such as constructed wetlands, flocculation and biosorption. In these biological processes, the enrichment mechanisms, accumulation and adsorption of radionuclides were used.
ii. The evaporation, electro dialysis, reverse osmosis, ion exchange, extraction and active carbon adsorption were used to separate the radionuclides by the separation, enrichment and sorption processes.
iii. The targeted radionuclides were separated by chemical reactions, such as electrolysis, chemical reduction and chemical precipitation.

Zemskova et al. [18, 19] introduced the M (M—Cu, Ni) potassium ferrocyanide chitosan-based biosorbents for the caesium and cerium radionuclide. This material can then separate the caesium radionuclide from the alkaline solution by co-precipitation methods. This radioactive metal is effectively adsorbed on the polymer matrix of chitosan. The Cu and Ni ions on the chitosan matrix promote the biosorbent performance of the suggested material. In this enhancement, the structural performance and chemical performances of chitosan biopolymer were changed. As a result, the adsorption performance of this material for radioactive metal was enhanced considerably (Figure 14.2).

In this research work [20], the chitosan-based biosorbents are classified into the following categories related to the metal-sorbent interactions: halogenated derivatives,

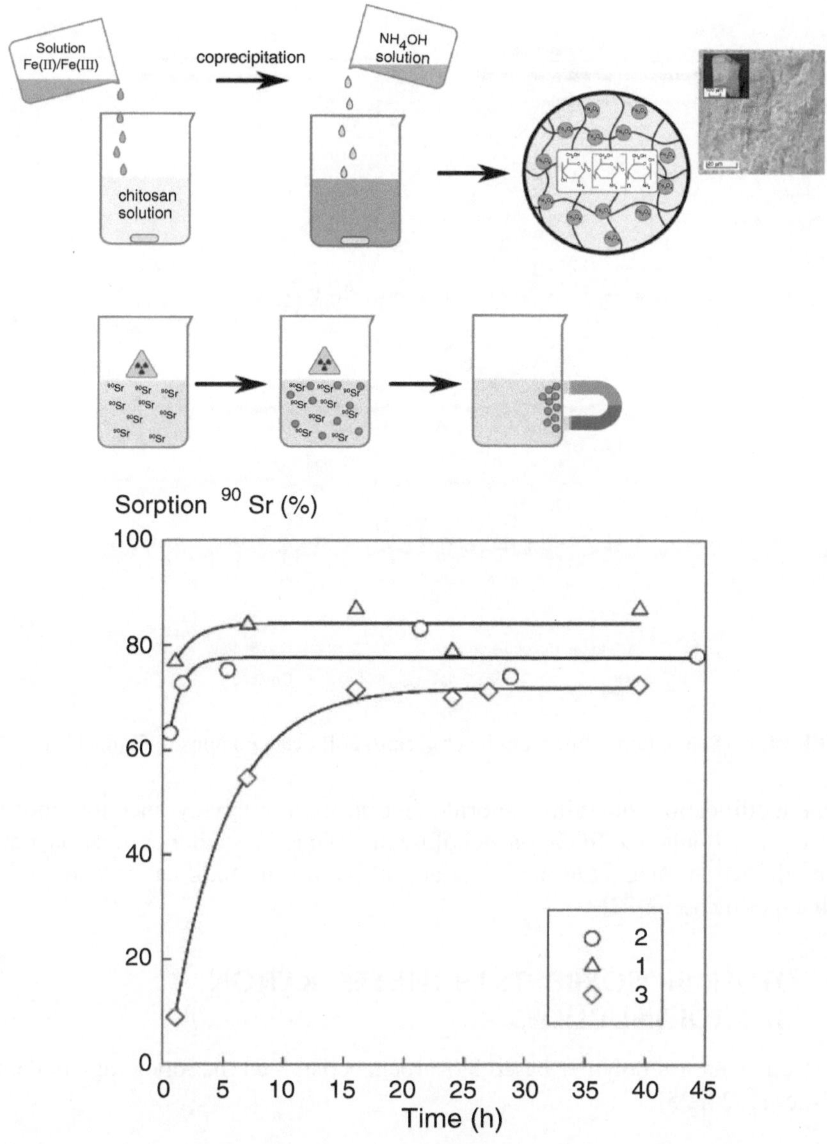

FIGURE 14.2 (a) Caesium and cerium radionuclides separation processes and (b) sorption kinetics of radionuclide on the surface of chitosan biosorbents [19]

chitosan-cyclodextrin conjugates, derivatives with 1,3 dicarbonyl compounds, chitosan graft sugars, chitosan graft copolymers, chitosan Ethylenediaminetetraacetic acid (EDTA)/ Diethylenetriaminepentaacetic acid (DTPA) complexes, chitosan crown ethers, derivatives of chitosan containing N, P and S as the heteroatom, templated chitosan and crosslinked chitosan (Figure 14.3). Ngah et al. [21] suggested several types of chitosan derivatives as biosorbents: bentonite modification, calcium

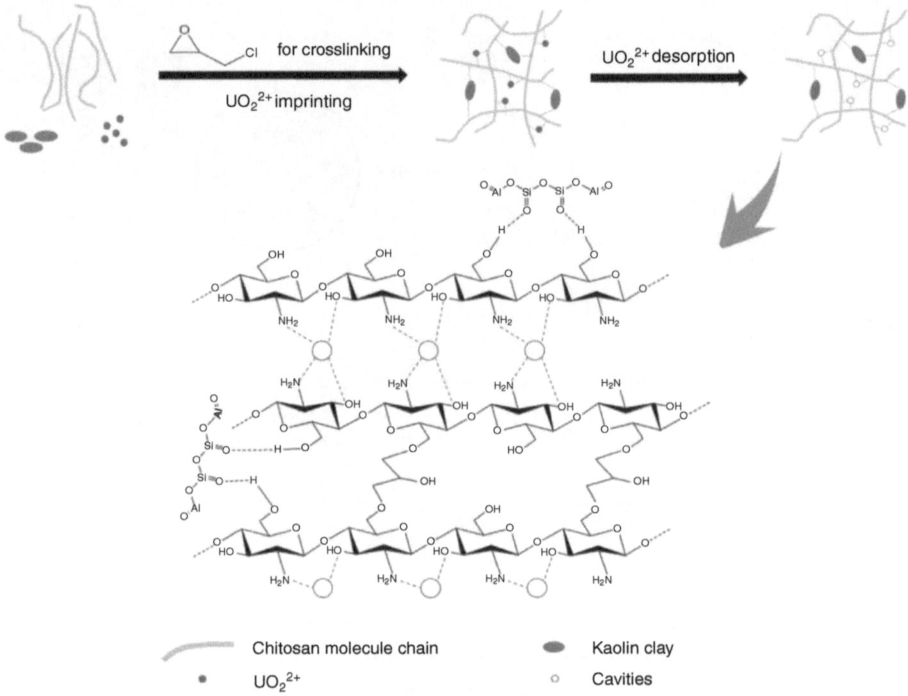

FIGURE 14.3 Separation of uranium by chitosan/kaolin clay composite foams [23]

alginate modification, polyvinyl chloride modification, polyvinyl alcohol modification, montmorillonite modification, cellulose modification, sand modification, cotton fibre modification, magnetite modification, oil palm ash modification and ceramic alumina modification [22].

14.3 OTHER BIOSORBENTS IN THE SEPARATION OF RADIONUCLIDES

The effective natural polymer-based biosorbents contained the following good characteristics [1, 24, 25]:

i. The natural polymer-based biosorbents have good mechanical, chemical, physical, diffusion, porosity and adsorption performances.
ii. These materials are easily recycled and regenerated materials after the biosorption of radionuclides with a strong alkaline and acidic medium.
iii. They can separate the radionuclides from the wastewater and other environmental examples at a high rate and effectively.
iv. The selectivity of these materials for the separation of radionuclides was high.
v. The sorption capacitance for radionuclides was relatively higher than the traditional methods.

vi. The sorption performance is very high and more stable in aggressive conditions: acidic, alkaline, hydraulic shock, high temperature, pH and low temperature.

Žukauskaitė et al. [26] suggested the new biosorbents named oak sawdust (*Quercus robur*) and moss (*Ptilium crista* – castrensis) for the separation of ^{137}Cs and plutonium isotopes from the aquatic system. It was found that

i. The rise of active functional groups is mainly responsible for the enhancement in the sorption capacitance of ^{137}Cs and plutonium isotopes. The hydroxyl and carboxyl groups are mainly responsible for the sorption of ^{137}Cs and plutonium isotopes on the surface and pore of oak sawdust (*Quercus robur*) and moss (*Ptilium crista* – castrensis). In this action, the radionuclides chemically or physically interacted with the functional groups (Figure 14.4).
ii. The increase in values of anion and cation on the sorbent promotes the sorption capacitance of this material.
iii. The pore value and area of the surface are important factors in the rise of sorption of capacitance.
iv. The sorption efficiency was over 80% for ^{137}Cs and plutonium isotopes, which means that over 80% of radionuclides in the wastewater solution were effectively removed from the examples.
v. The oak sawdust (*Quercus robur*) and moss (*Ptilium crista* – castrensis) biosorbents can be easily regenerated. Additionally, these biosorbents can remove radionuclides from aquatic systems in dynamic conditions.

Dulanská et al. [27] suggested an efficient biosorbent based on the wood-decay fungus *Fomes fomentarius* for the caesium radionuclides from the wastewater solution. It is reported that the sorption processes depend on the pH level and ionic strength. The suggested methodology was used in the Water Research Institute in Bratislava for the sorption of ^{137}Cs. The maximum adsorption capacitance of this biosorbent was 66 mg/g for caesium radionuclides.

El-khalafawy et al. [28] suggested the phalaris seeds peel powder (PSP) for the sorption of Cs(I) and Eu(III). This suggested biosorbents are low-cost and economically effective. In the sorption analysis of this biosorbent, the interference ions on the sorption of Eu(III) and Cs(I), the effect of media pH, initial metal concentration and contact time were investigated by the various spectroscopic, surface and theoretical methods. The obtained results suggested (Figure 14.5) that

i. The adsorption of Eu^{3+} and Cs^{+} on the surface of the biosorbent is endothermic, indicating that the adsorption of Eu^{3+} and Cs^{+} required additional energy. Additionally, their adsorption is spontaneously processed.
ii. The selectivity for the Eu^{3+} is higher than Cs^{+}.
iii. The separation indicator is lower than 1, confirming that the PSP is more effective for the sorption of Cs(I) and Eu(III).

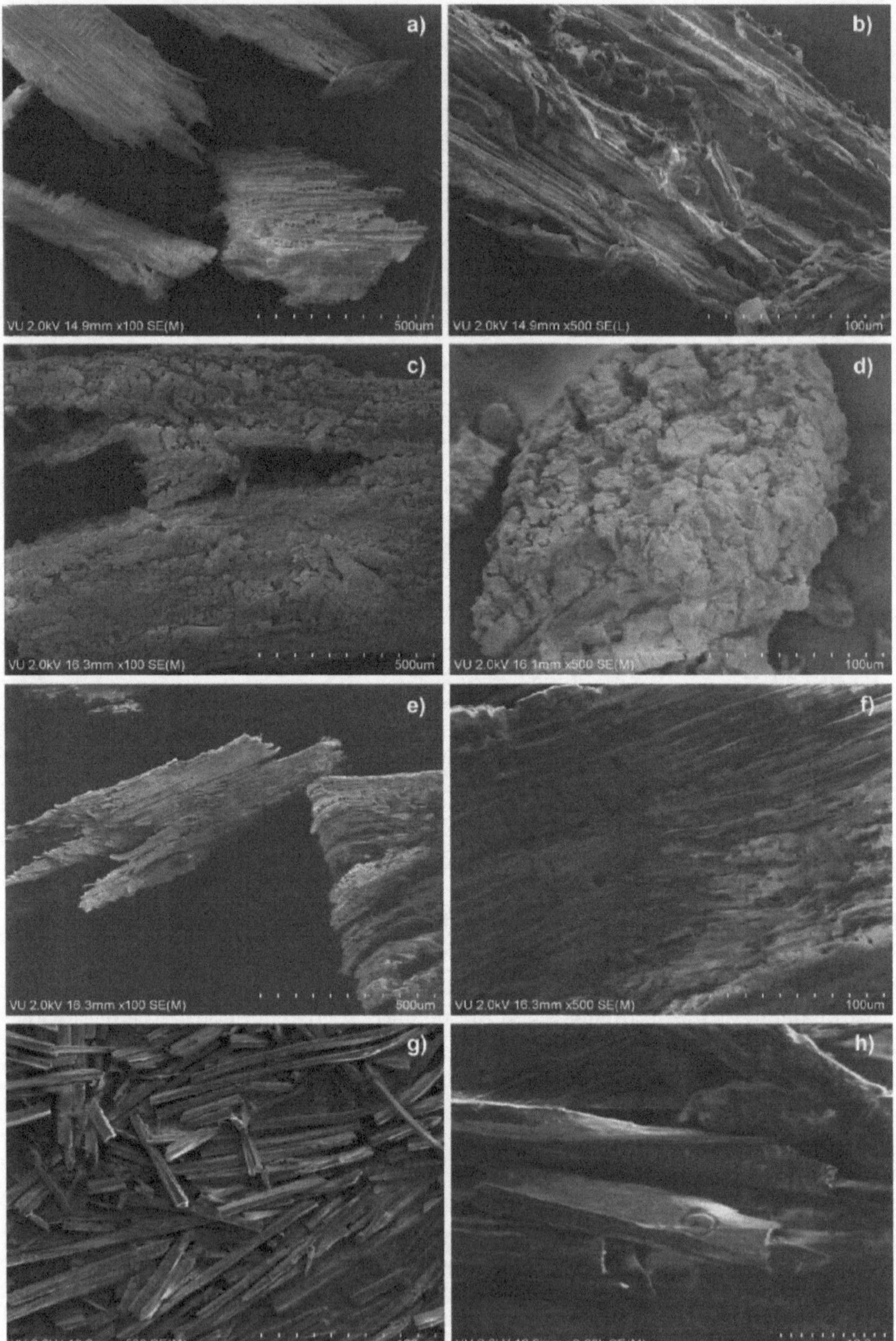

FIGURE 14.4 Scanning electron pictures before ×100 (a), ×500 (b) and after the functionalised with iron hydroxide ×100 (c), ×500 (d), after the carbonisation of sawdust ×100 (e), ×500 (f), additionally, carbonised sawdust functionalised with HCl ×500 (g), ×3000 (h) [26]

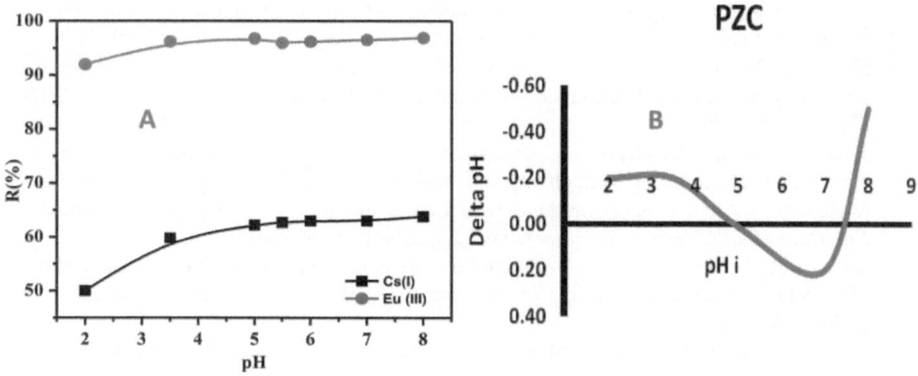

FIGURE 14.5 (a) Effectiveness of PSP for the sorption of Cs(I) and Eu(III) and (b) point of zero charge analysis [28]

14.4 CONCLUSION

In this book chapter, the application of biosorbents in the separation of radionuclides was reviewed and discussed with modern examples. Currently, the separation of radionuclides from environmental examples is an urgent task in modern engineering sciences. Chitosan-based biosorbents were new and green technology in the separation of radionuclides. These types of biosorbents were more efficient and low-cost materials in the separation of radionuclides. The preparation methods of chitosan-based materials were easy and did not require high-degree technologies. Chitosan is a natural product and green polymer. There are many research works related to the separation of radionuclides from wastewater and other environmental samples. In addition to this, the animal organ, plant parts, and algal, fungal and bacterial biosorbents were suggested as effective in the separation of radionuclides [29].

REFERENCES

1. Wang, J. and C. Chen, Chitosan-based biosorbents: Modification and application for biosorption of heavy metals and radionuclides. *Bioresource Technology*, 2014. **160**: p. 129–141.
2. Maksoud, M.I.A.A., et al., Insight on water remediation application using magnetic nanomaterials and biosorbents. *Coordination Chemistry Reviews*, 2020. **403**: p. 213096.
3. Kılınç, E., et al., A new method for the preconcentrations of U (VI) and Th (IV) by magnetized thermophilic bacteria as a novel biosorbent. *Analytical and Bioanalytical Chemistry*, 2021. **413**(4): p. 1107–1116.
4. Datta S, K. Radhapyari, N. Saha, S.K. Samanta, Recent trends in the application of biowaste for hazardous radioactive waste treatment, in *Environmental Sustainability and Industries*. 2022, Elsevier. p. 159–192.
5. Heidari, F., et al., Isolation of an efficient biosorbent of radionuclides (226Ra, 238U): Green algae from high-background radiation areas in Iran. *Journal of Applied Phycology*, 2017. **29**(6): p. 2887–2898.
6. Wang, J. and S. Zhuang, Removal of cesium ions from aqueous solutions using various separation technologies. *Reviews in Environmental Science and Bio/Technology*, 2019. **18**(2): p. 231–269.

7. Guptaa, N.K., et al., Biosorption-an alternative method for nuclear waste management: A critical. *Journal of Environmental Chemical Engineering*, 2018. **6**: p. 2159–2175.
8. Shutova, V.V., et al., Levan from Azotobacter vinelandii as a Component of Biosorbents for Heavy Metals and Radionuclides. *Applied Biochemistry and Microbiology*, 2021. **57**(1): p. 102–109.
9. Varala, S., et al., Desorption studies for the recovery of radionuclides (Th and Zr) and optimization using Taguchi mixed design L18 (2132)–A regeneration step for loaded biosorbent, general mathematical model for multistage operation. *Journal of Environmental Chemical Engineering*, 2017. **5**(6): p. 5396–5405.
10. Dai, Y., et al., Macroporous ion-imprinted chitosan foams for the selective biosorption of U (VI) from aqueous solution. *International Journal of Biological Macromolecules*, 2020. **164**: p. 4155–4164.
11. Zhuang, S. and J. Wang, Removal of cesium ions using nickel hexacyanoferrates-loaded bacterial cellulose membrane as an effective adsorbent. *Journal of Molecular Liquids*, 2019. **294**: p. 111682.
12. Alam, M.F., et al., Selective separation of radionuclides from environmental matrices using proprietary solid-phase extraction systems: A review. *Microchemical Journal*, 2022. **181**: p. 107637.
13. Fiaccabrino, D.E., P. Kunz, and V. Radchenko, Potential for production of medical radionuclides with on-line isotope separation at the ISAC facility at TRIUMF and particular discussion of the examples of 165Er and 155Tb. *Nuclear Medicine and Biology*, 2021. **94–95**: p. 81–91.
14. Kapashi E., M. Kapnisti, A. Dafnomili, F. Noli, Aloe Vera as an effective biosorbent for the removal of thorium and barium from aqueous solutions. *Journal of Radioanalytical and Nuclear Chemistry*. 2019 Jul 15. **321**: pp. 217–226.
15. Gamal, R., N.M. Sami, and H.S. Hassan, Assessment of modified Salvadora Persica for removal of 134Cs and 152+ 154Eu radionuclides from aqueous solution. *Environmental Science and Pollution Research*, 2022. **29**(2): p. 3072–3090.
16. Yin, Y., et al., Removal of strontium ions by immobilized Saccharomyces cerevisiae in magnetic chitosan microspheres. *Nuclear Engineering and Technology*, 2017. **49**(1): p. 172–177.
17. Wani, A.A., et al., Recent advances and future perspectives of polymer-based magnetic nanomaterials for detection and removal of radionuclides: A review. *Journal of Molecular Liquids*, 2022. **365**: p. 119976.
18. Zemskova, L., et al., Chitosan-based biosorbents: immobilization of metal hexacyanoferrates and application for removal of cesium radionuclide from aqueous solutions. *Journal of Sol-Gel Science and Technology*, 2019. **92**(2): p. 459–466.
19. Zemskova, L., et al., New chitosan/iron oxide composites: Fabrication and application for removal of Sr^{2+} radionuclide from aqueous solutions. *Biomimetics*, 2018. **3**(4): p. 39.
20. Varma, A.J., S.V. Deshpande, and J.F. Kennedy, Metal complexation by chitosan and its derivatives: A review. *Carbohydrate Polymers*, 2004. **55**(1): p. 77–93.
21. Ngah, W.S.W., L.C. Teong, and M.A.K.M. Hanafiah, Adsorption of dyes and heavy metal ions by chitosan composites: A review. *Carbohydrate Polymers*, 2011. **83**(4): p. 1446–1456.
22. Attia, L.A., M.A. Youssef, and O.A. Abdel Moamen, Feasibility of radioactive cesium and europium sorption using valorized punica granatum peel: Kinetic and equilibrium aspects. *Separation Science and Technology*, 2021. **56**(2): p. 217–232.
23. Yu, H., et al., Selective biosorption of U(VI) from aqueous solution by ion-imprinted honeycomb-like chitosan/kaolin clay composite foams. *International Journal of Biological Macromolecules*, 2022. **206**: p. 409–421.
24. Zhuang, S., et al., Adsorption of Co^{2+} and Sr^{2+} in aqueous solution by a novel fibrous chitosan biosorbent. *Science of the Total Environment*, 2022. **825**: p. 153998.

25. Gök, C., S. Aytas, and H. Sezer, Modeling uranium biosorption by Cystoseira sp. and application studies. *Separation Science and Technology*, 2017. **52**(5): p. 792–803.
26. Žukauskaitė, Z., et al., Separation of anthropogenic radionuclides from aqueous environment using raw and modified biosorbents. *Journal of Environmental Radioactivity*, 2022. **244–245**: p. 106829.
27. Dulanská, S., et al., Modified biosorbent wood-decay fungus Fomes fomentarius for pre-concentration of 137Cs in water samples. *Journal of Radioanalytical and Nuclear Chemistry*, 2018. **318**(3): p. 2493–2500.
28. El-khalafawy, A., D.M. Imam, and M.A. Youssef, Enhanced biosorption of europium and cesium ions from aqueous solution onto phalaris seed peel as environmental friendly biosorbent: Equilibrium and kinetic studies. *Applied Radiation and Isotopes*, 2022. **190**: p. 110498.
29. Gupta, N.K., et al., Biosorption-an alternative method for nuclear waste management: A critical review. *Journal of Environmental Chemical Engineering*, 2018. **6**(2): p. 2159–2175.

15 Biosorption of Volatile Organic Compounds

Godwin Anywar, Jane Namukobe, Patience Tugume, Jamilu Ssenku, Moses Okol, and Julius Mulindwa

15.1 WHAT ARE VOLATILE ORGANIC COMPOUNDS?

Volatile organic compounds (VOCs) are highly reactive hydrocarbons that are man-made and/or naturally occurring, toxic to humans and can be detrimental to the environment (Berenjian et al., 2012; Frezzini et al., 2019). VOCs are characterized by a low boiling point, high saturation vapor pressure, high volatility and are lighter than air, thus evaporate easily at room temperature (Anand et al., 2014; Pennerman et al., 2016; United States Environmental Protection Agency, 2017; Xiang et al., 2020). They therefore float into the air and travel long distances from the source of emission (Montero-Montoya et al., 2018).

VOCs are emitted as gases from certain solids or liquids by a wide range of products (Thurston, 2017). Because of their non-polar nature, VOCs are frequently used to dissolve, dilute or disperse materials that do not dissolve in water. They are therefore responsible for the characteristic smell of paints, pharmaceuticals, refrigerators, cosmetics, paints, dry-cleaning supplies, pesticides, building materials and furnishings (Anand and Mehendale, 2005). Other sources of VOCs include office equipment such as copiers and printers, correction fluids and carbonless copy paper, graphics and craft materials including glues and adhesives, permanent markers and photographic solutions.

15.1.1 Sources and Composition of Volatile Organic Compounds

The origin of VOCs is both anthropogenic and natural (Xiang et al., 2020). Many VOCs are generated by a variety of biological systems (Fisher et al., 2020), and anthropogenic activities emit pollutants into the environment which are difficult and expensive to eliminate (Gopinath et al., 2018). Natural origins of VOCs include wetlands, forests, oceans and volcanoes with the estimated global VOCs biogenic emission estimated at about 1,150 teragrams of carbon year (TgC/yr). Trees discharge about 75% of the atmospheric biogenic VOC (BVOC) globally through foliar emissions but also via microbial decomposition of organic material with an estimated flux of 750–1,000 TgC/yr (Korpi et al., 2009; Xi-Fu, 2016; Safieddine et al., 2017). In terrestrial systems, VOCs are generated for a variety of functions and mechanisms, including host defense, immune system stimulation, attractants for pollinators,

DOI: 10.1201/9781003366058-15

interkingdom signaling and intra- and inter-species communication (Loreto et al., 2014; Bailly & Weisskopf, 2017).

Anthropogenic activities that produce VOCs range from manufacturing industries, petrochemical industries, vehicular emissions and fuel storage, photochemical pollution and interior finishing (Atkinson, 2000; Kim et al., 2018; Zou et al., 2019). In the coastal environment, VOCs are released from the chemical degradation induced in particularly by heat and UV light (Lomonaco et al., 2020).

VOCs are made up of isoprene, monoterpenes and other reactive BVOCs (Wian, 2019). VOCs are typically industrial solvents, such as trichloroethylene (TCE); fuel oxygenates, such as methyl tert-butyl ether, or by-products, produced by chlorination in water treatment, such as chloroform. VOCs are often components of petroleum fuels, hydraulic fluids, paint thinners and dry-cleaning agents and are precursors for the formation of photochemical and haze smog (Thurston, 2017; Zhang et al., 2017b; Shen & Zhang, 2019).

While in the atmosphere, VOCs can be transformed by chemical reactions *in situ* especially by photo-oxidation leading to the formation of ozone (O_3) and organic aerosol that form secondary emissions (Palm et al., 2018). In particular, VOCs can react with nitrogen oxides to form ground-level O_3, or smog, which contributes to climate change (Aydin et al., 2012; Anand et al., 2014). The degradation of polyolefins generates VOCs belonging to the families of lactones, esters, ketones and carboxylic acids, with the consequent reduction of molecular weight (Gardette et al., 2013).

15.1.2 Challenges with Volatile Organic Compounds

Due to their wide use as solvents in many products, almost everyone is exposed to them mainly through environmental media by inhalation, ingestion and skin contact (Masih et al., 2016; Kim et al., 2019). When VOCs are discharged into the environment, they induce ecological and health hazards (Kumar & Ngueagni, 2022). Different VOCs are common ground-water contaminants (Thurston, 2017) and are persistent environmental pollutants (Frezzini et al., 2019; Okoro et al., 2022). VOCs are more concentrated in indoors than outdoors and therefore have short- and long-term adverse health effects. These include asthma, atopic dermatitis and neurologic problems, eye irritation headache, skin allergy, nose and throat discomfort, nausea and dizziness respiratory diseases, impaired concentration, short-term memory and can also damage the internal organs such as the liver and kidneys (Kandyala et al., 2010; Montero-Montoya et al., 2018; Shuai et al., 2018; Alford and Kumar, 2021).

Some VOCs induce cancer in animals, and some of them are suspected or known to be carcinogenic in humans (Loh, 2007). Generally, VOCs' effects on human health depends on the type of compound (from toxic to non-toxic which depends on carbon chain, degree of unsaturation and the presence of functional groups), the period of exposure and level of exposure. In particular, are amides which are sensitizers, aldehydes are irritants; hydrocarbons as cytogenic/mutagenic, teratogenic and many unsaturated, short-chain halocarbons are animal carcinogens (Anand, 2005; An et al., 2010; Li et al., 2013; Anand et al., 2014; Tsai, 2019). Specifically, some VOC such as toluene and xylene could cause serious neurosis (Wang et al., 2013; Niaz et al., 2015).

15.2 BIOSORPTION OF ORGANIC VOLATILE COMPOUNDS

Biosorption is a physicochemical process that involves the removal of organic or inorganic substances that may be gaseous, soluble or insoluble from solution by biological material (Gadd, 2009). It is a fast energy independent process that involves the use of live or dead microorganisms such as bacteria, fungi, algae or their components and agricultural wastes for the removal of potentially toxic pollutants like heavy metals, dye and polycyclic aromatic hydrocarbons (Gadd, 2009; Gopinath et al., 2018; Kumar & Ngueagni, 2022). Biosorption includes mechanisms such as absorption, adsorption, ion exchange, surface complexation and precipitation (Gadd, 2009).

Currently, there are diverse technologies that can be adopted to mitigate environmental pollution by VOCs. Such technologies include catalytic degradation (Shah & Li, 2019; Yang et al., 2019a,b), condensation (Zaitan et al., 2008) and biological filtration (Khan et al., 2016; Wu et al., 2017), aeration and biodegradation (Wang et al., 2014). However, most of these technologies are associated with drawbacks such as energy and chemical consumption, need for prolonged periods of time for effective treatment and are expensive in terms of equipment and maintenance (Premkumar et al., 2018). The adsorption-gathering technology exhibits a great promising because of its main advantages of low-cost, low-energy consumption and high removal efficiency (Liu et al., 2004; Yang et al., 2019a).

Biosorption has been considered a promising biotechnological approach to pollutant removal because of its efficiency, simplicity, similarity to conventional ion-exchange technology and availability of biomass (Mack et al. 2007; Gadd, 2009). Biosorption technology could reduce the operation costs by as much as 20–36% as compared to conventional systems (Ata et al., 2012; Vankar et al., 2013), such as activated carbon, which is the most used material to adsorb organic pollutants (Park et al., 2010b; Zhang et al., 2017a).

15.2.1 Factors Affecting Working of Biosorbents

A number of factors are affecting the biosorption equilibrium. These include the following:

i. Temperature: Inconsistent or conflicting results have been reported on how temperature affects biosorption equilibrium.
ii. pH: pH alters the chemical nature of the functional groups present in the biosorbent and hence the affinity of these groups toward metal ions.
iii. Biomass state and concentration: This determines the number of functional groups available at the adsorbent surface. This is also determined by the biosorbent state (dead or alive).
iv. Coexisting ions: The presence of other ions in solution directly affects the selectivity of the adsorbent and how effectively it will adsorb the pollutant of interest.
v. Contact time with the adsorbent: Increase in contact time leads to the increase in the percentage removal of the adsorbate by the adsorbent (Nwidi & Agunwamba, 2015; Samuel et al., 2020; Gupta et al., 2021).

15.2.2 Regeneration of Biosorbents

Biosorbents can be reused. After use, bisorbents can be regenerated through various methods such as filtration or centrifugation to separate it from the solution, followed by desorption by: (i) altering the pH alteration to desorb the adsorbate from the adsorbent; (ii) addition of chemicals which have functional groups that react with the adsorbate to desorb the adsorbate from the adsorbent and (iii) use of acids, bases, chelating groups such as ethylenediamine tetraacetic acid (EDTA) and organic solvents (Samuel et al., 2020; Gupta et al., 2021).

15.3 BIOSORPTION OF VOLATILE ORGANIC COMPOUNDS USING MICROORGANISMS

The discharge of VOCs in the environment has caught global attention, due only to their persistence, toxicity and mobility (Costa et al., 2012). This has been exacerbated by their widespread use and negative impact on all life forms in the environment. Microorganisms, especially heterotrophic bacteria and fungi, can be used to biodegrade environmental pollutants (Delhomenie & Heitz, 2005). This is attributed to their ability to detoxify organic and inorganic pollutants (Costa et al., 2015) and the fact that abiotic methods that have conventionally been used for the same purpose have demerits such as being expensive, environmentally unfriendly and dependent on the concentration of waste (Costa & Tavares, 2017). Although many biosorption studies have been performed on the ability of microorganisms to remove heavy metals from the environment, there have been limited studies on the removal of VOCs. However, different studies have showed the potential of microorganisms to bioremediate air and water environments.

Some technologies that have been employed in biosorption include biofilters, bioscrubbers and other bioreactor types (Estrada et al., 2012; Dobslaw et al., 2017). These methods have the potential to convert the pollutants into innocuous oxidation products with the help of microorganisms (Barbusinski et al., 2017). Due to many benefits of biofiltration, it has been utilized to eliminate gaseous pollutants from the air (Barbusinski et al., 2017). The following microorganisms have been utilized in debasement of VOCs by filtration: *Pseudomonas, Candida, Mycobacterium, Alcaligenes, Exophiala, Acinetobacter, Fusarium, Cladosporium, Rhodococcus, Aspergillus* and *Mucor* (Gopinath et al., 2018).

15.3.1 Biosorption of Volatile Organic Compounds by Fungi

According to Rybarczyk et al. (2021), *Candida albicans* and *Candida subhashii* efficiently removed cyclohexane and ethanol (95–99%) from air in a biotrickling filter. Microbial adsorbents synthesized from fungal cells of *Ophiostoma stenoceras* showed adsorption capacities for ethyl acetate, α-pinene and n-hexane at 620, 454 and 374 mg/g, respectively. These values were higher than those of other synthesized biosorbents (Cheng et al., 2020). This implies that fungal biosorbents have potential to remove VOC from the environment. Biological purification of off-gases using fungi is a cheap and ecofriendly way of neutralizing hydrophobic VOCs. Fungi exhibit higher effectiveness in VOCs removal than bacteria because of their ability to degrade hydrophobic VOC under extreme environmental conditions (Kennes & Veiga, 2004).

A symbiotic relationship between Cecropia trees and Azteca ants results in the emission of VOC in tree hollow structures for defense or communication (Guerrieri et al., 2009; Kigathi et al., 2019). However, these VOCs are toxic and could lead to anti-herbivory or intruder deterrence (Dobler et al., 2011; Agrawal et al., 2012) and may damage the eggs, larvae and pupae of the ants. Several species of Chaetothyriales have a high tolerance and assimilative capacity to such toxic volatile compounds at relatively high concentrations (Prenafeta-Boldú et al., 2018). Consequently, the Chaetothyriales have been used as biocatalysts in engineered air biofilters for the treatment of polluted air (Prenafeta-Boldú et al., 2008).

Synthetic dyes used in textile, paper, cosmetics, pharmaceutical and leather production are important environmental pollutants. Most of these dyes have complex structures rendering them recalcitrant to natural degradation processes (Abdallah & Taha, 2012). Various physicochemical approaches that have been used for the elimination of these dyes from the environment are costly thus biosorption is considered a cheaper alternative for treating dye-contaminated wastewater (Daneshvar et al., 2019). In a study by Bouras et al. (2021) *Aspergillus carbonarius* and *Penicillium glabrum* removed methylene blue from aqueous solutions at maximum biosorption capacities of 21.88 mg/g and 16.67 mg/g at 30°C. This suggests potential use of these fungi as biosorbents for the effective removal of methylene blue from dye wastewater. Other fungi such as *Trichoderma asperellum* (Marcharchand & Ting, 2017), *Aspergillus fumigatus* (Kabbout & Taha, 2014), *Trametes versicolor* (Casas et al. 2014), *Fusarium solani* (Hassan & Ai-Jawhari, 2015), *Coriolopsis* sp. and *Penicillium simplicissimum* (Chen & Ting, 2015) have also been reported as efficient dye-decolorizers. Yeast strains such as *Saccharomyces cerevisiae*, *Yarrowia lipolytica*, *Trichoderma lixii* and *Schizosaccharomyces pombe* have been used in the biosorption of dyes in the laboratory (Danouche et al., 2021). This is an indication of the potential of such yeast strains to be used on a larger scale for the removal of volatile dyes from the aquatic environment.

The fungus *O. stenoceras* captured and adsorbed various organic compounds in liquid and gas phases. Complete degradation of hexane, tetrahydrofuran and chlorobenzene took a shorter time compared to when the bacterium *Pseudomonas veronii* was used (Cheng et al., 2019). These findings are instrumental in exploring the special cell surface of the fungus in adsorption and bio-enhancement that could be utilized for treatment of organic contaminants using bacteria. Mycorrhizal fungi are also important in remediation of pollutants (Bouwer & Zehnder, 1993). For instance, the roots of *Lolium multiflorum* arbuscular mycorrhizal fungal hyphae translocated fluorene and phenanthrene (Gao et al., 2010). Additionally, *Polygonum aviculare* and its root-associated fungal strains effectively remediated soil polluted by petroleum (Mohsenzadeh et al., 2010).

Aspergillus flavus strain KRP1 was used for bioremediation of mercury-contaminated wastewater (Tanwer et al., 2022). Cells of lyophilized fungi, *O.*

stenoceras, after modification by aminomethylation purified VOC exhibiting higher adsorption capacities of 620, 454 and 374 mg/g for ethyl acetate, α-pinene and n-hexane, respectively, in comparison to synthesized biosorbents (Cheng et al., 2020). Several species of fungi including *Paecilomyces variotii*, *Paecilomyces*, *Phanerochaete chrysosporium*, *F. solani* and *Cladosporium sphaerospermum* were used for debasement of VOC (Behnamnia et al., 2009).

15.3.2 Biosorption of Volatile Organic Compounds by Bacteria

Pseudomonas spp. enhanced the rate of removal of VOC by potted *Spathiphyllum* sp. (Wolverton & Wolverton, 1993). This illustrated the role of microorganisms in enhancing the ability of potted plants to absorb VOC. Several endophytic and rhizospheric bacteria assisted plants in removing toxic compounds from soil (McGuinness & Dowling, 2009).

Pseudomonas, Proteus, Enterobacteriaceae, Alteromonas, Aeromonas and *Xanthomonas* are mercury resistant and are used for its bioremediation. *Vibrio fluvialis* exhibited a high bioremediation efficiency removing approximately 60% of mercury ions from industrial effluents. *Pseudomonas aeruginosa* showed the highest mercury bioremediating capacity of 62% for Hg^{2+} under laboratory conditions (Tanwer et al., 2022). These results indicate the potential of bacterial to efficiently clean up the environment and waste contaminated with mercury.

Bacillus subtilis efficiently decolorized Congo red dye in a short time in culture containing sucrose and ammonium chloride at temperature of 35°C and neutral pH 7.0 (Sarim et al., 2019). Immobilization of the bacteria on sodium alginate beads enhanced the percent decolorization indicating the potential to use such bacteria on a large scale to treat industrial effluents.

Biofiltration by *Hyphomicrobium* spp. associated with roots of ornamental house plants was clean indoor air of harmful VOC through degradation (Russell et al., 2014). This suggests that a range of ornamental houseplants that are in association with similar microbes could potentially be used as living biofilters to create healthier work and living spaces.

Bacteria, such as *Pseudomonas putida*, *Pseudomonas fluorescens*, *Rhodococcus fascians*, *Alcaligenes xylosoxidans*, *Burkholderia cepacia*, *Hyphomicrobium* spp., *Xanthobacter* spp., *Acinetobacter* spp., are used for the debasement of VOC. Briseno-Roa et al. (2006) noted that a biofilter achieved nearly 100% effectiveness at pollution levels of 0.5 g/m^3 or less while using bacterial populace. *P. putida* effectively evacuated VOC at 83% from an inlet contamination of 1 g/L (Gopinath et al., 2018).

Immobilized bacteria (*Pseudomonas* sp., *Acinetobacter* sp.) together with waste materials effectively removed benzene/toluene/ethylbenzene/xylenes/TCE/cis-1,2-dichloroethylene (BTEX/TCE/cis-DCE mixture) from the contaminated water (de Toledo et al., 2019). This offers potential for the utilization of microorganisms and waste materials in the filtration water of any contaminants.

15.3.3 Macro- and Microalgae Biosorption of Volatile Organic Compounds

Algae have been categorized into macroalgae which are represented by over 18,000 species of red, green and brown algae; and microalgae which are represented by over 20,000 species of green algae (*Chlorophyceae*), blue–green algae (*Cyanobacteria*/*Cyanophyceae*), yellow–green algae (*Xanthophyceae*), golden algae (*Chrysophyceae*), diatom (*Bacillariophyceae*) and other minor species (Yu et al., 2017; Abinandan et al., 2018). Despite the chemical (moisture, volatile matter, fixed carbon and ash composition) and biochemical (lipid, protein and carbohydrate composition) differences between the two groups both macroalgae and microalgae have similar applications. According to Lee et al. (2022), there are three feasible modes of application of both microalgae and macroalgae as adsorbents. These include the direct application as biosorbent with mechanical treatment, conversion via thermochemical conversion to charcoal-based adsorbents such as torrefied biomass, hydrochar, biochar and generation of activated carbon via activation techniques.

Macroalgae biosorption processes constitute a cost-effective and environmentally friendly alternative for the removal of petroleum-based soluble compounds (Flores-Chaparro et al., 2017). Macroalgae biomass in particular have achieved relevance in biosorption because of its lack of toxicity and large-scale availability (Rodriguez-Hernandez et al., 2017; Bilal et al., 2018; Elayadi et al., 2020). Due to the unique physicochemical properties of both macro- and microalgae in their adsorbent forms, the adsorption of hazardous pollutants was found to be highly effective. The process involved different mechanisms such as physisorption, chemisorption, ion exchange, complexation and others depending on the types of pollutants. Overall, both macroalgae and microalgae not only can be tailored into different forms of adsorbents based on the applications but their adsorption capacities are also far more superior compared to the conventional adsorbents. According to Anastopoulos and Kyzas (2015), algae have been used as biosorbent material and are 15.3% more effective than other kinds of biomass and 84.6% more than other microbial biosorbents.

Algae have received great attention as biosorbents due to their low nutrients' requirements, high sorption capacity, high surface area-to-volume ratio, lower production of sludge and their capacity for metal regeneration and recovery in the removal of pollutants from wastewater (Park et al., 2010a). Algal biomass can be used in live or dead form for the biosorption of VOCs (Zainith et al., 2021). This could be advantageous, since toxic pollutants may have no effect on such cells, making them relatively easy to handle (Vahabisani & An, 2021). Furthermore, dead cells do not need any further treatment or nutrition and can be deployed over many cycles. The composition of the cell wall of algae plays a significant role in the biosorption process. The microalgal cell wall consists mainly of polysaccharides, proteins (Decho & Herndl, 1995) and lipids, which offer several functional groups, such as $^{-}COOH$, ^{-}OH, $^{-}PO_3$, $-NH_2$, ^{-}SH and ^{-}CO, that can bind the VOCs. Due to the rapid biosorption capability, low cost and high efficiency, reusability, no toxic waste generation, algae could be an ideal and promising biosorbent.

15.4 BIOSORPTION OF VOLATILE ORGANIC COMPOUNDS BY FOOD WASTE MATERIALS

The production of low-cost adsorbents from agricultural food wastes is currently gaining popularity as novel bio-based products for environmental applications (Oliveira & Franca, 2008; Anastopoulos et al., 2022). It is estimated that one-third of the food produced globally for human consumption (about 1.3 billion tones) is lost or wasted each year along the food supply chain (Turon et al., 2014; Nicastro & Carillo, 2021). This figure is expected to rise if proper prevention polices are not put in place. The large amounts of food wastes generated make them a cheap raw material for the preparation of low-cost sorbents and are being thoroughly studied (Oliveira & Franca, 2008). The conventional waste management practices dispose or incinerate the wastes, contributing to environmental pollution while misusing biomass, a valuable resource with a great potential of reuse (Karić et al., 2022). Despite the inherent variability of the agricultural food wastes, they could evolve into an important industrial feedstock on account of its availability, versatility and sustainability, for the production of bio-based products (Turon et al., 2014). Among the bio-based products from agricultural wastes is the natural, environmental and economic adsorbent (Dai et al., 2018). Agricultural waste biomass may be used as: (i) an adsorbent in its original, raw form, following ambient drying and grinding; (ii) modified bio-based sorbents or (iii) a source material for the synthesis of activated carbon adsorbents through carbonization (Karić et al., 2022). The latter may be undesirable owing to not only the costs involved but also a large environmental footprint and its production associated with high raw resources consumption (Van Tran et al., 2022).

The use of raw agricultural wastes for adsorption of VOCs may be inefficient since they have low biosorption capacities compared to the synthetic commercial adsorbents (Karić et al., 2022). Thus, for efficient removal of VOC pollutants from the environment using agricultural wastes, pre-treatment is necessary. This has been done by adopting physical and chemical treatments using organic compounds, oxidizing agents, bases, organic and mineral acids (Ramrakhiani et al., 2016). According to Pokharel et al. (2020), chemical reagents such as calcium hydroxide, sodium carbonate, sodium hydroxide, nitric acid, hydrochloric acid, sulfuric acid, tartaric acid, citric acid, formaldehyde, methanol and hydrogen peroxide are commonly used to enhance pollutant-binding capacities of sorbents.

Agricultural wastes have loose and porous structures and are composed by cellulose, hemicellulose and lignin as major constituents and may possess other functional groups, including hydroxyl, aldehyde, carbonyl, carboxyl, phenolic and/or ether groups, which are able to interact with pollutants including the VOCs (Ali & Saeed, 2016; Dai et al., 2018). Due to the existence of various functional groups in agricultural wastes and thus the potential to bind with a wider range of VOC pollutants, they could be used in the environmental management of these pollutants generated from different sources.

Non-living or dead biomass like agricultural wastes such as husks, seeds, peels, active sludge, plant and wood wastes and stalks of different crops may be used as low-cost adsorbents for the removal of aliphatic and aromatic VOCs from wastewater

(Frezzini et al., 2019; Zainith et al., 2021). Banana peel for instance is an abundant and low-cost agricultural waste residue that showed a high adsorption capacity of phenolic compounds (Achak et al., 2009). These food waste materials usually contain high levels of cellulose, hemicellulose, lignin and protein and as such can constitute renewable natural resources for a plethora of inexpensive ecofriendly and sustainable materials (Oliveira & Franca, 2008).

Frezzini et al. (2019) investigated the adsorption capacities of three food waste materials (coffee waste, grape waste and lemon peel) with TCE, an aromatic VOC, and p-Xylene, an aliphatic Cl-VOCs. Adsorption equilibrium isotherms showed coffee waste to be the most efficient sorbent for the removal of both VOCs. This is probably due to the presence of lipids in coffee samples on which TCE and p-Xylene compounds are efficiently adsorbed, due to their lipophilic nature (Frezzini et al., 2019).

15.5 SYNTHETIC BIOSORBENTS AND THEIR EFFICACY IN REMOVING VOLATILE ORGANIC COMPOUND

Synthetic adsorbents are spherical crosslinked polymer particles with porous structures and no ion-exchange groups or functional groups. They adsorb organic compounds on the surface due to hydrophobic interaction between the compounds and the synthetic adsorbents. Synthetic carbonaceous media have been shown to be effective for treating VOC-contaminated water (Gallup et al., 1996).

Synthetic adsorbents have also found application in separation of extracts from plants such as herbal drugs and natural pigments, fermented antibiotics such as penicillins and cephalosporins, peptides such as insulin, proteins, vitamins and nutraceuticals. The surface of every biosorbent is characterized by the presence of functional groups such as amino, carboxyl, amide, carboxylate, thio-ether, thiols, sulfhydryl, imidazole, phenolic, phosphate and hydroxide groups. Pollutants ionically interact with these charged groups and adsorb onto the adsorbent surface (Samuel et al., 2020). Table 15.1 shows the main types of synthetic biosorbents and their characteristics

15.5.1 Advantages and Limitations of Using Synthetic Biosorbents

The advantages of this approach over conventional techniques such as ion exchange, coagulation and membrane separation include the following:

i. Scalability: They are scalable from analytical to industrial separations.
ii. Stability: They offer chemical stability under a wide range of conditions, such as pH.
iii. They are cost-effective due to their long shelf life and rejuvenation with severe conditions available.
iv. They require less sophisticated operation skill.
v. They generate limited sludge.
vi. Regeneration allows reuse of the adsorbent through multiple treatment cycles, with significant economic advantage (Chaukura et al., 2016; Anawar & Strezov, 2019; Baig et al., 2021).

TABLE 15.1
Types of synthetic biosorbents and their characteristics

Name of synthetic adsorbent	Characteristics	Examples
1. Polystyrenic adsorbents	Standard type of synthetic adsorbents. Made from poly(styrene-divinylbenzene)	HP20, HP20SS, HP21
2. Chemically modified polystyrenic adsorbent	Has bromine functionality attached to the poly(styrene-divinylbenzene) main chain. Addition of bromine increases the adsorption capacity and hydrophobicity of the chemically modified synthetic adsorbents Is applicable to expanded-bed adsorption or countercurrent operation.	SEPABEADS
3. Polymethacrylic adsorbent	Contains ester groups and is more polar than polystyrenic synthetic adsorbents Bind functional groups that can form hydrogen bonds better than polystyrene-based adsorbents.	CHP2MG, CHP2MGY

Source: Mitsubishi Chemical Corporation (2018).

However, synthetic biosorbents also have several challenges or limitations. These include difficulty in production and the fact that modification through cookery (synthesis/polymerization, crosslinking, template extraction) completely changes the product and hence its characteristics (Anawar & Strezov, 2019; Baig et al., 2021).

Most work on the use of modified or synthetic biosorbents has been done on removal of pollutants from wastewater as summarized in Table 15.3.

15.6 NANOADSORBENTS AND NANOCOMPOSITE ADSORBENTS

Synthesized, fabricated and upgraded nanoparticles have demonstrated efficient adsorbent properties e.g. in removal of wide range of heavy metals from wastewater. The modification of nanoadsorbent and development of nanocomposites increase the adsorption capacity and enhance the heavy metals removal. There are several types of adsorption mechanisms that explain the adherence of metal ions adsorbates to nanoadsorbents and nanocomposite adsorbents. These include surface adsorption, precipitation, electrostatic interaction and ion exchange (Baig et al., 2021; Nik-Abdul-Ghani & Alam, 2021). The categories of nanoadsorbents and their properties are shown in Table 15.3. These vary in their morphology, size and chemical properties.

15.6.1 Nanocomposite Adsorbents

Nanoadsorbents are modified by incorporating nanoparticles with polymer/metal/carbon to form nanocomposite adsorbents. Nanocomposites improve the adsorption performance and provide more specific interaction with the targeted contaminants.

TABLE 15.2
Modified/synthetic biosorbents

Country of use	Biosorbent	Type of modification	Application	Adsorption capacity q_t (mg/g)	References
China	Chitosan	Modified chitosan	Cu^{2+} removal from wastewater	65.8	Peng et al. (2014)
		Chitosan-coated cotton fibers	Removal of recovery of Hg (II)	98.9	Qu et al. (2009)
		Triethylene–tetramine grafted magnetic chitosan	Hg (II) sorption	1,044	Ge and Hua (2016)
			Adsorption of Pb (II) ion from aqueous solutions	370.63	Kuang et al. (2013)
Czech Republic	*Fomitopsis pinicola*	Acid impregnation (HCl)	Cr (VI) adsorption from WW	45.1	Pertile et al. (2021)
	Orange peel			34.1	
Egypt	*Scenedesmus obliquus* (Alga)	Alkaline (NaOH)	Removal of pharmaceutically active compound (PAC) – cefadroxil	68	Ali et al. (2018)
			Removal of PAC – ciprofloxacin	39	
India	Chitosan	Crosslinked chitosan-g-acrylonitrile	Adsorption of Cu (II) and Ni (II) ions	230.8	Ramya et al. (2011)
	Parthenium hysterophorus weed	Alkaline (NaOH)	Removal of ibuprofen	90.5	Mondal et al. (2016)
Peru	*Rumex acetosella*	Acid impregnation Sulfuric acid (H_2SO_4)	As^{3+}, Cd^{2+}, Pb^{2+} and Zn^{2+}	95.4, 109.89, 156.25, 119.05	Ligarda-Samanez et al. (2022)
Turkey	Waste apricot	Metal salt ($ZnCl_2$) impregnation	Removal of naproxen from wastewater	106.38	Önal et al. (2007)
Tunisia	Olive waste	Acid impregnation (H_3PO_4)	Removal of ibuprofen	12.6	Baccar et al. (2012)
			Removal of diclofenac	56.2	

TABLE 15.3
Nanoadsorbents and their properties

Type of nanosorbent	Examples	Properties
Carbon-based nanoadsorbents	Carbon nanotube (CNT), multiwalled CNT and graphene	
Metal oxide-based nanoadsorbents	Oxides of manganese zinc, nickel, iron, aluminum, titanium and magnesium	• They provide high removal capacity, high surface area and specific affinity toward heavy metal adsorption. • They range from 1 to 100 nm.
Polymer-based nanoadsorbents	Organic–inorganic hybrid polymers e.g. chitosan, dendrimers, cellulose nanoadsorbents	• Have stronger adsorption capacity, greater thermal stability and higher recyclability.

Source: Nik-Abdul-Ghani and Alam (2021).

Nanocomposites have many advantages in terms of low cost, stability, better mechanical properties, low energy consumption, susceptible to high temperature and harsh chemical environments (Nik-Abdul-Ghani & Alam, 2021).

REFERENCES

Abdallah, R., & Taha, S., 2012, Biosorption of methylene blue from aqueous solution by non-viable *Aspergillus fumigatus. Chemical Engineering Journal*, 195–196, 69–76. https://doi.org/10.1016/j.cej.2012.04.066

Abinandan, S., Subashchandrabose, S. R., Venkateswarlu, K., & Megharaj, M. (2018). Nutrient removal and biomass production: Advances in microalgal biotechnology for wastewater treatment. *Critical Reviews in Biotechnology*, 38(8), 1244–1260. https://doi.org/10.1080/07388551.2018.1472066

Achak, M., Hafidi, A., Ouazzani, N., Sayadi, S., & Mandi, L. (2009). Low cost biosorbent "banana peel" for the removal of phenolic compounds from olive mill wastewater: Kinetic and equilibrium studies. *Journal of Hazardous Materials*, 166(1), 117–125. https://doi.org/10.1016/j.jhazmat.2008.11.036

Agrawal, A. A., Petschenka, G., Bingham, R. A., Weber, M. G., & Rasmann, S. (2012). Toxic cardenolides: chemical ecology and coevolution of specialized plant–herbivore interactions. *New Phytologist*, 194, 28–45.

Alford, K. L., & Kumar, N. (2021). Pulmonary health effects of indoor volatile organic compounds-a meta-analysis. *International Journal of Environmental Research and Public Health*, 1578. https://doi.org/10.3390/ijerph18041578

Ali, A., & Saeed, K. (2016). Phenol removal from aqueous medium using chemically modified banana peels as low-cost adsorbent. *Desalination and Water Treatment*, 57(24), 11242–11254. https://doi.org/10.1080/19443994.2015.1041057

Ali, M. E. M., Abd El-Aty, A. M., Badawy, M. I., & Ali, R. K. (2018). Removal of pharmaceutical pollutants from synthetic wastewater using chemically modified biomass of green alga Scenedesmus obliquus. *Ecotoxicology and Environmental Safety*, 151, 144–152. https://doi.org/10.1016/j.ecoenv.2018.01.012

An, T., Wan, S., Li, G., Sun, L., & Guo, B. (2010). Comparison of the removal of ethanethiol in twin-biotrickling filters inoculated with strain RG-1 and B350 mixed microorganisms. *Journal of Hazardous Materials*, 183(1–3), 372–380.

Anand, S. S., & Mehendale, H. M. (2005). Volatile organic compounds (VOC). In: Philip Wexler (Ed.), *Encyclopedia of Toxicology*. Second Edition. Elsevier, pp. 450–455, ISBN 9780123694003, https://doi.org/10.1016/B0-12-369400-0/01015-2

Anand, S. S., Philip, B. K., & Mehendale, H. M. (2014). Volatile organic compounds. In: Philip Wexler (Eds.), *Encyclopedia of Toxicology*. Third Edition. Oxford: Academic Press, pp. 967–970.

Anastopoulos, I., Giannopoulos, G., Islam, A., Ighalo, J. O., Iwuchukwu, F. U., Pashalidis, I., Kalderis, D., Giannakoudakis, D. A., Nair, V., & Lima, E. C. (2022). Chapter13 - Potential environmental applications of helianthus annuus (sunflower) residue-based adsorbents for dye removal in (waste) waters. In: I. Anastopoulos, E. Lima, L. Meili, & D. Giannakoudakis (Eds.), *Biomass-Derived Materials for Environmental Applications*. Netherlands: Elsevier, pp. 307–318.

Anastopoulos, I., & Kyzas, G. Z. (2015). Progress in batch biosorption of heavy metals onto algae. *Journal of Molecular Liquids*, 209, 77–86. https://doi.org/10.1016/j.molliq.2015.05.023

Anawar, H., & Strezov, V. (2019). Synthesis of biosorbents from natural/agricultural biomass wastes and sustainable green technology for treatment of nanoparticle metals in municipal and industrial wastewater. In: *Emerging and Nanomaterial Contaminants in Wastewater*. Elsevier Inc. https://doi.org/10.1016/B978-0-12-814673-6.00004-8

Ata, A., Nalcaci, O. O., & Ovez, B. (2012). Macro algae *Gracilaria verrucosa* as a biosorbent: A study of sorption mechanisms. *Algal Research*, 1(2), 194–204. https://doi.org/10.1016/j.algal.2012.07.001

Atkinson, R. (2000). Atmospheric chemistry of VOCs and NOx. *Atmospheric Environment*, 34(12), 2063–2101. https://doi.org/10.1016/S1352-2310(99)00460-4

Aydin, B., Chan, N., & Malmiri H. J. (2012). Volatile organic compounds removal methods: A review. *American Journal of Biochemistry and Biotechnology*, 8(4), 220–229. https://doi.org/10.3844/ajbbsp.2012.220.229

Baccar, R., Sarrà, M., Bouzid, J., Feki, M., & Blánquez, P. (2012). Removal of pharmaceutical compounds by activated carbon prepared from agricultural by-product. *Chemical Engineering Journal*, 211–212, 310–317. https://doi.org/10.1016/j.cej.2012.09.099

Baig, U., Faizan, M., & Sajid, M. (2021). Effective removal of hazardous pollutants from water and deactivation of water-borne pathogens using multifunctional synthetic adsorbent materials: A review. *Journal of Cleaner Production*, 302. https://doi.org/10.1016/j.jclepro.2021.126735

Bailly, A., & Weisskopf, L. (2017). Mining the volatilomes of plant-associated microbiota for new biocontrol solutions. *Frontiers in Microbiology*, 8, 1638. https://doi.org/10.3389/fmicb.2017.01638

Barbusinski, K., Kalemba, K., Kasperczyk, D., Urbaniec, K., & Kozik, V. (2017). Biological methods for odor treatment–A review. *Journal of Cleaner Production*, 152, 223–241.

Behnamnia, M., Kalantari, K., & Ziaie, J. (2009). The effects of brassinosteroid on the induction of biochemical changes in *Lycopersicon esculentum* under drought stress. *Turkish Journal of Botany*, 33(6), 417–428.

Berenjian, A., Chan, N., & Malmiri, H. J. (2012). Volatile organic compounds removal methods: A review. *American Journal of Biochemistry and Biotechnology*, 8(4), 220–229. https://doi.org/10.3844/ajbbsp.2012.220.229

Bilal, M., Rasheed, T., Sosa-Hernández, J. E., Raza, A., Nabeel, F., & Iqbal, H. M. N. (2018). Biosorption: an interplay between marine algae and potentially toxic elements—A review. *Marine Drugs*, 16(2), 65.

Bouras, H. D., Isik, Z., Arikan, E. B., Yeddou, A. R., Bouras, N., Chergui, A.,… & Dizge, N. (2021). Biosorption characteristics of methylene blue dye by two fungal biomasses. *International Journal of Environmental Studies*, 78(3), 365–381.

Bouwer, E. J., & Zehnder, A. J. B. (1993). Bioremediation of organic compounds—Putting microbial metabolism to work. *Trends in Biotechnology*, 11, 360–367.

Briseno-Roa, L., Hill, J., Notman, S., Sellers, D., Smith, A. P., Timperley, C. M.,… & Griffiths, A. D. (2006). Analogues with fluorescent leaving groups for screening and selection of enzymes that efficiently hydrolyze organophosphorus nerve agents. *Journal of Medicinal Chemistry*, 49(1), 246–255.

Casas, N., Parella, T., Vincent, T., Caminal, G., & Sarra, M. (2014), Metabolites from the biodegradation of triphenyl methane dyes by Trametes versicolor or laccase. *Chemosphere*, 75, 1344–1349. https://doi.org/10.1016/j.chemosphere.2009.02.029

Chaukura, N., Gwenzi, W., Tavengwa, N., & Manyuchi, M. M. (2016). Biosorbents for the removal of synthetic organics and emerging pollutants: Opportunities and challenges for developing countries. *Environmental Development*, 19, 84–89. https://doi.org/10.1016/j.envdev.2016.05.002

Chen, S. H., & Ting, A. S. Y. (2015). Biodecolorization and biodegradation potential ofrecalcitrant triphenyl methane dyes by *Coriolopsis* sp. isolated from compost. *Journal of Environmental Management*, 150, 274–280. https://doi.org/10.1016/j.jenvman.2014.09.014

Cheng, Z., Feng, K. E., Su, Y., Ye, J., Chen, D., Zhang, S., Xiaomin Z., & Dionysiou, D. D. (2020). Novel biosorbents synthesized from fungal and bacterial biomass and their applications in the adsorption of volatile organic compounds. *Bioresource Technology*, 300, 122705. https://doi.org/10.1016/j.biortech.2019.122705

Cheng, Z., Zhang, X., Kennes, C., Chen, J., Chen, D., Ye, J.,… & Dionysiou, D. D. (2019). Differences of cell surface characteristics between the bacterium pseudomonas veronii and fungus ophiostoma stenoceras and their different adsorption properties to hydrophobic organic compounds. *Science of the Total Environment*, 650, 2095–2106.

Costa, F., Neto, M., Nicolau, A., & Tavares T. (2015). Biodegradation of diethylketone by Penicillium sp. and Alternaria sp.—A comparative study biodegradation of diethylketone by fungi. *Current Biochemical Engineering*, 2, 81–89. https://doi.org/10.2174/22127119016661408122225947

Costa, F., Quintelas, C., & Tavares, T. (2012). Kinetics of biodegradation of diethylketone by Arthrobacter viscosus. *Biodegradation*, 23(1), 81–92. https://doi.org/10.1007/s10532-011-9488-7

Costa, F., & Tavares, T. (2017). Sorption studies of diethylketone in the presence of Al^{3+}, Cd^{2+}, Ni^{2+} and Mn^{2+}, from lab-scale to pilot scale. *Environmental Technology*, 1–13. https://doi.org/10.1080/ 09593330.2016.1278462

Dai, Y., Sun, Q., Wang, W., Lu, L., Liu, M., Li, J., Yang, S., Sun, Y., Zhang, K., Xu, J., Zheng, W., Hu, Z., Yang, Y., Gao, Y., Chen, Y., Zhang, X., Gao, F., & Zhang, Y. (2018). Utilizations of agricultural waste as adsorbent for the removal of contaminants: A review. *Chemosphere*, 211, 235–253. https://doi.org/10.1016/j.chemosphere.2018.06.179

Daneshvar, E., Vazirzadeh, A., & Bhatnagar, A. (2019). Biosorption of methylene blue dye onto three different marine macroalgae: effects of different parameters on isotherm, kinetic and thermodynamic. *Iranian Journal of Science and Technology, Transactions A: Science* 43, 2743–2754. https://doi.org/10.1007/s40995-019-00764-8

Danouche, M., El Arroussi, H., Bahafid, W., & El Ghachtouli, N. (2021). An overview of the biosorption mechanism for the bioremediation of synthetic dyes using yeast cells. *Environmental Technology Reviews*, 10(1), 58–76.

de Toledo, R. A., Hin Chao, U., Shen, T., Lu, Q., Li, X., & Shim, H. (2019). Development of hybrid processes for the removal of volatile organic compounds, plasticizer, and pharmaceutically active compound using sewage sludge, waste scrap tires, and wood chips as sorbents and microbial immobilization matrices. *Environmental Science and Pollution Research*, 26(12), 11591–11604.

Decho, A. W., & Herndl, G. J. (1995). Microbial activities and the transformation of organic matter within mucilaginous material. *Science of the Total Environment*, 165(1), 33–42. https://doi.org/10.1016/0048-9697(95)04541-8

Delhomenie, M. C., & Heitz, M. (2005). Biofiltration of air: A review. *Critical Reviews in Biotechnology*, 25, 53–72. https://doi.org/10.1080/07388550590935814

Dobler, S., Petschenka, G., & Pankoke, H. (2011). Coping with toxic plant compounds—The insect'sperspectiveoniridoidglycosidesandcardenolides. *Phytochemistry*, 72, 1593–1604.

Dobslaw, D., Schulz, A., Helbich, S., Dobslaw, C., & Engesser, K. H. (2017). VOC removal and odor abatement by a low-cost plasma enhanced biotrickling filter process. *Journal of Environmental Chemical Engineering*, 5(6), 5501–5511.

Elayadi, F., Achak, M., Beniich, N., Belaqziz, M., & El Adlouni, C. (2020). Factorial design for optimizing and modeling the removal of organic pollutants from olive mill wastewater using a novel low-cost bioadsorbent. *Water, Air, & Soil Pollution*, 231(7), 351. http://doi.org/10.1007/s11270-020-04695-8

Estrada, J. M., Kraakman, N. B., Lebrero, R., & Muñoz, R. (2012). A sensitivity analysis of process design parameters, commodity prices and robustness on the economics of odour abatement technologies. *Biotechnology Advances*, 30(6), 1354–1363.

Fisher, C. L., Lane, P. D., Russell, M., Maddalena, R., & Lane, T. W. (2020). Low molecular weight volatile organic compounds indicate grazing by the marine rotifer brachionus plicatilis on the microalgae microchloropsis salina. *Metabolites*, 10(9), 361. https://doi.org/10.3390/metabo10090361

Flores-Chaparro, C. E., Chazaro Ruiz, L. F., Alfaro de la Torre, M. C., Huerta-Diaz, M. A., & Rangel-Mendez, J. R. (2017). Biosorption removal of benzene and toluene by three dried macroalgae at different ionic strength and temperatures: Algae biochemical composition and kinetics. *Journal of Environmental Management*, 193, 126–135. https://doi.org/10.1016/j.jenvman.2017.02.005

Frezzini, M. A., Massimi, L., Astolfi, M. L., Canepari, S., & Giuliano, A. (2019). Food waste materials as low-cost adsorbents for the removal of volatile organic compounds from wastewater. *Materials* (Basel), 12(24). https://doi.org/10.3390/ma12244242

Gadd, G. M. (2009). Biosorption: critical review of scientific rationale, environmental importance and significance for pollution treatment. *Journal of Chemical Technology & Biotechnology*, 84(1), 13–28. https://doi.org/10.1002/jctb.1999

Gallup, D. L., Isacoff, E. G., ^ Smith, D.N., III (1996). Use of ambersorb® carbonaceous adsorbent for removal of BTEX compounds from oil-field produced water. *Environmental Progress*, 15, 197–203. https://doi.org/10.1002/ep.670150320

Gao, Y., Cheng, Z., Ling, W., & Huang, J. (2010). Arbuscular mycorrhizal fungal hyphae contribute to the uptake of polycyclic aromatic hydrocarbons by plant roots. *Bioresource Technology*, 101, 6895–6901.

Gardette, M., Perthue, A., Gardette, J.-L., Janecska, T., Földes, E., Pukánszky, B., & Therias, S. (2013). Photo- and thermal-oxidation of polyethylene: comparison of mechanisms and influence of unsaturation content. *Polymer Degradation and Stability*, 98(11), 2383–2390. https://doi.org/10.1016/j.polymdegradstab.2013.07.017

Ge, H., & Hua, T. (2016) Synthesis and characterization of poly (maleic acid)-grafted crosslinked chitosan nanomaterial with high uptake and selectivity for Hg (II) sorption. *Carbohydrate polymers*, 153, 246–252.

Gopinath, M., Pulla, R. H., Rajmohan, K. S., Vijay, P., Muthukumaran, C., & Gurunathan, B. (2018). Bioremediation of volatile organic compounds in biofilters. In: S. J. Varjani et al. (Eds.), *Bioremediation: Applications for Environmental Protection and Management*. Singapore: Springer, pp. 301–330.

Guerrieri, F. J., Nehring, V., Jørgensen, C. G., Nielsen, J., Galizia, C. G., d'Ettorre, P. (2009). Ants recognize foes and not friends. *Proceedings of the Royal Society of London. Series B Biological Sciences*, 276, 2461–2468.

Gupta, A., Sharma, V., Sharma, K., Kumar, V., Choudhary, S., & Mishra, P. K. (2021). A review of adsorbents for heavy metal decontamination: Growing approach to wastewater treatment. *Materials (Basel)*, 14(16), 4702. doi: 10.3390/ma14164702.

Hassan, I. F., & Ai-Jawhari, H. (2015). Decolorization of methylene blue and crystal violet by some filamentous fungi. *International Journal of Environmental Bioremediation & Biodegradation*, 3(2), 62–65.

Kabbout, R., & Taha, S. (2014). Biodecolorization of textile dye effluent by biosorption on fungal biomass materials. *Physics Procedia*, 55, 437–444. https://doi.org/10.1016/j.phpro.2014.07.063

Kandyala, R., Raghavendra, S. P., & Rajasekharan, S. T. (2010). Xylene: an overview of its health hazards and preventive measures. *Journal of Oral and Maxillofacial Pathology*, 14(1), 1–5. https://doi.org/10.4103/0973-029x.64299

Karić, N., Maia, A. S., Teodorović, A., Atanasova, N., Langergraber, G., Crini, G., Ribeiro, A. R. L., & Đolić, M. (2022). Bio-waste valorisation: Agricultural wastes as biosorbents for removal of (in)organic pollutants in wastewater treatment. *Chemical Engineering Journal Advances*, 9, 100239. https://doi.org/10.1016/j.ceja.2021.100239

Kennes, C., & Veiga, M. C. (2004). Fungal biocatalysts in the biofiltration of VOC-polluted air. *Journal of Biotechnology*, 113, 305–319. https://doi.org/10.1016/j.jbiotec.2004.04.037

Kigathi, R. N., Weisser, W. W., Reichelt, M., Gershenzon, J., & Unsicker, S. B. (2019). Plant volatile emission depends on the species composition of the neighboring plant community. *BMC Plant Biology*, 19, 58.

Kim, J., Lee, S. S., & Khim, J. (2019). Peat moss-derived biochars as effective sorbents for VOCs' removal in groundwater. *Environmental Geochemistry and Health*, 41(4), 1637–1646. https://doi.org/10.1007/s10653-017-0012-9

Kim, J. M., Kim, J. H., Lee, C. Y., Jerng, D. W., & Ahn, H. S. (2018). Toluene and acetaldehyde removal from air on to graphene-based adsorbents with microsized pores. *Journal of Hazardous Materials*, 344, 458–465. https://doi.org/10.1016/j.jhazmat.2017.10.038

Khan, A. M., Wick, L. Y., Harms, H., & Thullner, M. (2016). Biodegradation of vapor-phase toluene in unsaturated porous media: Column experiments. *Environmental Pollution*, 211, 325–331. https://doi.org/10.1016/j.envpol.2016.01.013

Korpi, A., Järnberg J., & Pasanen, A-L. (2009). Microbial volatile organic compounds. *Critical Reviews in Toxicology*, 39(2), 139–193, https://doi.org/10.1080/10408440802291497

Kuang, S. P., Wang, Z. Z., Liu, J., & Wu, Z. C. (2013). Preparation of triethylene-tetramine grafted magnetic chitosan for adsorption of Pb(II) ion from aqueous solutions. *Journal of Hazardous Materials*, 260, 210–219. https://doi.org/10.1016/j.jhazmat.2013.05.019

Kumar, P. S., & Ngueagni, P. T. (2022). Chapter 15- removal of volatile organic carbon and heavy metols through microbial approach. In: M. Shah, S. Rodriguez-Couto & J. Biswas (Eds.), *An Innovative Role of Biofiltration in Wastewater Treatment Plants* (WWTPs). Netherlands: Elsevier, pp. 285–308.

Lee, X. J., Ong, H. C., Ooi, J., Yu, K. L., Tham, T. C., Chen, W.-H., & Ok, Y. S. (2022). Engineered macroalgal and microalgal adsorbents: Synthesis routes and adsorptive performance on hazardous water contaminants. *Journal of Hazardous Materials*, 423, 126921. https://doi.org/10.1016/j.jhazmat.2021.126921

Li, G., Zhang, Z., Sun, H., Chen, J., An, T., & Li, B. (2013). Pollution profiles, health risk of VOCs and biohazards emitted from municipal solid waste transfer station and elimination by an integrated biological-photocatalytic flow system: A pilot-scale investigation. *Journal of Hazardous Materials*, 250, 147–154.

Ligarda-Samanez, C. A., Choque-Quispe, D., Palomino-Rincón, H., Ramos-Pacheco, B. S., Moscoso-Moscoso, E., Huamán-Carrión, M. L., Peralta-Guevara, D. E., Obregón-Yupanqui, M. E., Aroni-Huamán, J., Bravo-Franco, E. Y., Palomino-Rincón, W., & de la Cruz, G. (2022). Modified polymeric biosorbents from Rumex acetosella for the removal of heavy metals in wastewater. *Polymers*, 14(11). https://doi.org/10.3390/polym14112191

Liu, H. L., Chen, B. Y., Lan, Y. W., & Cheng, Y. C. (2004). Biosorption of Zn(II) and Cu(II) by the indigenous Thiobacillus thiooxidans. *Chemical Engineering Journal*, 97(2–3), 195–201. https://doi.org/10.1016/S1385-8947(03)00210-9

Lomonaco, T., Manco, E., Corti, A., La Nasa, J., Ghimenti, S., Biagini, D., Di Francesco, F., Modugno, F., Ceccarini, A., Fuoco, R., & Castelvetro, V. (2020). Release of harmful volatile organic compounds (VOCs) from photo-degraded plastic debris: A neglected source of environmental pollution. *Journal of Hazardous Materials*, 394, 122596. https://doi.org/10.1016/j.jhazmat.2020.122596

Loreto, F., Dicke, M., Schnitzler, J. P., & Turlings, T. C. (2014). Plant volatiles and the environment. *Plant, Cell & Environment*, 37(8), 1905–1908. http://doi.org/10.1111/pce.12369

Mack, C., Wilhelmi, B., Duncan, J. R., & Burgess, J. E. (2007). Biosorption of precious metals. *Biotechnology Advances*, 25(3), 264–271. https://doi.org/10.1016/j.biotechadv.2007.01.003

Marcharchand, S., & Ting, A. S. Y. (2017). Trichoderma asperellum cultured in reduced concentrations of synthetic medium retained dye decolourization efficacy. *Journal of Environmental Management*, 203, 542–549. https://doi.org/10.1016/j.jenvman.2017.06.068

Masih, A., Lall, A. S., Taneja, A., & Singhvi, R. (2016). Inhalation exposure and related health risks of BTEX in ambient air at different microenvironments of a terai zone in north India. *Atmospheric Environment*, 147, 55–66. https://doi.org/10.1016/j.atmosenv.2016.09.067

McGuinness, M., & Dowling, D. (2009). Plant-associated bacterial degradation of toxic organic compounds in soil. *International Journal of Environmental Research and Public Health*, 6, 2226–2247.

Mitsubishi Chemical Corporation. (2018). Diaion TM Technical Manual – Synthetic adsorbents: Diaion TM and Sepabeads TM.

Mohsenzadeh, F., Nasseri, S., Mesdaghinia, A., Nabizadeh, R., Zafari, D., Khodakaramian, G., & Chehregani, A. (2010). Phytoremediation of petroleum-polluted soils: Application of polygonum aviculare and its root-associated (penetrated) fungal strains for bioremediation of petroleum-polluted soils. *Ecotoxicology and Environmental Safety*, 73, 613–619.

Mondal, S., Aikat, K., & Halder, G. (2016). Biosorptive uptake of ibuprofen by chemically modified Parthenium hysterophorus derived biochar: Equilibrium, kinetics, thermodynamics and modeling. *Ecological Engineering*, 92, 158–172. https://doi.org/10.1016/j.ecoleng.2016.03.022

Montero-Montoya, R., López-Vargas, R., & Arellano-Aguilar, O. (2018). Volatile organic compounds in air: Sources, distribution, exposure and associated illnesses in children. *Annals of Global Health*, 84(2), 225–238. http://doi.org/10.29024/aogh.910

Niaz, K., Bahadar, H., Maqbool, F., & Abdollahi, M. (2015). A review of environmental and occupational exposure to xylene and its health concerns. *EXCLI Journal*, 14, 1167–1186. http://doi.org/10.17179/excli2015–623

Nicastro, R., & Carillo, P. (2021). Food loss and waste prevention strategies from farm to fork. *Sustainability*, 13(10), 5443. https://doi.org/10.3390/su13105443

Nik-Abdul-Ghani, J., & Alam. (2021). The role of nanoadsorbents and nanocomposite adsorbents in the removal of heavy metals from wastewater: A review and prospect. *Pollution*, 7(1), 153–179. https://doi.org/10.22059/poll.2020.307069.859

Nwidi, I., & Agunwamba, J. (2015). Selection of biosorbents for biosorption of three heavy metals in a flow-batch reactor using removal efficiency as parameter. *Nigerian Journal of Technology*, 34(2), 406. https://doi.org/10.4314/njt.v34i2.27

Okoro, H. K., Pandey, S., Ogunkunle, C. O., Ngila, C. J., Zvinowanda, C., Jimoh, I., Lawal, I. A., Orosun, M. M., & Adeniyi, A. G. (2022). Nanomaterial-based biosorbents: Adsorbent for efficient removal of selected organic pollutants from industrial wastewater. *Emerging Contaminants*, 8, 46–58. https://doi.org/10.1016/j.emcon.2021.12.005

Oliveira, L. S., & Franca, A. S. (2008). Low-cost adsorbents from agri-food wastes. In: L. V. Greco and M. N. Bruno (Eds.), *Food Science and Technology: New Research*, Hauppauge, New York: Nova Science Publishers, Inc., pp. 171-209.

Önal, Y., Akmil-Başar, C., & Sarici-Özdemir, Ç. (2007). Elucidation of the naproxen sodium adsorption onto activated carbon prepared from waste apricot: Kinetic, equilibrium and thermodynamic characterization. *Journal of Hazardous Materials*, 148(3), 727–734. https://doi.org/10.1016/j.jhazmat.2007.03.037

Palm, B. B., de Sá, S. S., Day, D. A., Campuzano-Jost, P., Hu, W., Seco, R., Sjostedt, S.J., Park, J.-H., Guenther, A. B., Kim, S., Brito, J., Wurm, F., Artaxo, P., Thalman, R., Wang, J., Yee, L. D., Wernis, R., Isaacman-VanWertz, G., Goldstein, A. H., Liu, Y., Springston, S. R., Souza, R., Newburn, M. K., Alexander, M. L., Martin, S. T., & Jimenez, J. L. (2018). Secondary organic aerosol formation from ambient air in an oxidation flow reactor in central Amazonia. *Atmospheric Chemistry and Physics*, 18, 467–493.

Park, D., Yun, Y.-S., & Park, J. M. (2010a). The past, present, and future trends of biosorption. *Biotechnology and Bioprocess Engineering*, 15(1), 86–102. https://doi.org/10.1007/s12257-009-0199-4

Park, K.-H., Balathanigaimani, M. S., Shim, W.-G., Lee, J.-W., & Moon, H. (2010b). Adsorption characteristics of phenol on novel corn grain-based activated carbons. *Microporous and Mesoporous Materials*, 127(1), 1–8. https://doi.org/10.1016/j.micromeso.2009.06.032

Peng, S., Meng, H., Ouyang, Y., & Chang, J. (2014). Nanoporous magnetic cellulose-chitosan composite microspheres: Preparation, characterization, and application for Cu(II) adsorption. *Industrial and Engineering Chemistry Research*, 53(6), 2106–2113. https://doi.org/10.1021/ie402855t

Pennerman, K. K., AL-Maliki, H. S., Lee, S., & Bennett J. W. (2016). Chapter 7 – Fungal volatile organic compounds (VOCs) and the genus aspergillus. In: Vijai Kumar Gupta, *New and Future Developments in Microbial Biotechnology and Bioengineering*, Elsevier, pp. 95–115, ISBN 9780444635051. https://doi.org/10.1016/B978-0-444-63505-1.00007-5

Pertile, E., Dvorský, T., Václavík, V., & Heviánková, S. (2021). Use of different types of biosorbents to remove cr (Vi) from aqueous solution. *Life*, 11(3). https://doi.org/10.3390/life11030240

Pokharel, A., Acharya, B., & Farooque, A. (2020). Biochar-assisted wastewater treatment and waste valorization. *Applications of Biochar for Environmental Safety*, 19, DOI: 10.5772/intechopen.92288.

Premkumar, M. P., Thiruvengadaravi, K. V., Senthil Kumar, P., Nandagopal, J., & Sivanesan, S. (2018). Eco-friendly treatment strategies for wastewater containing dyes and heavy metals. In: T. Gupta, A. K. Agarwal, R. A. Agarwal, & N. K. Labhsetwar (Eds.), *Environmental Contaminants: Measurement, Modelling and Control*. Singapore: Springer Singapore, pp. 317–360.

Prenafeta-Boldú, F.X., de Hoog, G.S., & Summerbell, R.C. (2018). Fungal communities in hydrocarbon degradation. In: T. J. McGenity (Ed.), *Microbial Communities Utilizing Hydrocarbons and Lipids: Members, Metagenomics and Ecophysiology*. Cham: Springer International Publishing, pp. 1–36.

Prenafeta-Boldú, F.X., Illa, J., van Groenestijn, J., & Flotats, X. (2008). Influence of synthetic packing materials on the gas dispersion and biodegradation kinetics in fungal air biofilters. *Applied Microbiology and Biotechnology*, 79, 319–327.

Qu, R., Sun, C., Ma, F., Zhang, Y., Ji, C., Xu, Q., Wang, C., & Chen, H. (2009). Removal and recovery of Hg(II) from aqueous solution using chitosan-coated cotton fibers. *Journal of Hazardous Materials*, 167(1–3), 717–727. https://doi.org/10.1016/j.jhazmat.2009.01.043

Ramya, R., Sankar, P., Anbalagan, S., & Sudha, P. N. (2011). Adsorption of Cu (II) and Ni (II) ions from metal solution using crosslinked chitosan-g-acrylonitrile copolymer. *International Journal of Environmental Sciences*, 1(6), 1323.

Ramrakhiani, L., Ghosh, S., & Majumdar, S. (2016). Surface modification of naturally available biomass for enhancement of heavy metal removal efficiency, upscaling prospects, and management aspects of spent biosorbents: a review. *Applied Biochemistry and Biotechnology*, 180(1), 41–78. https://doi.org/10.1007/s12010-016-2083-y

Rodriguez-Hernandez, M. C., Flores-Chaparro, C. E., & Rangel-Mendez, J. R. (2017). Influence of dissolved organic matter and oil on the biosorption of BTEX by macroalgae in single and multi-solute systems. *Environmental Science and Pollution Research*, 24(26), 20922–20933. https://doi.org/10.1007/s11356-017-9672-3

Russell, J. A., Hu, Y., Chau, L., Pauliushchyk, M., Anastopoulos, I., Anandan, S., & Waring, M. S. (2014). Indoor-biofilter growth and exposure to airborne chemicals drive similar changes in plant root bacterial communities. *Applied and Environmental Microbiology*, 80(16), 4805–4813.

Rybarczyk, P., Marycz, M., Szulczyński, B., et al. (2021). Removal of cyclohexane and ethanol from air in biotrickling filters inoculated with Candida albicans and Candida subhashii. *Archives of Environmental Protection*. https://doi.org/10.24425/aep.2021.136445

Safieddine, S. A., Heald, C. L., & Henderson, B. H. (2017). The global non-methane reactive organic carbon budget: a modeling perspective: Global reactive organic carbon budget. *Geophysical Research Letters*, 44, 3897–3906.

Samuel P. N. K. J., Soosai, M. R., Ganesh Moorthy, I., & Sankar, K. (2020). Material and process selection for biosorption. In: *Handbook of Environmental Chemistry*. Vol. 104. Springer Science and Business Media Deutschland GmbH, pp. 241–259. https://doi.org/10.1007/698_2020_507

Sarim, K. M., Kukreja, K., Shah, I., & Choudhary, C. K. (2019). Biosorption of direct textile dye Congo red by Bacillus subtilis HAU-KK01. *Bioremediation Journal*, 23(3), 185–195.

Shah, K. W., & Li, W. (2019). A review on catalytic nanomaterials for volatile organic compounds VOC removal and their applications for healthy buildings. *Nanomaterials*, 9(6), 910.

Shen, Y., & Zhang, N. (2019). Facile synthesis of porous carbons from silica-rich rice husk char for volatile organic compounds (VOCs) sorption. *Bioresource Technology*, 282, 294–300. https://doi.org/10.1016/j.biortech.2019.03.025

Shuai, J., Kim, S., Ryu, H., Park, J., Lee, C. K., Kim, G.-B., Ultra, V. U., & Yang, W. (2018). Health risk assessment of volatile organic compounds exposure near Daegu dyeing industrial complex in South Korea. *BMC Public Health*, 18(1), 528. https://doi.org/10.1186/s12889-018-5454-1

Tanwer, N., Bumbra, P., Khosla, B., & Laura, J. S. (2022). Mercury pollution and its bioremediation by microbes. In: *Microbes and Microbial Biotechnology for Green Remediation*. Elsevier, pp. 651–664.

Thurston, G. D. (2017). Outdoor air pollution: Sources, atmospheric transport, and human health effects. In: Stella R. Quah (Ed.,), *International Encyclopedia of Public Health*. Second Edition.Academic Press, pp. 367–377. https://doi.org/10.1016/B978-0-12-803678-5.00320-9

Tsai, W.-T. (2019). An overview of health hazards of volatile organic compounds regulated as indoor air pollutants. *Reviews on Environmental Health*, 34(1), 81–89. https://doi.org/10.1515/reveh-2018-0046

Turon, X., Venus, J., Arshadi, M., Koutinas, M., Lin, C. S., & Koutinas, A. (2014). Food waste and byproduct valorization through bio-processing: Opportunities and challenges. *BioResources*, 9(4), 5774–5777.

United States Environmental Protection Agency (2017). Technical overview of volatile organic compounds. https://www.epa.gov/indoor-air-quality-iaq/technical-overview-volatile-organic-compounds Last updated on MARCH 14, 2023

Vahabisani, A., & An, C. (2021). Use of biomass-derived adsorbents for the removal of petroleum pollutants from water: A mini-review. *Environmental Systems Research*, 10(1), 25. https://doi.org/10.1186/s40068-021-00229-1

Van Tran, S., Nguyen, K. M., Nguyen, H. T., Stefanakis, A. I., & Nguyen, P. M. (2022). Chapter 27 - Food processing wastes as a potential source of adsorbent for toxicant removal from water. In: A. Stefanakis & I. Nikolaou (Eds.), *Circular Economy and Sustainability*. Amsterdam, Netherlands: Elsevier, pp. 491–507.

Vankar, P. S., Sarswat, R., Dwivedi, A. K., & Sahu, R. S. (2013). An assessment and characterization for biosorption efficiency of natural dye waste. *Journal of Cleaner Production*, 60, 65–70. https://doi.org/10.1016/j.jclepro.2011.09.021

Wang, H., Nie, L., Li, J., Wang, Y., Wang, G., Wang, J., & Hao, Z. (2013). Characterization and assessment of volatile organic compounds (VOCs) emissions from typical industries. *Chinese Science Bulletin*, 58(7), 724–730. https://doi.org/10.1007/s11434-012-5345-2

Wang, M (2019). Study of volatile organic compounds (VOC) in the cloudy atmosphere: Air/droplet partitioning of VOC. *Earth Sciences*. Université Clermont Auvergne. 2019CLFAC080.

Wang, N., Zhang, Y., Zhu, F., Li, J., Liu, S., & Na, P. (2014). Adsorption of soluble oil from water to graphene. *Environmental Science and Pollution Research*, 21(10), 6495–6505. https://doi.org/10.1007/s11356-014-2504-9

Wolverton, B. C., & Wolverton, J. D. (1993). Plants and soil microorganisms: Removal of formaldehyde, xylene, and ammonia from the indoor environment. *Journal of Mississippi Academy of Sciences*, 38, 11–15.

Wu, C., Xu, P., Xia, Y., Li, W., Li, S., & Wang, X. (2017). Microbial compositions and metabolic interactions in one- and two-phase partitioning airlift bioreactors treating a complex VOC mixture. *Journal of Industrial Microbiology and Biotechnology*, 44(9), 1313–1324. http://doi.org/10.1007/s10295-017-1955-7

Xiang, W., Zhang, X., Chen, K., Fang, J., He, F., Hu, X., Tsang, D. C. W., Ok, Y. S., & Gao, B. (2020). Enhanced adsorption performance and governing mechanisms of ball-milled biochar for the removal of volatile organic compounds (VOCs). *Chemical Engineering Journal*, 385, 123842. https://doi.org/10.1016/j.cej.2019.123842

Xi-Fu. (2016). Indoor microbial volatile organic compound (MVOC) levels and associations with respiratory health, sick building syndrome (SBS), and allergy. In: C. Viegas, A. C. Pinheiro, R. Sabino, S. Viegas, J. Brandão, C. Veríssimo (Eds.), *Environmental Mycology in Public Health*. Academic Press, pp. 387–395, https://doi.org/10.1016/B978-0-12-411471-5.00022

Yang, F., Gao, S., Ding, Y., Tang, S., Chen, H., Chen, J., Liu, J., Yang, Z., Hu, X., & Yuan, A. (2019a). Excellent porous environmental nanocatalyst: Tactically integrating size-confined highly active MnOx in nanospaces of mesopores enables the promotive catalytic degradation efficiency of organic contaminants. *New Journal of Chemistry*, 43(48), 19020–19034. https://doi.org/10.1039/C9NJ05092B

Yang, F., Shao, B., Liu, X., Gao, S., Hu, X., Xu, M., Wang, Y., Zhou, S., & Kong, Y. (2019b). Nanosheet-like Ni-based metasilicate towards the regulated catalytic activity in styrene oxidation via introducing heteroatom metal. *Applied Surface Science*, 471, 822–834. https://doi.org/10.1016/j.apsusc.2018.12.074

Yu, K. L., Show, P. L., Ong, H. C., Ling, T. C., Chi-Wei Lan, J., Chen, W.-H., & Chang, J.-S. (2017). Microalgae from wastewater treatment to biochar – Feedstock preparation and conversion technologies. *Energy Conversion and Management*, 150, 1–13. https://doi.org/10.1016/j.enconman.2017.07.060

Zainith, S., Saxena, G., Kishor, R., & Bharagava, R. N. (2021). Chapter 20 - Application of microalgae in industrial effluent treatment, contaminants removal, and biodiesel production: Opportunities, challenges, and future prospects. In: G. Saxena, V. Kumar, & M. P.

Shah (Eds.), *Bioremediation for Environmental Sustainability*. Amsterdam, Netherlands: Elsevier, pp. 481–517.

Zaitan, H., Bianchi, D., Achak, O., & Chafik, T. (2008). A comparative study of the adsorption and desorption of o-xylene onto bentonite clay and alumina. *Journal of Hazardous Materials*, 153(1), 852–859. https://doi.org/10.1016/j.jhazmat.2007.09.070

Zhang, X., Gao, B., Creamer, A. E., Cao, C., & Li, Y. (2017a). Adsorption of VOCs onto engineered carbon materials: A review. *Journal of Hazardous Materials*, 338, 102–123. https://doi.org/10.1016/j.jhazmat.2017.05.013

Zhang, X., Gao, B., Zheng, Y., Hu, X., Creamer, A. E., Annable, M. D., & Li, Y. (2017b). Biochar for volatile organic compound (VOC) removal: Sorption performance and governing mechanisms. *Bioresource Technology*, 245, 606–614. https://doi.org/10.1016/j.biortech.2017.09.025

Zou, W., Gao, B., Ok, Y. S., & Dong, L. (2019). Integrated adsorption and photocatalytic degradation of volatile organic compounds (VOCs) using carbon-based nanocomposites: A critical review. *Chemosphere*, 218, 845–859. https://doi.org/10.1016/j.chemosphere.2018.11.175

16 Nano-biosorbents and Their Applications in Electrochemical Sensing and Adsorptive Removal of Environmental Pollutants

Shreanshi Agrahari, Ankit Kumar Singh, Ravindra Kumar Gautam, and Ida Tiwari

16.1 INTRODUCTION

Pollution has enhanced people's desire for a better life. Since the past decade, industrialization has accelerated to meet demand (Khan et al. 2019). Modern technology and a growing range of industries improve people's lives (Sahu et al. 2021). Rising pollution from these businesses has had harmful consequences. Multiple sectors throughout the globe strive to fulfil public demand (Karri, Shams, and Sahu 2018). Food, fibre, textile, paper, nuclear power plant, pharmaceutical, oil refinery, construction, canning, tanneries, paper, pulp mills, and breweries (Khan et al. 2019; Werkneh and Rene 2019). These industries produce toxins, byproducts, and secondary pollutants that are released into the environment. Non-biodegradable pollutants harm people, plants, and animals. Often, tainted wastewater causes plants to develop slowly, produce few fruits or vegetables, and die (Agarwal, Gupta, and Agarwal 2019). Contaminated industrial waste also causes infections in animals, harming livestock. In humans, pollutants cause mutagenesis and cancer. Cholera, diarrhea, renal illnesses, skin irritations, and asthma are caused by industrial pollutants (Prabakar et al. 2018). Given the dangers of industrial effluents and the illnesses they cause, researchers are searching for effective wastewater cleansing solutions.

Coagulation/flocculation (Dotto et al. 2019), precipitation (Navamani Kartic, Aditya Narayana, and Arivazhagan 2018), electrolytic processes (Aquino et al. 2017), and advanced electrochemical processes (Khan et al. 2019) are popular approaches for the removal of environmental contaminants (*c.f.* Figure 16.1). Despite their popularity, these approaches have limits. These processes produce secondary pollutants, which are expensive, time-consuming, and inefficient. To address these issues,

DOI: 10.1201/9781003366058-16

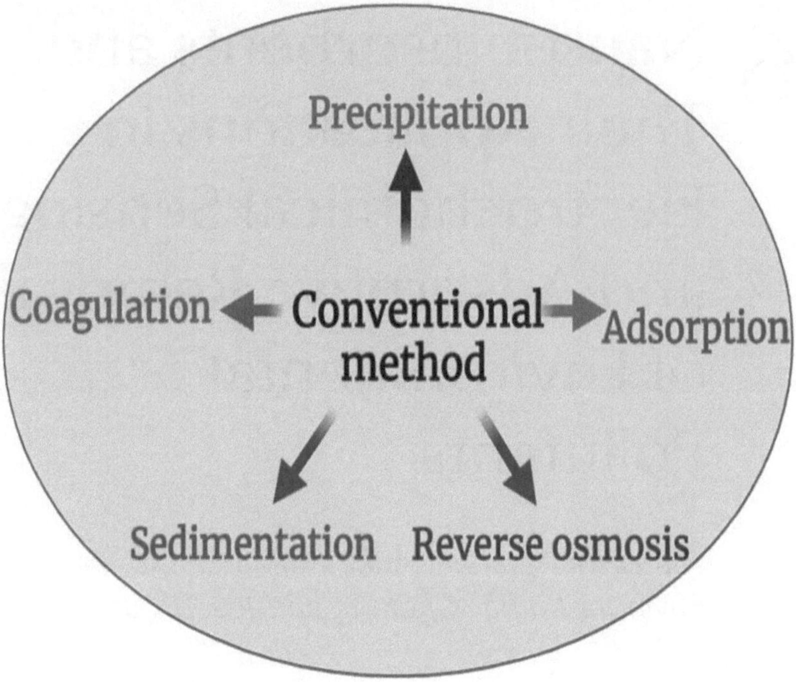

FIGURE 16.1 Several conventional techniques used for detection of environmental contaminants

scientists are developing greener methods for decontaminating waste (Navamani Kartic, Aditya Narayana, and Arivazhagan 2018; Collivignarelli et al. 2019).

Nanotechnology is a new field of science and technology (Mahamadi 2019). The nanoparticle size range is 1–100 nm, which lies in the macroscopic and molecular bulk materials (Khan et al. 2019). Nanoparticles exhibit physical, chemical, and electrical characteristics. Biosorption performance may be increased by introducing nanomaterials, which have small size, big surface-to-volume ratio, large surface areas, high reactivity, and greater sorption capacity.

Biosorbents are biomaterials that absorb metals from water. Biosorbents are microbial or plant biomass. Biosorbent formulation requires the proper biomass. Biosorption is predominantly cell surface sequestration; hence, cell wall alteration might affect metal ion binding. The effectiveness of biosorption may be increased by adding nanomaterials by capturing their beneficial qualities, such as tiny size, big surface-to-volume ratio, large surface areas, high reactivity, and greater sorption capacity (Singh et al. 2021). Nano-biosorbents are better than traditional biosorbents in removing pollutants (Kaushik and Malik 2009). The nano-biosorbents are biological components with one dimension and a size range of less than 100 nm. Nano-biosorbents are nontoxic, have a high sorption capacity, are sensitive even at low concentrations (ppb), and are desorbable, increasing their reusability (*c.f.* Figure 16.2). Other aspects that promote nano-biosorbent efficacy include

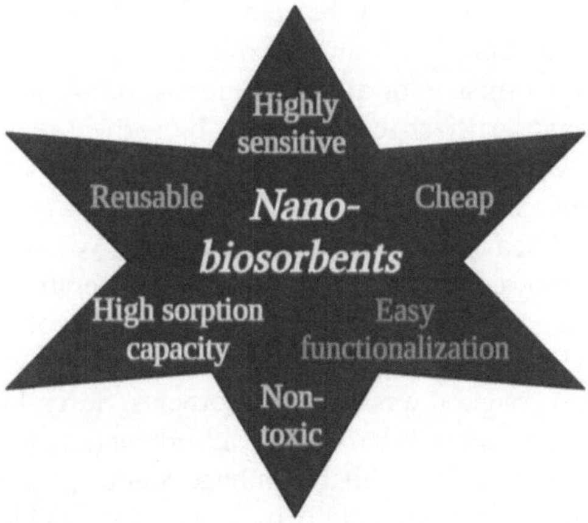

FIGURE 16.2 Advantages of nano-biosorbents

cost-effectiveness, selectivity, and functionalization (Singh et al. 2022). Due to its usefulness, nano-biosorbents are used to eliminate environmental contaminants.

This chapter examines nanosystem as a multidisciplinary science with the goal of designing smart-sensing technologies on combining with non-bioabsorbent for detecting environmental contaminants. It explores various immobilization processes for nano-biosorbents. Moreover, various electrochemical techniques using nano-biosorbents for analysis pollutants are also examined.

16.2 NANO-BIOSORBENTS

Inorganic or organic molecules from diluted water solutions may be abiotically adsorbed and concentrated by living or dead biomass via a process known as biosorption (Demirbas 2008). To remove substances from water bodies, a method known as biosorption is used. Biosorbents are materials generated from nonliving biomass. Heavy metals, rare earth elements, radionuclides, and metalloids are "traditionally" removed by biosorption, although research and applications have expanded to include the removal of organics, such as dye (Kaushik and Malik 2009). High-value proteins, steroids, pharmaceuticals, and antibiotics are among the substances that biosorption has been suggested to recover. The mechanism behind microbial biosorption of heavy metals and related elements was discovered after decades of biosorption research. Adsorption, ion exchange, chelation, complexation, and micro-precipitation are all involved (Khan et al. 2019). These rely on metal kind and ionic form, microbial biomass metal binding sites, and external environmental variables. Biosorbents are made from biomass with high metal-binding ability. Investigated biomass types include bacterial, cyanobacterial, fungal, algal, plant, and animal (chitosan) (Kaushik and Malik 2009).

Biosorbents may also be made from nonliving microbes. Biosorption may extract heavy metals from wastewater based on the metal-binding capabilities of biological components. By passively binding with nonliving microorganisms like bacteria, fungi, and yeasts, heavy metals may be removed by biosorption (Volesky 2001). Biosorption offers a number of advantages over conventional treatment methods, including low cost, high efficacy for diluted solutions, little chemical and/or biological sludge, no additional fertilizer requirements, biosorbent regeneration, and metal recovery (Mahamadi 2019). Proteins, polysaccharides, and phenolic compounds, in particular, which include functional groups that may attach to metal ions, are responsible for the sorption of heavy metals onto biomaterials (Volesky 2001).

Several biomasses showed excellent removal. pH, temperature, and adsorbent dosage impact biosorption capabilities. Industrial use is still far off (Vilar, Botelho, and Boaventura 2007). The majority of biomasses can remove heavy metals from solutions, however not all of them are good wastewater adsorbents. Several important properties must be stated to make the material an industrial adsorbent (Lesmana et al. 2009) which includes high adsorption capacity, available in huge number, low economic value and alternate usage. Attached metals are readily retrieved, whereas biosorbent is reusable.

Many biosorbents and/or alternative adsorbents have a high capacity for adsorption, with some of them being superior than adsorbents that are easily available in markets (Macaskie 1990). This suggests that the majority of biosorbents and/or other adsorbents have promised for commercial use. In nature, there are several low-binding biomasses. Their capacity for adsorption may be improved by pretreatment or chemical modification. Acid, alkali, or other oxidizing and organic chemicals are added during chemical modification. Boiling, heating, autoclaving, and freeze-drying are all used to modify material physically. Chemical activation approaches are undesirable because they lose the advantages of being economical and environmentally friendly. Chemical waste handling is expensive and presents significant challenges (Volesky 2001).

Large markets exist for inexpensive biosorbents. Metal-bearing effluents are a problem in the mining and ore-processing industry, electroplating and metal finishing, smelting, and tanneries (Ali et al. 2020). Biosorption's potential is huge. Cheaper biosorbents might create new environmental markets inaccessible to ion-exchange resins due to their exorbitant cost for clean-up operations. These factors show biosorption's economic feasibility and promise for heavy metal removal/recovery. Microbial and plant biomass and other biological processes provide interesting alternatives to existing heavy metal treatment approaches (Agarwal, Gupta, and Agarwal 2019). New biological technologies may supplement existing treatments but not replace them.

Metal recovery may be an extra source of money from a necessary water treatment. Biosorption is a well-researched method for removing and recovering metals from aqueous solutions. Biosorption can remove metals across a wide pH and temperature range, has fast adsorption and desorption kinetics, and is low-cost to operate. Biosorbent may be made using cheap growing medium or as an industrial byproduct (Ahluwalia and Goyal 2007). Biosorption is more cost-effective than its nearest opponent, ion exchange, and can be converted simply. Heavy metal recovery reduces costs. Knowing that hundreds of biosorbents may bind diverse contaminants, enough study has been done on biomaterials to comprehend biosorption (Jain, Varshney, and Srivastava 2017; Aziz et al. 2020). Through ongoing study, particularly on pilot and full-scale biosorption processes, biosorption technology may become more useful and appealing than presently employed technologies (Sartaj et al. 2020).

Biosorption technique may remove toxic metals cost-effectively and reduce hazardous waste sludges. Metals may be easily recovered and recycled by biosorption. Biosorbent materials have metal-binding sites throughout (granules, fibres) (Sharma and Bhattacharya 2017). Biosorbents are an excellent new wastewater treatment option due to their high metal-collecting capability, cheap cost, and frequent reuse. Bacterial biosorbents are tiny, low-density, weak particles. Despite having a high capacity for biosorption, achieving equilibrium quickly, being inexpensive to operate, and having effective particle mass transfer. They frequently have issues with solid–liquid separation, biomass swelling, regenerating/reusing, and column pressure loss (Tran et al. 2019).

Several researchers reported methods for synthesizing nano-biosorbent material (*c.f.* Figure 16.3), including chemical, physical, microwave-assisted, ball mill, precipitation, sol–gel, green synthesis, and biological synthesis (Ali 2012; Kanamarlapudi, Chintalpudi, and Muddada 2018). These techniques were used to remove contaminants from water sources, including synthetic colors, dangerous compounds, heavy metals, and biological waste. Green chemistry is the backbone of the production of nano-biosorbents, and the precursor is the key to classification. Precursors for nano-biosorbents are listed include waste, biological precursors, and plant extracts (Tran et al. 2019).

Fruit peels, sugarcane debris, sawdust, and other agricultural wastes were all used as reducing and stabilizing agents for nano-biosorbent (Ali 2012). In most studies, nano-biosorbent was synthesized utilizing flowers, roots, latex, leaves, etc. Algae, fungi, and bacteria are biological antecedents for biosorbent, which removes toxins from water sources. According to some studies, nano-biosorbent may be made

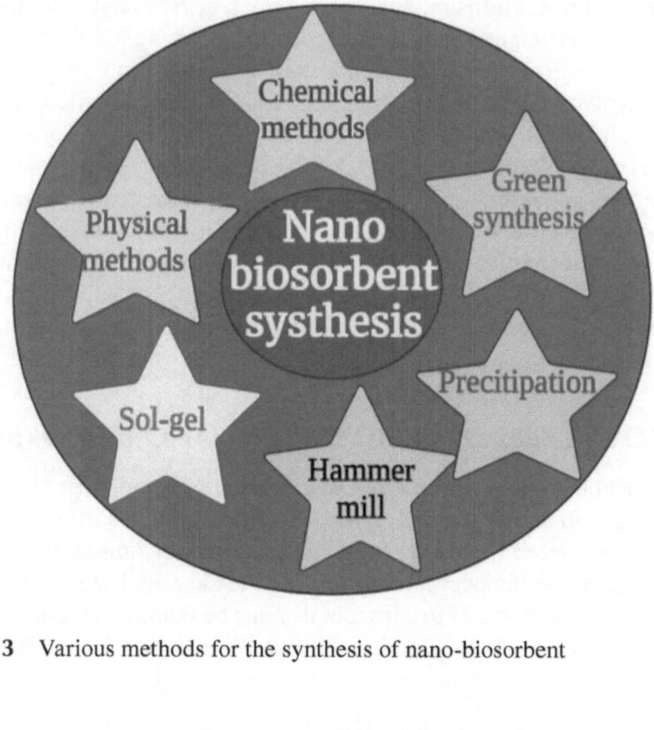

FIGURE 16.3 Various methods for the synthesis of nano-biosorbent

utilizing chitosan, dextrin, and other chemical precursors, as well as oil from natural sources. Bioadsorbents have gained popularity over time. Agricultural waste, marine debris, microbial biosorbents, industrial byproducts, and minerals are common bioadsorbents (Vilar, Botelho, and Boaventura 2007). These biomaterials have great biosorption capability, are available in large numbers, are inexpensive, and are easily desorbable and reusable.

Nano-biosorbent morphology and shape affect properties and performance. Bottom-up and top-down nanoadsorbent synthesis (Hegazi 2013). Top-down processes include ball milling, reactive milling, and mechanical alloying. Bottom-up nanoparticle production is modern. Sol–gel, molecular self-assembly, and physical/chemical vapor deposition yield nanoadsorbents. Several innovative nanosorbents have been synthesized and employed to remove pollutants from sewage, including silica/Fe_3O4 nanoparticles, nanometals, and nanometal oxides zeolites, magnetic nanoparticles s like activated charcoal/Fe_3O_4 nanoparticles and bentonite/$CuFe_2O_4$ nanocomposite, carbon nanotubes (CNTs) (Vojoudi et al. 2017). Demonstrates sorption capacity, efficiency, and ideal conditions for removing contaminants from wastewater (El-Ashtoukhy, Amin, and Abdelwahab 2008).

Nano-biosorption involves creating nano-biosorbents, or biomaterials having nanometer-scale components (Ali et al. 2020). Nanoscale characteristics of nano-biosorbents fluctuate significantly. Nano-biosorbents have more surface-to-volume than microbiosorbents. Nano-biosorbents combine biosorption with nanotechnology. Nano-biosorbents bind pollutants such as heavy metals, dyes, pesticides, and inorganic ions in wastewater (Aziz et al. 2020). Biosorption enables convenient availability of low-cost sorbents, which is a benefit over other approaches. Biological byproducts may be utilized as sorbents including starch, cellulose, lignin, chitin, chitosan, sugarcane bagasse, rice husk, etc. (Sartaj et al. 2020). Biosorption efficiently removes suspended compounds from dilute aqueous solutions. This process's great sensitivity removes ppm to ppb pollutants. Biosorption involves electrostatic contact, ion exchange, microprecipitation, complexation, chelation, coordination, etc. Nanotechnology improves biosorption characteristics (Wang and Pang 2020). Nano-biosorbents have improved pollutant removal capacity. The advantages of nano-biosorbents include their high surface-area-to-volume ratios, renewability, biocompatibility, high natural abundance, better mechanical characteristics, optical transparency, sustainability, adaptable surface chemistry, and environmental inertness. Nano-biosorbents are widely employed to decontaminate wastewater, assuring clean water and a green environment (Aquino et al. 2017).

16.3 VARIOUS TYPES OF IMMOBILIZED NANO-BIOSORBENTS

The sorbent must be employed as a fixed or extended bed and must not cause a significant pressure drop across the bed in order to function as a sorption system. This has to be processed, sized, pelleted, chemically altered, or immobilized. These are intended to enhance metal-specific binding sites inside a bed reactor (Gadd 2008). The ability of microbial biomass to sorb metals must be maintained during a continuous industrial process, and it must be immobilized (Singh and Tiwari 2020).

Free cells need high hydrostatic pressures to flow because they have weak mechanical properties and small particle sizes. Tensions biomass disintegration by using the cellular immobilization (Siddiquee, Rovina, and Azad 2015). Immobilized biomass provides reusability, high biomass loading, and minimum clogging in continuous flow systems. Biomass solidification structures generate the proper size, mechanical strength, and stiffness. Chemical engineering unit activities need porosity. Biomass immobilization uses many approaches. Inert supports, trapping in polymeric matrix, covalent attachment to vector chemicals, or cell cross-linking are several literature-based biosorption techniques (Volesky 2007).

Heavy metals are non-degradable, persistent contaminants that pollute aquatic systems. Treatment might vary from a major procedure for heavily contaminated industrial waste to a polishing step for eliminating trace quantities (Dodbiba, Ponou, and Fujita 2015). Standard methods for treating wastewater include oxidation/reduction, ion exchange, and precipitation as hydroxides/sulfides. In order to purify polluted water, microbial or plant biomass may be utilised to passively bind metals. Biosorption provides cost-effective industrial wastewater control. In prolonged contact with a metal-containing fluid, living biomass may bioaccumulate metals intracellularly (Szilva et al. 1998). Living and nonliving biomass can biosorb, however only live biomass can bioaccumulate. Biosorption may occur through complexation, coordination, chelatation of metals, ion exchange, adsorption, and inorganic microprecipitation (Veglio' and Beolchini 1997; Tran et al. 2019).

To improve metal binding capacity and comprehend biosorption, many techniques have been explored to alter microbial cell walls. Physical and chemical treatments may be used to adjust regrown biomass' metal-binding characteristics. Heating/boiling, freezing/thawing, drying, and lyophilisation are physical treatments (Gadd 1990; Vojoudi et al. 2017).

Biosorbent drying precedes the biosorption/desorption cycle in several studies. Drying algae's new biomass improves storage and heavy metal absorption. The drying procedure caused shrinkage and porosity loss in the sea algae Sargassum sp. Morphological examination of Sargassum sp. algae revealed pore holes filled with cellular material (Veglio' and Beolchini 1997). Biomass type determines the biosorption technology utilised. Some plants and components don't need immobilisation (Hegazi 2013). Small single cells in contactors must immobilise. Cross-linking and trapping in polymeric matrices are common biosorbent formulation processes. The right technology and operating parameters must be chosen for an economically successful treatment. Several approaches may be used to immobilized nano-biosorbents suited for development applications, including entrapment, cross-linking, bacteria, and fungi (*c.f.* Figure 16.4).

16.3.1 Immobilization Through Cross-Linking

A cross-linker forms stable cellular aggregates. This method is used to immobilize algae. Formaldehyde, glutaric dialdehyde, divinylsulfone, and formaldehyde–urea mixtures are examples of cross-linkers (Veglio' and Beolchini 1997). *Saccharomyces cerevisiae* cell walls were used to make two types of magnetic biosorbents, and they were compared with nonmagnetic cells. Both of these magnetic biosorbents were found to bind Cu^{2+} to 225 and 50 m mol g^{-1}, Cd^{2+} to 90 and 25 m mol g^{-1}, and Ag^{+} to 80 and 45 m mol g^{-1}. 400, 125, and 75 m mol g^{-1} for nonmagnetic cell walls (Patzak et al. 1997). The sorption capacity is almost always decreased by cross-linking, and *Azolla filiculoides* immobilized with epichlorohydrin is no exception (Fogarty and Tobin 1996).

16.3.2 Immobilization Through Entrapment

Cells are immobilized in matrices. Insoluble Ca-alginate has been utilized to recover metals by live and non-viable cells. Polyacrylamide, polysulfone, polyethylenimine, and polyhydroxyethylmethacrylate are also utilized (Szilva et al. 1998). Immobilized materials are soft gel particles. Polysulfone and polyethylenimine immobilized proteins were the strongest. When yeast cell wall envelopes were immobilized in a silica matrix, many new biosorbents were produced (Szilva et al. 1998).

16.3.3 Immobilization Using Microorganisms

The immobilization of microorganisms in a polymeric matrix has greater potential, with advantages such as particle size control, biomass regeneration and reuse, simple separation of biomass and effluent, high biomass loading, and minimal clogging under continuous flow conditions (Veglio' and Beolchini 1997). Alginate, carrageenan, agar, agarose, and chitosan, an amino polysaccharide made from chitin, were used in experiments together with synthetic polymer such as acrylamide, polyurethane, polyvinyl alcohol, and resins (Bohumil Volesky 2007). Several researchers immobilized bacterial biomass for metal biosorption. Important biosorbent immobilization matrices include sodium alginate, polysulfone, polyacrylamide, and polyurethane. Immobilized biomass' environmental applicability depends on the immobilization matrix. The polymeric matrix influences the biosorbent particle's mechanical strength and chemical resistance (Veglio' and Beolchini 1997).

During immobilization, it's important to prevent mass transfer limits and additional process costs (Bai and Abraham 2003). After immobilization, the biomass is generally kept inside the matrix, hence mass transfer resistance determines biosorption rate. Mass transfer resistance delays equilibrium, but a proper immobilization matrix should enable all active binding sites to access the solution, albeit slowly. Immobilizing biomass increases process costs. Biosorption is typically marketed as a cost-effective alternative to other established methods (Veglio' and Beolchini 1997). Immobilizing biomass for biosorption increases process costs, although it's typically essential for real-world applications (Vijayaraghavan, Lee, and Yun 2008).

Biomass weakens and mushes when it comes into touch with water. In column applications, the small size and low density of free cells offer issues that result in significant pressure losses. Applications for microbial biosorption favor biomass that is immobilized or palletized (Wang and Chen 2009). Biomass may be immobilized using alginate, polyacrylamide, polyvinyl alcohol, polysulfone, silica gel, cellulose, and glutaraldehyde. It would be possible to create a biosorbent material with the ideal size, mechanical strength, stiffness, and porosity by immobilizing biomass in solid structures. Entrapment was used to immobilize *Phanerochaete chrysosporium* within Ca-alginate beads in order to extract Hg(II) and Cd(II) ions from aqueous solution. With 10 mM HCl, the alginate–fungus beads could be recovered to up to 97% of their original state (Arica et al. 2003). By entrapment, carboxymethylcellulose was used to immobilize Trametes versicolor mycelia, and their greatest biosorption capacities were 1.51 mmol Cu, 0.85 mmol Pb, and 1.33 mmol Zn per g of dry biosorbent, respectively (Bayramoğlu, Bektaş, and Arica 2003). To remove metal ions from aqueous solutions, *Monilesaurus rouxii* biomass was put onto a column and immobilized in polysulfone. For single component metal solutions, the removal capacities for Pb, Cd, Ni, and Zn were 4.06, 3.76, 0.36, and 1.36 mg g^{-1}, respectively (Yan and Viraraghavan 2001). Cd, Ni, and Zn capacities were 0.36, 0.31, and 0.4 mg g^{-1} in a multi-component metal solution. *Aspergillus niger* was immobilized in polysulfone beads to extract Cd, Cu, Pb, Ni. Cd, Cu, Pb, and Ni removal capacities were 3.60, 2.89, 10.05, and 1.08 mg g^{-1}. Polysulfone is an amorphous, hard, heat-resistant, and chemically stable thermoplastic (Yan and Viraraghavan 2001). Important application criteria include choosing suitable and reasonably priced support materials for biosorbent immobilization as well as strengthening reuse procedures and properties of immobilized biomaterials including mechanical intensity, chemical stability, and porous ratio (Wang and Chen 2009).

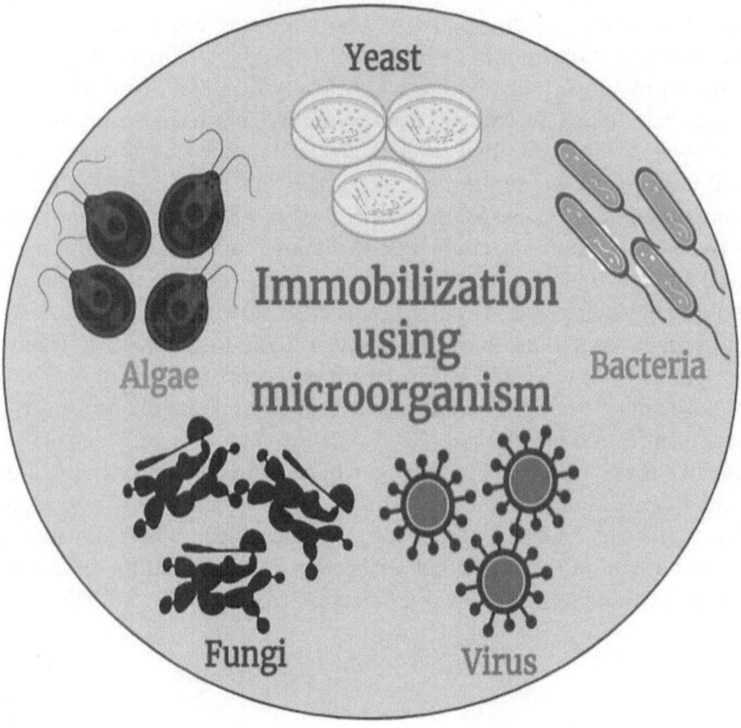

FIGURE 16.4 Various microorganisms for the immobilization of nano-biosorbent

16.4 CONVENTIONAL TECHNIQUES FOR THE REMOVAL OF ENVIRONMENTAL CONTAMINANTS

Physical, chemical, and biological methods remove contaminants from environment. Chemical precipitation, electrocoagulation, filtration, ion exchange, electrochemical treatment, membrane technology, adsorption on activated carbon, zeolite, and evaporation are conventional ways for eliminating contaminants from environment (Volesky and Naja 2007). Chemical precipitation is unsuccessful and creates enormous amounts of difficult-to-treat sludge. Ion exchange, membrane technologies, and activated carbon adsorption are too costly to treat significant amounts of low-concentration heavy metal water and wastewater (Karri, Shams, and Sahu 2018). Developing and implementing cost-effective metal removal/recovery processes improves industrial processing operations' competitiveness. Due to disadvantages and the need for more cost-efficient and effective procedures, alternative separation technologies have been developed. Recently, passive treatment techniques have gained popularity (Bai and Abraham 2003). These use geochemical and biological processes to enhance water quality with minimum upkeep. Biological removal may remove heavy metals from wastewater using microorganisms (fungi, algae, bacteria), plants (living or dead), and biopolymers. Biosorption provides benefits over traditional approaches. Metal biosorption research concentrate on anthropogenic point sources of metal releases for practical reasons. Biosorption removes metals selectively across a wide pH and temperature range, with fast adsorption and desorption kinetics and cheap capital and operating costs. Biosorbent may be made from cheap growing media or industrial byproducts. Different metal cations may be removed by biosorbents with a wide variety of metal affinities (Khan et al. 2019). A nanobentonite-intercalated carboxymethyl chitosan composite (nZVI@nBent-CMC) was developed (Eltaweil et al. 2021). Anionic Congo red (CR) and cationic crystal violet (CV) were eliminated from the final composite. For CR and CV, the nano-biosorbent possesses sorption capacities of 884.95 mg g^{-1} and 505.05 mg g^{-1}, respectively. The pseudo-second-order kinetics model and Langmuir isotherm are followed by the thermodynamically spontaneous sorption process. A heavy metal-removing bio-composite was also prepared (Yildirim, Baran, and Acay 2020). The fungal-extract-based nano-biosorbent was utilized to remove Ni(II) and Cu(II) from water. fungal-extract-based (FE-CB) bio-nanosorbent removed 7.18 mg g^{-1} Cu(II) and 8.50 mg g^{-1} Ni(II). The findings demonstrated that the process is spontaneous, endothermic, and beneficial.

Nano-bioabsorbent is used as an electrode material in electrochemical biosensing, energy storage, and pollutant removal from the environment (dyes, hazardous metal ions) (Vijayaraghavan, Lee, and Yun 2008). Further, electrochemical sensors have recently gained attention for their quick reaction, simple operation, great sensitivity, superior selectivity, and real-time in situ detection (Agrahari et al. 2022).

16.5 VARIOUS ELECTROCHEMICAL TECHNIQUES USING NANO-BIOSORBENTS FOR THE DETECTION OF ENVIRONMENTAL CONTAMINANTS

Electrochemical analysis measures conductivity between electrodes, current, potential, impedance, and field effect. Galvanostatic (with current control) and potentiostatic (with potential control) methods are examples of dynamic techniques. Three controls are polarography, voltammetry, and ammeter. The electrode in electrochemical biosensors may function as an ammeter, potentiometer, conductivity converter, and impedimetry converter. Electrochemical measurement instruments consist of a detector and two control electrodes (work). The biosensor's control electrode is Ag/AgCl and the detecting electrode is a transducer. The voltage of the control (reference) electrode is altered in relation to the detecting electrode. The potential of the control electrodes must be constant. An auxiliary electrode is used when the current through the reference electrode changes in potential. In three-electrode systems, a current is generated between the auxiliary and detection electrodes, and the potential of the detector electrode is regulated by the reference electrode.

16.5.1 Potentiometry

The potential difference between two floating electrodes is measured using potentiometry with almost no current flowing. The voltage between two electrodes shows sample structure in potentiometry. This technique often yields data on the ion concentration (activity) in the sample. Nagana et al. used sawdust from *Triplochiton scleroxylon* (Ayous) by drying with Reactive Red 120 (RR120) and Reactive Black 5 (RB5) (Ngana et al. 2021). Functionalized sawdust used as GC modifiers increased Pb detection (II). After optimizing experimental conditions, sub-nanomolar detection limits were established (1.3 nM, 0.7 nM, and 0.4 nM for glassy carbon modified *Ayous* sawdust (GC/SA) , GC/SA-RB5, and GC/SA-RR120).

16.5.2 Amperometry

The amperometric method assesses the current intensity brought on by the oxidation or reduction of electroactive species during an electrochemical reaction. Amprometry, which is more sensitive than potentiometric biosensors, relates emission current intensity and electroactive chemical concentration (which makes or consumes electrons). Fozing Mekeuo et al. study designs and evaluate a lignocellulosic electrochemical sensor for pesticide analysis. (Fozing Mekeuo et al. 2022). To improve the sorption of the amperometric sensor, maleic anhydride was grafted on Ayous sawdust and mixed with CNTs to create a conductive composite. The sensor resulted in sensitivity of 2.61 ± 0.08 and limit of detection of 0.04 μM.

16.5.3 Voltammetry

It measures current-induced voltage changes between two electrodes. Voltammetry is an ammeter subclass. Voltammetry types vary by voltage (direct current or alternating currenta) and application (scan, pulse, square wave). The current between the working and auxiliary electrodes is monitored.

The graphic shows flow and prospective modifications. Voltammetry scan speeds must be tuned while scanning potential. How potential is applied differs between voltammetric methods. Using *Bougainvillea spectabilis* as a precursor, (Veeramani et al. 2017) manufacture graphene sheet-like porous activated carbon with a high specific surface area. The catechin sensor's sensitivity, linear range, and detection limit were 7.2 $\mu A\ \mu M^{-1}\ cm^{-2}$, 4–368 μM, and 0.67 μM, respectively. Single-walled CNTs have special properties that are shown by the electrocatalytic oxidation of three to four-dihydroxyphenylacetic acid (DOPAC) at a glassy carbon electrode modified with single walled carbon nanotubes SWNTs. DOPAC at this SWNT-modified electrode demonstrates diffusion-controlled electrochemistry. Peak current increased linearly from 1.0×10^{-6} to 1.2×10^{-4} M with DOPAC concentration, and 4.0×10^{-7} M was the detection limit (Wang et al. 2002).

16.5.4 Impedimetric

Electrochemical impedance spectroscopy studies electron transport and diffusion in electrochemical processes. Impedance is the complicated resistance in a resistor, capacitor, and inductor circuit. Modeling electrochemical events as resistor, capacitor, and inductor at the soluble electrode interface. Using a circuit's impedance spectrum, electrode surface events may be studied. Resistance is measured by applying a modest sinusoidal voltage at frequency w. Nyquist plots the resultant spectrum as imagined resistance to actual resistance. High-frequency semicircular component is associated to electron transmission, whereas low-frequency linear part is related to emission. By changing the electrode's current resistance, specific surface interactions may be observed.

Conductivity subset of spectroscopic impedance. Electrochemical biosensors translate biological information into electronic signals using electrodes. Extensive research has been done on electrochemical sensors, and some of these sensors have been commercialized and used in clinical, industrial, environmental, and agricultural domains. Electrochemical biosensors have a biological detector. Electrochemical biosensors integrate electrochemical analysis with biological activity. The analyte interacts with the bioreceptor, producing an electrical signal proportional to the analyte concentration. Electrochemical conversion is beneficial since many analytes lack strong fluorescence, and tagging molecules is problematic. By integrating electrochemical sensitivity with bioreceptor selectivity, fluorescence biosensor detection limits are achieved. Co-immobilized acetylcholinesterase and choline oxidase. Enzymes were immobilized using a hybrid entrapment/surface attachment technique. Entrapped and surface-immobilized enzymes detected 10–500 and 10–250 ppb aldicarb, respectively (Kok, Bozoglu, and Hasirci 2002).

16.6 CHALLENGES IN THE ANALYSIS OF ENVIRONMENTAL POLLUTANTS USING NANO-BIOSORBENT

Nanotechnology is useful for contaminant detection. There are several immobilization approaches, but none can be utilized to build nanobiosensors using nano-absorbent. Research is needed to overcome these limitations and offer improved methodologies with high sensitivity, selectivity, and robustness. No evidence links metal resistance to high biosorption. These biomass types have been used in metal biosorption investigations. Even though certain cultures have high metal biosorption, their biomass is not easily used. It would have to be specially farmed, making it uneconomical. For environmental applications, the cost-effectiveness of biosorbent materials is crucial, while synthetic ion exchange resins may also effectively bind metals. However, their high price makes normal wastewater treatment uneconomical. The consistency of industrial biomass must be changed to provide biosorbent materials for large-scale sorbing equipment. The typical consistency is dry cake or powder, moist "mud," or both. It has to be processed into tiny grains in order to withstand sorption conditions.

16.7 CONCLUSIONS AND FUTURE ASPECTS

Researchers and scientists are exploring novel, cost-effective, and green ways to remove toxins due to increased pollution. Nano-biosorbents are used to remove pollutants from the environment. This chapter introduces nano-biosorbents and their use in waste treatment using electrochemical techniques. Biosorption provides cheap operating costs, little chemical usage, no needs for nutrients or disposal of biological or inorganic sludge, high efficiency at low metal concentrations, and no metal toxicity problems. Bio-based components are abundant, non-toxic, biocompatible, renewable, and easily modifiable. Using these materials' beneficial features and nanotechnology, nano-biosorbents are created. Nano-biosorbents have been utilized to remove toxins from wastewater and are potential tools for a clean environment. Biosorption has many traits with ion-exchange technology, and biosorbents might be regarded as direct rivals of ionex resins despite a shorter life cycle and poorer selectivity. High cost restricts ion use exchange's. Not all companies generating metal-bearing effluents can afford such extensive treatment, thus most settle for simple decontamination. The gathered information supports commercial biosorption techniques. Huge marketplaces already exist and may develop as global laws get tougher and metal demand rises. Future attempts to increase biosorbent selectivity and shelf life, biosorption model dependability and performance, and pilot size demonstrations should support large-scale applications. Biosorption might be used in future separation technologies with renewable biosorbents supplementing traditional techniques in hybrid or integrated systems.

16.7.1 Author Contributions

SA: Methodology, Conceptualization, Visualization, Writing – original draft; **AKS:** Investigation, Writing – Review & Editing; **RKG:** Writing – Review & Editing, Data curation, Formal analysis **IT:** Supervision, Validation.

ACKNOWLEDGMENTS

The Scheme for Promotion of Academic and Research Collaboration SPARC (Scheme Number-6019) is acknowledged. SA (Chem./2019-2020/RET-2/Sept-19-term/1/975) and A K S (Chem./2018-19/RET/Sept.18-term/1/4809) are grateful to UGC for their research fellowships. Illustrations are made using BioRender.com, and therefore, the service provider is greatly acknowledged.

REFERENCES

Agarwal, Prashant, Ritika Gupta, and Neeraj Agarwal. 2019. "Advances in Synthesis and Applications of Microalgal Nanoparticles for Wastewater Treatment." *Journal of Nanotechnology* 2019. https://doi.org/10.1155/2019/7392713.

Agrahari, Shreanshi, Ravindra Kumar Gautam, Ankit Kumar Singh, and Ida Tiwari. 2022. "Nanoscale Materials-Based Hybrid Frameworks Modified Electrochemical Biosensors for Early Cancer Diagnostics: An Overview of Current Trends and Challenges." *Microchemical Journal* 172 (PB): 106980. https://doi.org/10.1016/j.microc.2021.106980.

Ahluwalia, Sarabjeet Singh, and Dinesh Goyal. 2007. "Microbial and Plant Derived Biomass for Removal of Heavy Metals from Wastewater." *Bioresource Technology* 98 (12): 2243–2257. https://doi.org/10.1016/j.biortech.2005.12.006.

Ali, Imran. 2012. "New Generation Adsorbents for Water Treatment." *Chemical Reviews* 112 (10): 5073–5091. https://doi.org/10.1021/cr300133d.

Ali, Nisar, Salah Uddin, Adnan Khan, Saraf Khan, Sana Khan, Nauman Ali, Hamayun Khan, Hammad Khan, and Muhammad Bilal. 2020. "Regenerable Chitosan-Bismuth Cobalt Selenide Hybrid Microspheres for Mitigation of Organic Pollutants in an Aqueous Environment." *International Journal of Biological Macromolecules* 161: 1305–1317. https://doi.org/10.1016/j.ijbiomac.2020.07.132.

Aquino, José M., Douglas W. Miwa, Manuel A. Rodrigo, and Artur J. Motheo. 2017. "Treatment of Actual Effluents Produced in the Manufacturing of Atrazine by a Photo-Electrolytic Process." *Chemosphere* 172: 185–192. https://doi.org/10.1016/j.chemosphere.2016.12.154.

Arica, M. Yakup, Çiğdem Arpa, Bülent, Kaya, S. Bektaş, Adil Denizli, and Ö Genç. 2003. "Comparative Biosorption of Mercuric Ions from Aquatic Systems by Immobilized Live and Heat-Inactivated Trametes Versicolor and Pleurotus Sajur-Caju." *Bioresource Technology* 89 (2): 145–154. https://doi.org/10.1016/S0960-8524(03)00042-7.

Aziz, Aisha, Nisar Ali, Adnan Khan, Muhammad Bilal, Sumeet Malik, Nauman Ali, and Hamayun Khan. 2020. "Chitosan-Zinc Sulfide Nanoparticles, Characterization and Their Photocatalytic Degradation Efficiency for Azo Dyes." *International Journal of Biological Macromolecules* 153: 502–512. https://doi.org/10.1016/j.ijbiomac.2020.02.310.

Bai, R. Sudha, and T. Emilia Abraham. 2003. "Studies on Chromium(VI) Adsorption-Desorption Using Immobilized Fungal Biomass." *Bioresource Technology* 87 (1): 17–26. https://doi.org/10.1016/S0960-8524(02)00222-5.

Bayramoğlu, Gülay, Sema Bektaş, and M. Yakup Arica. 2003. "Biosorption of Heavy Metal Ions on Immobilized White-Rot Fungus Trametes Versicolor." *Journal of Hazardous Materials* 101 (3): 285–300. https://doi.org/10.1016/S0304-3894(03)00178-X.

Collivignarelli, Maria Cristina, Alessandro Abbà, Marco Carnevale Miino, and Silvestro Damiani. 2019. "Treatments for Color Removal from Wastewater: State of the Art." *Journal of Environmental Management* 236 (February): 727–745. https://doi.org/10.1016/j.jenvman.2018.11.094.

Demirbas, Ayhan. 2008. "Heavy Metal Adsorption onto Agro-Based Waste Materials: A Review." *Journal of Hazardous Materials* 157 (2–3): 220–229. https://doi.org/10.1016/j.jhazmat.2008.01.024.

Dodbiba, Gjergj, Josiane Ponou, and Toyohisa Fujita. 2015. "Biosorption of Heavy Metals." *Microbiology for Minerals, Metals, Materials and the Environment* 409–426. https://doi.org/10.4018/978-1-5225-8903-7.ch077.

Dotto, Juliana, Márcia Regina Fagundes-Klen, Márcia Teresinha Veit, Soraya Moreno Palácio, and Rosangela Bergamasco. 2019. "Performance of Different Coagulants in the Coagulation/Flocculation Process of Textile Wastewater." *Journal of Cleaner Production* 208: 656–665. https://doi.org/10.1016/j.jclepro.2018.10.112.

El-Ashtoukhy, El Sayed Zakaria, N. K. Amin, and Ola Abdelwahab. 2008. "Removal of Lead (II) and Copper (II) from Aqueous Solution Using Pomegranate Peel as a New Adsorbent." *Desalination* 223 (1–3): 162–173. https://doi.org/10.1016/j.desal.2007.01.206.

Eltaweil, Abdelazeem S., Ashraf M. El-Tawil, Eman M. Abd El-Monaem, and Gehan M. El-Subruiti. 2021. "Zero Valent Iron Nanoparticle-Loaded Nanobentonite Intercalated Carboxymethyl Chitosan for Efficient Removal of Both Anionic and Cationic Dyes." *ACS Omega* 6 (9): 6348–6360. https://doi.org/10.1021/acsomega.0c06251.

Fogarty, Robert V., and John M. Tobin. 1996. "Fungal Melanins and Their Interactions with Metals." *Enzyme and Microbial Technology* 19 (4): 311–317. https://doi.org/10.1016/0141-0229(96)00002-6.

Fozing Mekeuo, Ghislaine Ariane, Christelle Despas, Charles Péguy Nanseu-Njiki, Alain Walcarius, and Emmanuel Ngameni. 2022. "Preparation of Functionalized Ayous Sawdust-Carbon Nanotubes Composite for the Electrochemical Determination of Carbendazim Pesticide." *Electroanalysis* 34 (4): 667–676. https://doi.org/10.1002/elan.202100262.

Gadd, Geoffrey M. 1990. "Heavy Metal Accumulation by Bacteria and Other Microorganisms." *Experientia* 46 (8): 834–840. https://doi.org/10.1007/BF01935534.

Gadd, Geoffrey M. 2008. "Accumulation and Transformation of Metals by Microorganisms." *Biotechnology: Second, Completely Revised Edition* 10–12: 225–264. https://doi.org/10.1002/9783527620999.ch9k.

Hegazi, Hala Ahmed. 2013. "Removal of Heavy Metals from Wastewater Using Agricultural and Industrial Wastes as Adsorbents." *HBRC Journal* 9 (3): 276–282. https://doi.org/10.1016/j.hbrcj.2013.08.004.

Jain, Priyanka, Shilpa Varshney, and Shalini Srivastava. 2017. "Synthetically Modified Nano-Cellulose for the Removal of Chromium: A Green Nanotech Perspective." *IET Nanobiotechnology* 11 (1): 45–51. https://doi.org/10.1049/iet-nbt.2016.0036.

Kanamarlapudi, Sri Lakshmi Ramya Krishna, Vinay Kumar Chintalpudi, and Sudhamani Muddada. 2018. "Application of Biosorption for Removal of Heavy Metals from Wastewater." *Biosorption*. https://doi.org/10.5772/intechopen.77315.

Karri, Rama Rao, Shahriar Shams, and Jaya Narayan Sahu. 2018. *4-Overview of Potential Applications of Nano-Biotechnology in Wastewater and Effluent Treatment. Nanotechnology in Water and Wastewater Treatment: Theory and Applications*. Elsevier Inc. https://doi.org/10.1016/B978-0-12-813902-8.00004-6.

Kaushik, Prachi, and Anushree Malik. 2009. "Fungal Dye Decolourization: Recent Advances and Future Potential." *Environment International* 35 (1): 127–141. https://doi.org/10.1016/j.envint.2008.05.010.

Khan, Adnan, Nisar Ali, Muhammad Bilal, Sumeet Malik, Syed Badshah, and Hafiz M.N. Iqbal. 2019. "Engineering Functionalized Chitosan-Based Sorbent Material: Characterization and Sorption of Toxic Elements." *Applied Sciences (Switzerland)* 9 (23). https://doi.org/10.3390/app9235138.

Kok, Fatma N., Faruk Bozoglu, and Vasif Hasirci. 2002. "Construction of an Acetylcholinesterase-Choline Oxidase Biosensor for Aldicarb Determination." *Biosensors and Bioelectronics* 17 (6–7): 531–539. https://doi.org/10.1016/S0956-5663(02)00009-X.

Lesmana, Sisca O., Novie Febriana, Felycia E. Soetaredjo, Jaka Sunarso, and Suryadi Ismadji. 2009. "Studies on Potential Applications of Biomass for the Separation of Heavy Metals from Water and Wastewater." *Biochemical Engineering Journal* 44 (1): 19–41. https://doi.org/10.1016/j.bej.2008.12.009.

Macaskie, Lynne E. 1990. "An Immobilized Cell Bioprocess for the Removal of Heavy Metals from Aqueous Flows." *Journal of Chemical Technology & Biotechnology* 49 (4): 357–379. https://doi.org/10.1002/jctb.280490408.

Mahamadi, Courtie. 2019. "Will Nano-Biosorbents Break the Achilles' Heel of Biosorption Technology?" *Environmental Chemistry Letters* 17 (4): 1753–1768. https://doi.org/10.1007/s10311-019-00909-6.

Navamani Kartic, Dhayabaran, B. C.H. Aditya Narayana, and M. Arivazhagan. 2018. "Removal of High Concentration of Sulfate from Pigment Industry Effluent by Chemical Precipitation Using Barium Chloride: RSM and ANN Modeling Approach." *Journal of Environmental Management* 206: 69–76. https://doi.org/10.1016/j.jenvman.2017.10.017.

Ngana, Beaufils Ngatchou, Patrick Marcel Tchekwagep Seumo, Lionel Magellan Sambang, Gustave Kenne Dedzo, Charles Peguy Nanseu-Njiki, and Emmanuel Ngameni. 2021. "Grafting of Reactive Dyes onto Lignocellulosic Material: Application for Pb(II) Adsorption and Electrochemical Detection in Aqueous Solution." *Journal of Environmental Chemical Engineering* 9 (1): 104984. https://doi.org/10.1016/j.jece.2020.104984.

Patzak, M., P. Dostalek, R. V. Fogarty, I. Safarik, and J. M. Tobin. 1997. "Development of Magnetic Biosorbents for Metal Uptake." *Biotechnology Techniques* 11 (7): 483–487. https://doi.org/10.1023/A:1018453814472.

Prabakar, Desika, Subha Suvetha K., Varshini T. Manimudi, Thangavel Mathimani, Gopalakrishnan Kumar, Eldon R. Rene, and Arivalagan Pugazhendhi. 2018. "Pretreatment Technologies for Industrial Effluents: Critical Review on Bioenergy Production and Environmental Concerns." *Journal of Environmental Management* 218: 165–180. https://doi.org/10.1016/j.jenvman.2018.03.136.

Sahu, J. N., Rama Rao Karri, Hossain M. Zabed, Shahriar Shams, and Xianghui Qi. 2021. "Current Perspectives and Future Prospects of Nano-Biotechnology in Wastewater Treatment." *Separation and Purification Reviews* 50 (2): 139–158. https://doi.org/10.1080/15422119.2019.1630430.

Sartaj, Seema, Nisar Ali, Adnan Khan, Sumeet Malik, Muhammad Bilal, Menhad Khan, Nauman Ali, Sajjad Hussain, Hammad Khan, and Sabir Khan. 2020. "Performance Evaluation of Photolytic and Electrochemical Oxidation Processes for Enhanced Degradation of Food Dyes Laden Wastewater." *Water Science and Technology* 81 (5): 971–984. https://doi.org/10.2166/wst.2020.182.

Sharma, S., and A. Bhattacharya. 2017. "Drinking Water Contamination and Treatment Techniques." *Applied Water Science* 7 (3): 1043–1067. https://doi.org/10.1007/s13201-016-0455-7.

Siddiquee, Shafiquzzaman, Kobun Rovina, and Sujjat Al Azad. 2015. "Heavy Metal Contaminants Removal from Wastewater Using the Potential Filamentous Fungi Biomass: A Review." *Journal of Microbial & Biochemical Technology* 07 (06): 384–393. https://doi.org/10.4172/1948-5948.1000243.

Singh, Ankit Kumar, Ravindra Kumar Gautam, Shreanshi Agrahari, Jyoti Prajapati, and Ida Tiwari. 2022. "Graphene Oxide Supported Fe 3 O 4-MnO 2 Nanocomposites for Adsorption and Photocatalytic Degradation of Dyestuff: Ultrasound Effect, Surfactants

Role and Real Sample Analysis." *International Journal of Environmental Analytical Chemistry,* 1–27. https://doi.org/10.1080/03067319.2022.2091930.

Singh, Ankit Kumar, Nandita Jaiswal, Ravindra Kumar Gautam, and Ida Tiwari. 2021. "Development of G-C3N4/Cu-DTO MOF Nanocomposite Based Electrochemical Sensor towards Sensitive Determination of an Endocrine Disruptor BPSIP." *Journal of Electroanalytical Chemistry* 887 (April): 115170. https://doi.org/10.1016/J.JELECHEM.2021.115170.

Singh, Ankit Kumar, and Ida Tiwari. 2020. *Nanomaterial Synthesis and Mechanism for Enzyme Immobilization: Part II.* https://doi.org/10.1007/978-981-13-9333-4_8.

Szilva, János, Gabriela Kuncová, Milan Patzák, and Pavel Dostálek. 1998. "The Application of a Sol-Gel Technique to Preparation of a Heavy Metal Biosorbent from Yeast Cells." *Journal of Sol-Gel Science and Technology* 13 (1–3): 289–294. https://doi.org/10.1023/a:1008659807522.

Tran, Hai Nguyen, Hoang Chinh Nguyen, Seung Han Woo, Tien Vinh Nguyen, Saravanamuthu Vigneswaran, Ahmad Hosseini-Bandegharaei, Jörg Rinklebe, et al. 2019. "Removal of Various Contaminants from Water by Renewable Lignocellulose-Derived Biosorbents: A Comprehensive and Critical Review." *Critical Reviews in Environmental Science and Technology* 49 (23): 2155–2219. https://doi.org/10.1080/10643389.2019.1607442.

Veeramani, Vediyappan, Mani Sivakumar, Shen Ming Chen, Rajesh Madhu, Hatem R. Alamri, Zeid A. Alothman, Md Shahriar A. Hossain, et al. 2017. "Lignocellulosic Biomass-Derived, Graphene Sheet-like Porous Activated Carbon for Electrochemical Supercapacitor and Catechin Sensing." *RSC Advances.* https://doi.org/10.1039/c7ra07810b.

Veglio', F., and F. Beolchini. 1997. "Removal of Metals by Biosorption: A Review." *Hydrometallurgy* 44 (3): 301–316. https://doi.org/10.1016/s0304-386x(96)00059-x.

Vijayaraghavan, K., Min Woo Lee, and Yeoung Sang Yun. 2008. "Evaluation of Fermentation Waste (Corynebacterium Glutamicum) as a Biosorbent for the Treatment of Nickel(II)-Bearing Solutions." *Biochemical Engineering Journal* 41 (3): 228–233. https://doi.org/10.1016/j.bej.2008.04.019.

Vilar, Vítor J.P., Cidália M.S. Botelho, and Rui A.R. Boaventura. 2007. "Modeling Equilibrium and Kinetics of Metal Uptake by Algal Biomass in Continuous Stirred and Packed Bed Adsorbers." *Adsorption* 13 (5–6): 587–601. https://doi.org/10.1007/s10450-007-9029-1.

Vojoudi, Hossein, Alireza Badiei, Shahriyar Bahar, Ghodsi Mohammadi Ziarani, Farnoush Faridbod, and Mohammad Reza Ganjali. 2017. "A New Nano-Sorbent for Fast and Efficient Removal of Heavy Metals from Aqueous Solutions Based on Modification of Magnetic Mesoporous Silica Nanospheres." *Journal of Magnetism and Magnetic Materials* 441: 193–203. https://doi.org/10.1016/j.jmmm.2017.05.065.

Volesky, Bohumil. 2001. "Detoxification of Metal-Bearing Effluents: Biosorption for the next Century." *Hydrometallurgy* 59 (2–3): 203–216. https://doi.org/10.1016/S0304-386X(00)00160-2.

Volesky, Bohumil. 2007. "Biosorption and Me." *Water Research* 41 (18): 4017–4029. https://doi.org/10.1016/j.watres.2007.05.062.

Volesky, Bohumil, and G. Naja. 2007. "Biosorption Technology: Starting up an Enterprise." *International Journal of Technology Transfer and Commercialisation* 6 (2/3/4): 196. https://doi.org/10.1504/ijttc.2007.017806.

Wang, Huisong, and Guangchang Pang. 2020. "Baked Bread Enhances the Immune Response and the Catabolism in the Human Body in Comparison with Steamed Bread." *Nutrients* 12 (1): 1–15. https://doi.org/10.3390/nu12010001.

Wang, Jianlong, and Can Chen. 2009. "Biosorbents for Heavy Metals Removal and Their Future." *Biotechnology Advances* 27 (2): 195–226. https://doi.org/10.1016/j.biotechadv.2008.11.002.

Wang, Jianxiu, Meixian Li, Zujin Shi, Nanqiang Li, and Zhennan Gu. 2002. "Electrocatalytic Oxidation of Norepinephrine at a Glassy Carbon Electrode Modified with Single Wall Carbon Nanotubes." *Electroanalysis* 14 (3): 225–230. https://doi.org/10.1002/1521-4109(200202)14:3<225::AID-ELAN225>3.0.CO;2-I.

Werkneh, Adhena Ayaliew, and Eldon R. Rene. 2019. "Applications of Nanotechnology and Biotechnology for Sustainable Water and Wastewater Treatment." *Energy, Environment, and Sustainability*, 405–430. https://doi.org/10.1007/978-981-13-3259-3_19.

Yan, Guangyu, and T. Viraraghavan. 2001. "Heavy Metal Removal in a Biosorption Column by Immobilized M. Rouxii Biomass." *Bioresource Technology* 78 (3): 243–249. https://doi.org/10.1016/S0960-8524(01)00020-7.

Yildirim, Ayfer, M. Firat Baran, and Hilal Acay. 2020. "Kinetic and Isotherm Investigation into the Removal of Heavy Metals Using a Fungal-Extract-Based Bio-Nanosorbent." *Environmental Technology and Innovation* 20: 101076. https://doi.org/10.1016/j.eti.2020.101076.

Index

Note: Bold page numbers refer to tables; *italic* page numbers refer to figures.